T0335935

LECTURE NOTES ON
Algebraic Structure of Lattice-Ordered Rings

LECTURE NOTES ON
Algebraic Structure of
Lattice-Ordered Rings

Jingjing Ma

University of Houston-Clear Lake, USA

World Scientific

NEW JERSEY · LONDON · SINGAPORE · BEIJING · SHANGHAI · HONG KONG · TAIPEI · CHENNAI

Published by

World Scientific Publishing Co. Pte. Ltd.

5 Toh Tuck Link, Singapore 596224

USA office: 27 Warren Street, Suite 401-402, Hackensack, NJ 07601

UK office: 57 Shelton Street, Covent Garden, London WC2H 9HE

Library of Congress Cataloging-in-Publication Data
Ma, Jingjing.
    Lecture notes on algebraic structure of lattice-ordered rings / Jingjing Ma (University of Houston-Clear Lake, USA).
        pages cm
    Includes bibliographical references and index.
    ISBN 978-981-4571-42-5 (hardcover : alk. paper)
    1. Lattice ordered rings. 2. Algebra.   I. Title.
    QA247.M24 2014
    511.3'3--dc23

                                2013051147

British Library Cataloguing-in-Publication Data
A catalogue record for this book is available from the British Library.

Image credit: "Forma Perfecta" designed by Ekaterina Lukasheva.
Folded and photographed by Inna Lupanova.

Printed in Singapore

To Li, Cheng, and Elisa

Preface

This book is an introduction to the theory of lattice-ordered rings. It is suitable for graduate and advanced undergraduate students who have finished an abstract algebra class. It can also be used as a self-study book for one who is interested in the area of lattice-ordered rings.

The book mainly presents some foundations and topics in lattice-ordered rings. Since we concentrate on lattice orders, most results are stated and proved for such structures, although some of results are true for partially ordered structures. This book considers general lattice-ordered rings. However I have tried to compare results in general lattice-ordered rings with results in f-rings. Actually a lot of research work in general lattice-ordered rings is to generalize the results of f-rings. I have also tried to make the book self-contained and to give more details in the proofs of the results. Because of elementary nature of the book, some results are given without proofs. Certainly references are given for those results.

Chapter 1 consists of background information on lattice-ordered groups, vector lattices, and lattice-ordered rings and algebras. Those results are basic and fundamental. An important structure theory on lattice-ordered groups and vector lattices presented in Chapter 1 is the structure theory of lattice-ordered groups and vector lattices with a basis. Chapter 2 presents algebraic structure of lattice-ordered algebras with a distributive basis, which is a basis in which each element is a distributive element. Chapter 3 concentrates on positive derivations of lattice-ordered rings. This topic hasn't been systematically presented before and I have tried to present most of the important results in this area. In Chapter 4, some topics of general lattice-ordered rings are considered. Section 4.1 consists of some characterizations of lattice-ordered matrix rings with the entrywise order over lattice-ordered rings with positive identity element. Section 4.2 gives the algebraic structure of lattice-ordered rings with positive cycles. In general lattice-ordered rings, f-elements often play important roles on their structures. In Section 4.3 we present some result along this line. Section 4.4 is about extending lattice orders in an Ore domain to its quotient ring. In Section 4.5 we consider how to generalize results on lattice-ordered matrix algebras over totally ordered fields to lattice-ordered matrix algebras over totally ordered integral domains. Section 4.6 consists of some results on lattice-ordered rings in which the identity element may not be positive. In Section 4.7, all lattice orders on 2×2 upper triangular matrix algebras over a totally ordered field are constructed, and some results are given for higher dimension triangular matrix algebras. Finally in Chapter 5, properties and structure of ℓ-ideals of lattice-ordered rings with a positive identity elements are presented.

I would like to thank Dr. K.K. Phua, the Chairman and Editor-in-Chief of World Scientific Publishing, for inviting me to write this lecture notes volume. I also want to express my thanks to my colleague Ms. Judy Bergman, University of Houston-Clear Lake, who has kindly checked English usage and grammar of the book. I will certainly have full responsibility for mistakes in the book, and hopefully they wouldn't give the reader too much trouble to understand its mathematical contents.

Jingjing Ma

Houston, Texas, USA
December 2013

Contents

Chapter 1

Introduction to ordered algebraic systems

In this chapter, we introduce various ordered algebraic systems and present some basic and important properties of these systems.

1.1 Lattices

For a nonempty set A, a binary relation $\leq$ on A is called a *partial order* on A if the following properties are satisfied.

(1) (reflexivity) $a \leq a$ for all $a \in A$,
(2) (antisymmetry) $a \leq b$, $b \leq a$ implies $a = b$ for all $a, b \in A$,
(3) (transitivity) $a \leq b$, $b \leq c$ implies $a \leq c$, for all $a, b, c \in A$.

The set A under a partial order $\leq$ is called a *partially ordered set*. One may write $b \geq a$ to denote $a \leq b$, and $a < b$ (or $b > a$) to mean that $a \leq b$ and $a \neq b$. If either $a \leq b$ or $b \leq a$, then a and b are called *comparable*, otherwise a and b are called *incomparable*. A partial order $\leq$ on a set A is called a *total order* if any two elements in A are comparable. In the case that $\leq$ is a total order, A is called a *totally ordered set* or a *chain*. Suppose that two partial orders, $\leq$ and $\leq'$, are defined on the same set A. Then we say that $\leq'$ is an *extension* of $\leq$ if, for all $a, b \in A$, $a \leq b$ implies $a \leq' b$.

A partial order $\leq$ on A induces a partial order on any nonempty subset B of A, that is, for any $a, b \in B$, define $a \leq b$ in B if $a \leq b$ with respect to the original partial order of A. The induced partial order on B is denoted by the same symbol $\leq$.

For a subset B of a partially ordered set A an *upper bound* (*lower bound*) of B in A is an element $x \in A$ ($y \in A$) such that $b \leq x$ ($b \geq y$) for each $b \in B$. We may simply denote that $x \in A$ ($y \in A$) is an upper (lower) bound of B by $B \leq x$ ($B \geq y$). B is called *bounded* in A if B has both an upper

1

bound and a lower bound in A. The set of all upper (lower) bounds of B in A is denoted by $U_A(B)$ $(L_A(B))$. If $B = \emptyset$, where $\emptyset$ denotes empty set, then $U_A(B) = L_A(B) = A$. An element $u \in B$ $(v \in B)$ is called the *least element (greatest element)* of B if $u \le b$ $(v \ge b)$ for each $b \in B$. A subset B of a partially ordered set may not have a least (greatest) element, but if there exists one, then it is unique since partial orders are antisymmetric. An element $w \in B$ $(z \in B)$ is called a *minimal element (maximal element)* in B if for any $b \in B$, $b \le w$ $(b \ge z)$ implies $b = w$ $(b = z)$, that is, no element in B is strictly less (greater) than w (z). A subset of a partially ordered set may contain more than one minimal or maximal element.

Suppose that L is a partially ordered set with a partial order $\le$. The $\le$ is called a *lattice order* and L is called a *lattice* under $\le$ if for any $a, b \in L$, the set $U_L(\{a, b\})$ has the least element and the set $L_L(\{a, b\})$ has the greatest element, namely, for any $a, b \in L$, the subset $\{a, b\}$ has the least upper bound and greatest lower bound that are denoted respectively by

$$a \vee b \ \text{ and } a \wedge b$$

$a \vee b$ is also called the *sup* of a and b, and $a \wedge b$ is also called the *inf* of a and b. A nonempty subset B of a lattice L is called a *sublattice* of L if for any $a, b \in B$, $a \vee b, a \wedge b \in B$. A lattice L is called *distributive* if for all $a, b, c \in L$,

$$a \vee (b \wedge c) = (a \vee b) \wedge (a \vee c) \ \text{ and } \ a \wedge (b \vee c) = (a \wedge b) \vee (a \wedge c),$$

and L is called *complete* if each subset of L has both an inf and a sup in L. In a lattice L, for any $a, b, c \in L$, by the definition of least upper bound and greatest lower bound, we have

$$a \vee (b \vee c) = (a \vee b) \vee c \ \text{ and } \ a \wedge (b \wedge c) = (a \wedge b) \wedge c.$$

This is true for any finitely many elements in L, and hence we just use $a_1 \vee \cdots \vee a_n$ and $a_1 \wedge \cdots \wedge a_n$ to denote the sup and inf of $a_1, \cdots, a_n$, respectively.

The following is an example that illustrates some concepts defined above. More examples may be found in the exercises of this chapter.

Example 1.1. For a given set A, let $P_A = \{B \mid B \text{ is a subset of } A\}$ be the power set of A. For two subsets B, C of A, define $B \le C$ if $B \subseteq C$, where "$B \subseteq C$" means that B is a subset of C. Then $\le$ is actually a lattice order and for any $B, C \in P_A$, $B \vee C = B \cup C$ and $B \wedge C = B \cap C$. Clearly $\emptyset$ is the least element of P_A and A is the greatest element of P_A. Moreover, P_A is a distributive and complete lattice (Exercise 3).

If A contains more than one element, then P_A is not a totally ordered set since for two different elements $a, b \in A$, the sets $\{a\}$ and $\{b\}$ are not comparable. Also the subset $B = \{\{a\}, \{b\}\}$ of P_A has no least and greatest element, and each element in B is a minimal element and a maximal element since $\{a\}$ and $\{b\}$ are not comparable.

This is a suitable place to state Zorn's lemma, which is equivalent to Axiom of Choice. For the proof and other equivalent forms of the lemma, see [Steinberg (2010)].

Theorem 1.1 (Zorn's Lemma). *Let A be a nonempty partially ordered set. If each subset of A which is a chain has an upper bound in A, then A contains a maximal element.*

1.2 Lattice-ordered groups and vector lattices

In this section we introduce partially ordered groups, lattice-ordered groups, vector lattices, and consider some basic properties of those ordered algebraic systems. We will always use addition to denote group operation although it may not be commutative. Certainly for a vector lattice, the addition on it is commutative.

1.2.1 Definitions, examples, and basic properties

Definition 1.1. A *partially ordered group* G is a group and a partially ordered set under a partial order $\leq$ such that G satisfies the following monotony law: for any $a, b \in G$,

$$a \leq b \;\Rightarrow\; c + a \leq c + b \text{ and } a + c \leq b + c \text{ for all } c \in G.$$

A partially ordered group G is a *lattice-ordered group* (ℓ-*group*) if the partial order is a lattice order, and G is a *totally ordered group* (*o-group*) if the partial order is a total order.

In a partially ordered group G, an element g is called *positive* if $g \geq 0$, where 0 is the identity element of G, and g is called *strictly positive* if $g > 0$. The set $G^+ = \{g \in G \mid g \geq 0\}$ is called the *positive cone* of G, and define $-G^+ = \{g \in G \mid -g \in G^+\} = \{g \in G \mid g \leq 0\}$, which is called *negative cone* of G. G^+ is a normal subsemigroup of G containing 0, but no other element along with its inverse, as shown in the following result. From the following two theorems, positive cones characterize partially ordered groups.

Theorem 1.2. *For a partially ordered group G, the positive cone G^+ satisfies the following three conditions:*

(1) $G^+ + G^+ \subseteq G^+$,
(2) $g + G^+ + (-g) \subseteq G^+$, for all $g \in G$,
(3) $G^+ \cap -G^+ = \{0\}$.

Proof. (1) Let $g, f \in G^+$. Then $0 \leq f \leq g + f$, so $0 \leq g + f$. Thus $g + f \in G^+$.

(2) Let $f \in G^+$. Then $0 = g + (-g) \leq g + f + (-g)$, so $g + f + (-g) \in G^+$.

(3) Clearly $0 \in G^+ \cap -G^+$. Suppose that $g \in G^+ \cap -G^+$. Then $g \geq 0$ and $-g \geq 0$, so $g \geq 0$ and $g \leq 0$, and hence $g = 0$. $\qquad\square$

Theorem 1.3. *Let G be a group and P be a subset of G which satisfies the following three conditions:*

(1) $P + P \subseteq P$,
(2) $g + P + (-g) \subseteq P$ for all $g \in G$,
(3) $P \cap -P = \{0\}$, where $-P = \{g \in G \mid -g \in P\}$.

For any $a, b \in G$, define $a \leq b$ if $b - a \in P$. Then $\leq$ is a partial order on G and G becomes a partially ordered group with the positive cone P.

Proof. For any $a \in G$, $a - a = 0 \in P$ implies $a \leq a$, so $\leq$ is reflexive. Suppose that for $a, b \in G$, $a \leq b$ and $b \leq a$, then $b - a, a - b \in P$, so $b - a \in P$ and $b - a = -(a - b) \in -P$. Thus $b - a = 0$ by (3), and hence $a = b$, so $\leq$ is antisymmetric. Now assume that $a \leq b$ and $b \leq c$ for $a, b, c \in G$. Then $b - a, c - b \in P$, so by (1) $c - a = (c - b) + (b - a) \in P$. Thus $a \leq c$, so $\leq$ is transitive. Suppose that $a \leq b$ for $a, b \in G$ and $g \in G$. Then from $b - a \in P$ and (2),

$$(g + b) - (g + a) = g + (b - a) + (-g) \in P,$$

so $g + a \leq g + b$. Also

$$(b + g) - (a + g) = b + g - g - a = b - a \in P,$$

so $a + g \leq b + g$. Therefore G is a partially ordered group with respect to the partial order $\leq$. Clearly $G^+ = \{g \in G \mid g \geq 0\} = P$. $\qquad\square$

Theorem 1.4. *Suppose that G is a partially ordered group with the positive cone P.*

(1) G is an ℓ-group if and only if $G = \{a - b \mid a, b \in P\}$ and P is a lattice under the induced partial order from G.

(2) G is a totally ordered group if and only if $G = P \cup -P$.

Proof. (1) Suppose that G is an ℓ-group. For $g \in G$, let $f = g \wedge 0$. Then $-f \in P$ and $g - f \in P$. Since $g = (g - f) - (-f)$, $G = \{a - b \mid a, b \in P\}$. It is clear that for any $a, b \in P$, $a \vee b, a \wedge b \in P$. Conversely, suppose that $G = \{a - b \mid a, b \in P\}$ and P is a lattice with respect to the induced partial order from G. For any $g \in G$, let $g = x - y$, $x, y \in P$. Suppose that $z = x \vee y \in P$. We claim that $g \vee 0 = z - y$ in G. It is clear that $z - y \geq 0, g$. Suppose that $u \in G$ and $u \geq g, 0$. Then $u + y \geq x, y$ and $u + y \in P$, so $u + y \geq z$. Then it follows that $u \geq z - y$, and hence $g \vee 0 = z - y$ in G. Similarly to show that $g \wedge 0$ exists in G. Generally for any $g, f \in G$, it is straightforward to check that

$$g \vee f = [(g - f) \vee 0] + f \text{ and } g \wedge f = [(g - f) \wedge 0] + f$$

(Exercise 5). Therefore G is a lattice, so G is an ℓ-group.

(2) If $G = P \cup -P$, then for any $g, f \in G$, either $g - f \in P$ or $-P$, and hence $g \geq f$ or $g \leq f$. Thus G is a total order. The converse is clear. □

A partially ordered group is called *directed* if each element is a difference of two positive elements. An ℓ-group is directed by Theorem 1.4(1). However a partially ordered group which is directed may not be an ℓ-group as shown in Example 1.2(3). A partially ordered group G is said to be *Archimedean* if for any $a, b \in G^+$, $na \leq b$ for all $n \in \mathbb{Z}^+$ implies $a = 0$, where $\mathbb{Z}^+$ is the set of all positive integers.

In this book we often use notation (G, P) to denote a partially ordered group or an ℓ-group with the positive cone P.

We illustrate partially ordered groups and ℓ-groups by a few examples. P will always denote the positive cone of a partially ordered group.

Example 1.2.

(1) Let G be the additive group of $\mathbb{Z}$ or $\mathbb{Q}$, or $\mathbb{R}$ with the usual order between real numbers. Then G is an Archimedean totally ordered group.

(2) Consider the group direct product $\mathbb{R} \times \mathbb{R}$. Let (x, y) belong to P if either $y > 0$ or $y = 0$ and $x \geq 0$. Then $\mathbb{R} \times \mathbb{R}$ is a totally ordered group which is not Archimedean since for any $n \in \mathbb{Z}^+$, $n(1, 0) \leq (0, 1)$.

(3) Consider $\mathbb{R} \times \mathbb{R}$ again. Define $(x, y) \in P$ if $x > 0$ and $y > 0$, or $(x, y) = (0, 0)$. Then $\mathbb{R} \times \mathbb{R}$ is an Archimedean partially ordered group but not an ℓ-group. For instance, $(1, 0)$ and $(0, 0)$ have no least upper

bound. We leave the verification of this fact as an exercise to the reader (Exercise 6). We note that for any $(x, y) \in \mathbb{R} \times \mathbb{R}$, $(x, y) = (x, 0) + (0, y)$, and $(x, 0), (0, y)$ are either positive or negative, so (x, y) can be written as a difference of two positive elements. Thus this partially ordered group is directed.

Since in this book, we concentrate on lattice orders, in the following we only prove some basic properties of ℓ-groups.

Theorem 1.5. *Let G be an ℓ-group.*

(1) For all $a, b, c, d \in G$, $c + (a \vee b) + d = (c + a + d) \vee (c + b + d)$, $c + (a \wedge b) + d = (c + a + d) \wedge (c + b + d)$.

(2) For all $a, b \in G$, $-(a \vee b) = (-a) \wedge (-b)$, $-(a \wedge b) = (-a) \vee (-b)$.

(3) As a lattice, G is distributive.

(4) For all $a, b \in G$, $a - (a \wedge b) + b = a \vee b$. If G is commutative, then $a + b = (a \wedge b) + (a \vee b)$, for all $a, b \in G$.

(5) If $na \geq 0$ for some positive integer n, then $a \geq 0$.

(6) If $x, y_1, \cdots, y_n$ are positive elements such that $x \leq y_1 + \cdots + y_n$, then $x = x_1 + \cdots + x_n$ for some positive elements $x_1, \cdots, x_n$ with $x_i \leq y_i, i = 1, \cdots, n$.

(7) If $x, y_1, \cdots, y_n$ are positive elements, then $x \wedge (y_1 + \cdots + y_n) \leq (x \wedge y_1) + \cdots + (x \wedge y_n)$.

Proof.   (1) From $a \vee b \geq a, b$, we have $c + (a \vee b) + d \geq (c + a + d), (c + b + d)$, so

$$c + (a \vee b) + d \geq (c + a + d) \vee (c + b + d).$$

On the other hand, $(c + a + d), (c + b + d) \leq (c + a + d) \vee (c + b + d)$ implies

$$a, b \leq -c + (c + a + d) \vee (c + b + d) + (-d),$$

and hence

$$a \vee b \leq -c + (c + a + d) \vee (c + b + d) + (-d).$$

Therefore $c + (a \vee b) + d \leq (c + a + d) \vee (c + b + d)$. We conclude that $c + (a \vee b) + d = (c + a + d) \vee (c + b + d)$. Similarly we have $c + (a \wedge b) + d = (c + a + d) \wedge (c + b + d)$.

(2) We have

$$a, b \leq a \vee b \Rightarrow -(a \vee b) \leq -a, -b \Rightarrow -(a \vee b) \leq -a \wedge -b,$$

and

$$-a \wedge -b \leq -a, -b \Rightarrow a, b \leq -(-a \wedge -b) \Rightarrow a \vee b \leq -(-a \wedge -b),$$

so $-a \wedge -b \le -(a \vee b)$. Therefore $-(a \vee b) = -a \wedge -b$. Similarly $-(a \wedge b) = -a \vee -b$.

(3) For $a, b, c \in G$, we show that $a \wedge (b \vee c) = (a \wedge b) \vee (a \wedge c)$. Let $d = b \vee c$. Then $a \wedge b \le a \wedge d$ implies $0 \le (a \wedge d) - (a \wedge b)$. Since

$$-d + (a \wedge d) = (-d + a) \wedge 0 \le (-b + a) \wedge 0 = -b + (a \wedge b),$$

we have $0 \le (a \wedge d) - (a \wedge b) \le d - b$. Similarly, $0 \le (a \wedge d) - (a \wedge c) \le d - c$. Thus

$$
\begin{aligned}
0 &\le [(a \wedge d) - (a \wedge b)] \wedge [(a \wedge d) - (a \wedge c)] \\
&\le (d - b) \wedge (d - c) \\
&= d + (-b \wedge -c) \\
&= d - d \\
&= 0,
\end{aligned}
$$

so $[(a \wedge d) - (a \wedge b)] \wedge [(a \wedge d) - (a \wedge c)] = 0$. Hence $(a \wedge d) - [(a \wedge b) \vee (a \wedge c)] = 0$, that is, $a \wedge (b \vee c) = (a \wedge b) \vee (a \wedge c)$.

The distributive property $a \vee (b \wedge c) = (a \vee b) \wedge (a \vee c)$ can be proved by replacing each element in $a \wedge (b \vee c) = (a \wedge b) \vee (a \wedge c)$ with its additive inverse.

(4) From (1) and (2),

$$a - (a \wedge b) + b = a + (-a \vee -b) + b = b \vee a = a \vee b,$$

and if G is commutative, it is clear that $a + b = (a \vee b) + (a \wedge b)$.

(5) By (1) and mathematical induction,

$$n(a \wedge 0) = na \wedge (n-1)a \wedge \cdots \wedge a \wedge 0.$$

Since $na \ge 0$, we have $na \wedge 0 = 0$, so

$$n(a \wedge 0) = (n-1)a \wedge \cdots \wedge a \wedge 0 = (n-1)(a \wedge 0).$$

Adding the inverse of $(n-1)(a \wedge 0)$ to both sides, we get $a \wedge 0 = 0$ and hence $a \ge 0$.

(6) Suppose that $x \le y_1 + y_2$. Let $x_1 = x \wedge y_1$ and $x_2 = -x_1 + x$. Then $x = x_1 + x_2$, $0 \le x_1 \le y_1$, and

$$0 \le x_2 = -x_1 + x = (-x \vee -y_1) + x = 0 \vee (-y_1 + x) \le y_2.$$

Generally, $x \le y_1 + \cdots + y_n$ implies $x = x_1 + x_1'$ with $0 \le x_1 \le y_1$ and $0 \le x_1' \le y_2 + \cdots + y_n$ by previous argument. Continuing this process or using mathematical induction, we will arrive at $x = x_1 + \cdots + x_n$ with $0 \le x_i \le y_i$ for $i = 1, \cdots, n$.

(7) By (6) $x \wedge (y_1 + \cdots + y_n) = z_1 + \cdots + z_n$, where $0 \le z_i \le y_i$ for $i = 1, \cdots, n$. Then each $z_i \le z_1 + \cdots + z_n \le x$, so $z_i \le x \wedge y_i$, and hence $x \wedge (y_1 + \cdots + y_n) \le (x \wedge y_1) + \cdots + (x \wedge y_n)$. $\qquad \square$

Two strictly positive elements a, b of an ℓ-group G are called *disjoint* if $a \wedge b = 0$. A subset $\{a_1, \cdots, a_n\}$ of G is called disjoint if each element in it is strictly positive and $a_i \wedge a_j = 0$ for any $i \neq j$.

Theorem 1.6. *Let G be an ℓ-group and $a, b, c, a_1, \cdots, a_n \in G$.*

(1) If a and b are disjoint, and $c \geq 0$, then $a \wedge (b + c) = a \wedge c$.
(2) If $a \wedge b = a \wedge c = 0$, then $a \wedge (b + c) = 0$.
(3) If $\{a_1, \cdots, a_n\}$ is a disjoint set, then $a_1 \vee \cdots \vee a_n = a_1 + \cdots + a_n$. In particular, if $a \wedge b = 0$, then $a + b = a \vee b = b \vee a = b + a$, that is, disjoint elements commute.

Proof.   (1) Since $a + c \geq a$,

$$a \wedge c = a \wedge ((a \wedge b) + c) = a \wedge [(a + c) \wedge (b + c)] = a \wedge (b + c).$$

(2) follows from (1).

(3) By (2), $(a_1 + \cdots + a_{n-1}) \wedge a_n = 0$, so

$$(a_1 + \cdots + a_{n-1}) \vee a_n = a_1 + \cdots + a_{n-1} + a_n$$

by Theorem 1.5(4). Continuing this process or using mathematical induction, we arrive at $a_1 \vee \cdots \vee a_{n-1} \vee a_n = a_1 + \cdots + a_{n-1} + a_n$.   □

Let G be an ℓ-group. For $g \in G$, the *positive part* g^+, the *negative part* g^- and the *absolute value* $|g|$ are defined as follows.

$$g^+ = g \vee 0, \quad g^- = (-g) \vee 0, \quad |g| = g \vee (-g).$$

Since $g + g^- = g + (-g \vee 0) = 0 \vee g = g^+$, $g = g^+ - g^-$.

Theorem 1.7. *Let G be an ℓ-group and $f, g \in G$.*

(1) $|g| = g^+ + g^-$.
(2) $g^+ \wedge g^- = 0$.
(3) If $f \wedge g = 0$, then $f = (f - g)^+$ and $g = (f - g)^-$.
(4) $ng^+ = (ng)^+$, $ng^- = (ng)^-$, and $n|g| = |ng|$ for any positive integer n.
(5) $|f + g| \leq |f| + |g| + |f|$. If G is commutative, then $|f + g| \leq |f| + |g|$.

Proof.   (1) $|g| \geq g, -g$ implies $2|g| \geq 0$. By Theorem 1.5(5), $|g| \geq 0$. Then by Theorem 1.5(1), we have

$$
\begin{aligned}
g^+ + g^- &= (g \vee 0) + (-g \vee 0) \\
&= [(g \vee 0) + (-g)] \vee (g \vee 0) \\
&= 0 \vee (-g) \vee g \vee 0 \\
&= 0 \vee |g| \\
&= |g|.
\end{aligned}
$$

(2) Since G is a distributive lattice,

$$g^+ \wedge g^- = (g \vee 0) \wedge (-g \vee 0) = (g \wedge -g) \vee 0 = -|g| \vee 0 = 0,$$

by Theorem 1.5(2) and (3).

(3) If $f \wedge g = 0$, then $-f \vee -g = 0$, and

$$f = f + 0 = f + (-f \vee -g) = 0 \vee (f - g) = (f - g)^+$$

and $g = (g - f)^+ = (f - g)^-$.

(4) By (2) and Theorem 1.6(2), $ng^+ \wedge ng^- = 0$ (Exercise 7). Since disjoint elements commute, $-g^- + g^+ = g^+ - g^-$, so

$$(ng)^+ = (n(g^+ - g^-))^+ = (ng^+ - ng^-)^+ = ng^+$$

by (3). Then $ng^- = n(-g)^+ = (-ng)^+ = (ng)^-$, and

$$n|g| = n(g^+ + g^-) = ng^+ + ng^- = (ng)^+ + (ng)^- = |ng|.$$

(5) Since $|f|, |g| \geq 0$, $f + g \leq |f| + |g| \leq |f| + |g| + |f|$ and

$$-(f + g) = (-g) + (-f) \leq |g| + |f| \leq |f| + |g| + |f|,$$

so

$$|f + g| = (f + g) \vee -(f + g) \leq |f| + |g| + |f|.$$

From the above argument, if G is commutative, then $|f + g| \leq |f| + |g|$. $\square$

A subset C of an ℓ-group G is called *convex* if for all $g \in G$ and $c, d \in C$, $c \leq g \leq d$ implies $g \in C$. A convex ℓ-subgroup of G is a subgroup of G which is convex and a sublattice of G. Clearly G and $\{0\}$ are convex ℓ-subgroups of G, and the intersection of a family of convex ℓ-subgroups of G is a convex ℓ-subgroup of G. For a subset X of G, the intersection of all convex ℓ-subgroups containing X is the smallest convex ℓ-subgroup that contains X, which is called the convex ℓ-subgroup *generated* by X and denoted by $C_G(X)$ or just $C(X)$.

One method of constructing convex ℓ-subgroups is by using a polar that is defined as follows. For a subset X of an ℓ-group G, the *polar* of X is

$$X^\perp = \{a \in G \mid |a| \wedge |x| = 0, \forall x \in X\}$$

and the *double polar* of X is $X^{\perp\perp} = (X^\perp)^\perp$. Clearly $X \subseteq X^{\perp\perp}$ and $X^{\perp\perp\perp} = X^\perp$ (Exercise 8). If $X = \{x\}$, then $X^\perp$ and $X^{\perp\perp}$ are denoted by $x^\perp$ and $x^{\perp\perp}$.

Theorem 1.8. *Let G be an ℓ-group.*

(1) A subgroup H of G is a convex ℓ-subgroup of G if and only if for any $a \in H, x \in G$, $|x| \leq |a|$ implies $x \in H$.

(2) For each subset X of G, $X^{\perp}$ is a convex ℓ-subgroup of G.

(3) $C(X) = \{g \in G \mid |g| \leq |x_1| + \cdots + |x_n| \text{ for some } x_1, \cdots, x_n \in X\}$.

(4) The subgroup of G generated by a family of convex ℓ-subgroups is a convex ℓ-subgroup of G.

Proof. (1) Suppose that H is a convex ℓ-subgroup of G and $|x| \leq |a|$ for some $a \in H$ and $x \in G$. Since H is a sublattice of G, $a \in H$ implies $a^+, a^- \in H$, and hence $|a| = a^+ + a^- \in H$. Then that H is convex implies $|x| \in H$, so $x^+, x^- \in H$ by the convexity of H again. Hence $x = x^+ - x^- \in H$.

Conversely, let H be a subgroup with the given property. Let $a, b \in H$ and $x \in G$ such that $a \leq x \leq b$. Then $0 \leq x - a \leq b - a \in H$, so $x - a \in H$, and $x = (x - a) + a \in H$. Thus H is convex. Let $a, b \in H$. Then $(b - a)^+ \leq |b - a|$ implies $(b - a)^+ \in H$, so $a \vee b = (b - a)^+ + a \in H$. Similarly $a \wedge b \in H$. Therefore H is a sublattice of G, and hence H is a convex ℓ-subgroup of G.

(2) Let $a, b \in X^{\perp}$. By Theorem 1.7(5) and Theorem 1.6(2), for any $x \in X$,

$$|a - b| \wedge |x| \leq (|a| + |-b| + |a|) \wedge |x| = (|a| + |b| + |a|) \wedge |x| = 0,$$

so $|a - b| \wedge |x| = 0$. Thus $a - b \in X^{\perp}$, that is, $X^{\perp}$ is a subgroup of G. Then it is clear that $X^{\perp}$ is a convex ℓ-subgroup by (1).

(3) Let

$$H = \{g \in G \mid |g| \leq |x_1| + \cdots + |x_n| \text{ for some } x_1, \cdots, x_n \in X\}$$

and $a, b \in H$. By Theorem 1.7(5) again, $|a - b| \leq |a| + |b| + |a|$, so $a - b \in H$, that is, H is a subgroup of G. Then by (1), H is a convex ℓ-subgroup of G. Clearly $X \subseteq H$ and any convex ℓ-subgroup of G containing X contains H. Hence $C(X) = H$.

(4) Let $\{C_i \mid i \in I\}$ be a family of convex ℓ-subgroups of G and C be the subgroup of G generated by $\{C_i \mid i \in I\}$. Suppose that $|g| \leq |c|$ for some $g \in G$ and $c \in C$. Let $c = \sum_{j=1}^{n} c_j$ with $c_j \in \cup\{C_i \mid i \in I\}$. Then by Theorem 1.7(5), $|g|$ is less than or equal to a sum of elements from $\cup\{C_i^+ \mid i \in I\}$, so since $g^+, g^- \leq |g|$, by Theorems 1.5(6), g^+, g^- can be written as a sum of elements from $\cup\{C_i^+ \mid i \in I\}$, so $g = g^+ - g^- \in C$. Thus by (1), C is a convex ℓ-subgroup of G. □

For an ℓ-group G, we use $\mathcal{C}(G)$ to denote the set of all convex ℓ-subgroups of G and partially order $\mathcal{C}(G)$ by set inclusion. It is well known and not hard to show that the set of all subgroups of a group is a lattice under set inclusion. For any two subgroups A and B, $A \wedge B = A \cap B$ and $A \vee B$ is the subgroup generated by $A \cup B$.

Theorem 1.9. *Let G be an ℓ-group. $C(G)$ is a complete distributive sublattice of the lattice of subgroups of G. Moreover, if A, $\{A_i \mid i \in I\}$ are convex ℓ-subgroups of G, then $A \cap (\vee_{i \in I} A_i) = \vee_{i \in I}(A \cap A_i)$.*

Proof. The intersection of a family of convex ℓ-subgroups is a convex ℓ-subgroup, and by Theorem 1.8(4), the subgroup generated by a family of convex ℓ-subgroups is also a convex ℓ-subgroup, so $C(G)$ is a complete sublattice of the lattice consisting of all subgroups of G.

Suppose that $A, A_1, A_2 \in C(G)$. We show that

$$A \vee (A_1 \cap A_2) = (A \vee A_1) \cap (A \vee A_2).$$

Let C, C_1, and C_2 be the subgroup generated by $A \cup (A_1 \cap A_2)$, $A \cup A_1$, and $A \cup A_2$ respectively. Since

$$A \cup (A_1 \cap A_2) = (A \cup A_1) \cap (A \cup A_2),$$

$C \subseteq C_1 \cap C_2$. Let $g \in C_1 \cap C_2$. By Theorem 1.8(3), we have

$$|g| \leq |x_1| + \cdots + |x_n| \text{ and } |g| \leq |y_1| + \cdots + |y_m|$$

for some $x_i \in A \cup A_1$ and $y_j \in A \cup A_2$. Then

$$|g| \leq (|x_1| \wedge |y_1| + \cdots + |x_1| \wedge |y_m|) + \cdots + (|x_n| \wedge |y_1| + \cdots + |x_n| \wedge |y_m|)$$

by Theorem 1.5(7). If $x_i \in A$ or $y_j \in A$, then $|x_i| \wedge |y_j| \in A$. Otherwise $x_i \in A_1$ and $y_j \in A_2$ implies $|x_i| \wedge |y_j| \in A_1 \cap A_2$. Thus each term $|x_i| \wedge |y_j| \in A \cup (A_1 \cap A_2)$, so $g \in C$ by Theorem 1.8(3). Hence $C_1 \cap C_2 \subseteq C$. Therefore $C = C_1 \cap C_2$, that is, $A \vee (A_1 \cap A_2) = (A \vee A_1) \cap (A \vee A_2)$.

Finally it is clear that $\vee_{i \in I}(A \cap A_i) \subseteq A \cap (\vee_{i \in I} A_i)$. If $g \in A \cap (\vee_{i \in I} A_i)$, then $|g| \leq |c_1| + \cdots + |c_n|$ with $c_k \in \cup_{i \in I} A_i$. By Theorem 1.5(6), $|g| = g_1 + \cdots + g_n$ with $0 \leq g_k \leq |c_k|$ for $k = 1, \cdots, n$. Then each $g_k \leq |g|$, so $g_k \leq |g| \wedge |c_k| \in A \cap A_{i_k}$ for some A_{i_k} implies $g_k \in A \cap A_{i_k}$. It follows that $g \in \vee_{i \in I}(A \cap A_i)$, and hence we also have $A \cap (\vee_{i \in I} A_i) \subseteq \vee_{i \in I}(A \cap A_i)$. Therefore $A \cap (\vee_{i \in I} A_i) = \vee_{i \in I}(A \cap A_i)$. $\square$

Let G be an ℓ-group and $\{C_i \mid i \in I\}$ be a family of convex ℓ-subgroups of G. G is call a *direct sum* of $\{C_i \mid i \in I\}$, denoted by $G = \oplus_{i \in I} C_i$, if G is generated by $\{C_i \mid i \in I\}$ and $C_i \cap C_j = \{0\}$ for any $i, j \in I$ with $i \neq j$.

Theorem 1.10. *Let G be an ℓ-group. Suppose that G is a direct sum of a family of convex ℓ-subgroups $\{C_i \mid i \in I\}$.*

(1) If $c_1 + \cdots + c_n = 0$, where $c_i \in C_{k_i}$ and $k_1, \cdots, k_n$ are distinct, then each $c_i = 0$.

(2) Each element $0 \neq a \in G$ can be uniquely written as $a = c_1 + \cdots + c_n$ with $0 \neq c_i \in C_{k_i}$ and $k_1, \cdots, k_n$ are distinct. Moreover $a \geq 0$ if and only if each $c_i \geq 0$.

Proof. (1) If $c_1 + \cdots + c_n = 0$, then $c_1 = -c_n - \cdots - c_2$ implies that
$$c_1 \in C_{k_1} \cap (C_{k_2} \vee \cdots \vee C_{k_n}).$$
By Theorem 1.9,
$$c_1 \in (C_{k_1} \cap C_{k_2}) \vee \cdots \vee (C_{k_1} \cap C_{k_n}) = \{0\}.$$
Thus $c_1 = 0$. Similarly $c_2 = \cdots = c_n = 0$.

(2) For $0 \leq a \in C_i$ and $0 \leq b \in C_j$ with $i \neq j$, since $C_i \cap C_j = \{0\}$, $a \wedge b = 0$, so $a + b = b + a$ by Theorem 1.6(3). Thus elements in C_i and C_j commute (Exercise 9). It follows then that each $a \in G$ with $a \neq 0$ can be written as $a = c_1 + \cdots + c_n$ with $0 \neq c_i \in C_{k_i}$ and $k_1, \cdots, k_n$ are distinct. The uniqueness follows from (1).

Clearly if each $c_i \geq 0$, then $a \geq 0$. Suppose that $a = c_1 + \cdots + c_n \geq 0$. Then $-c_1 \leq c_2 + \cdots + c_n \in C_{k_2} \vee \cdots \vee C_{k_n}$ implies $(-c_1)^+ \leq (c_2 + \cdots + c_n)^+ \in C_{k_2} \vee \cdots \vee C_{k_n}$, so $(-c_1)^+ \in C_{k_2} \vee \cdots \vee C_{k_n}$. Then by Theorem 1.9, we have
$$(-c_1)^+ \in (C_{k_1} \cap C_{k_2}) \vee \cdots \vee (C_{k_1} \cap C_{k_n}) = \{0\},$$
so $(-c_1)^+ = 0$, and hence $c_1 \geq 0$. Similarly $c_2 \geq 0, \cdots, c_n \geq 0$. $\square$

Let G be an ℓ-group and N be a normal convex ℓ-subgroup of G. Define the relation on the quotient group G/N by
$$x + N \leq y + N \text{ if } x \leq y + z \text{ for some } z \in N.$$
The relation is well-defined since if $x_1 + N = x + N$ and $y_1 + N = y + N$, then $x = x_1 + c$ and $y = y_1 + d$ for some $c, d \in N$, so $x = x_1 + c \leq y + z = y_1 + (d + z)$ implies $x_1 \leq y_1 + (d + z - c)$ with $d + z - c \in N$. Thus $x_1 + N \leq y_1 + N$.

It is clear that the relation defined above is reflexive and transitive. Suppose that $x + N \leq y + N$ and $y + N \leq x + N$ for some $x, y \in G$. Then $x \leq y + z$ and $y \leq x + w$ for some $z, w \in N$, so $-y + x \leq z \in N$ and $-x + y \leq w$ implies that
$$|-y + x| = (-y + x) \vee (-x + y) \leq z \vee w \in N.$$
It follows that $-y + x \in N$, and hence $x + N = y + N$, that is, the relation is also antisymmetric. Therefore it is a partial order on G/N.

Theorem 1.11. G/N *is an ℓ-group with respect to the partial order defined above.*

Proof. Suppose that $x + N \leq y + N$ and $z + N \in G/N$. Then $x \leq y + a$ for some $a \in N$. Since $z + x \leq (z + y) + a$,

$$(z + N) + (x + N) \leq (z + N) + (y + N),$$

and since

$$x + z \leq y + a + z = (y + z) + (-z + a + z)$$

with $-z + a + z \in N$,

$$(x + N) + (z + N) \leq (y + N) + (z + N).$$

Hence G/N is a partially ordered group.

We show next that

$$(x + N) \vee (y + N) = (x \vee y) + N \text{ and } (x + N) \wedge (y + N) = (x \wedge y) + N$$

for any $x, y \in G$. Clearly $x + N, y + N \leq (x \vee y) + N$. Let $x + N, y + N \leq z + N$ for some $z \in G$. Then $x \leq z + a$ and $y \leq z + b$ for some $a, b \in N$, so $-z + x \leq a$ and $-z + y \leq b$ and $(-z + x) \vee (-z + y) \leq a \vee b \in N$. Then it follows that $x \vee y \leq z + (a \vee b)$, and hence $(x \vee y) + N \leq z + N$. Therefore $(x + N) \vee (y + N) = (x \vee y) + N$. Similarly $(x + N) \wedge (y + N) = (x \wedge y) + N$. Hence G/N is an ℓ-group. □

The ℓ-group G/N with the lattice order defined above is called the *quotient ℓ-group* of G by N.

Let G and H be ℓ-groups. A group homomorphism $f : G \to H$ is called an *ℓ-homomorphism* if f also preserves sup and inf, namely, for any $a, b \in G$,

$$f(a \vee b) = f(a) \vee f(b) \text{ and } f(a \wedge b) = f(a) \wedge f(b).$$

For example, for an ℓ-group G and a normal convex ℓ-subgroup N, it is easy to check that the group homomorphism $\varphi : G \to G/N$ defined by $\varphi(a) = a + N$ is an ℓ-homomorphism called the *projection* (Exercise 11). An *ℓ-isomorphism* is a group isomorphism that preserves sup and inf. If there exists an ℓ-isomorphism between two ℓ-groups G and H, then they are called *ℓ-isomorphic* and denoted by $G \cong H$.

Theorem 1.12. *Let G and H be ℓ-groups and $f : G \to H$ be a group homomorphism. Then f is an ℓ-homomorphism if and only if $x \wedge y = 0$ $(x \vee y = 0) \Rightarrow f(x) \wedge f(y) = 0$ $(f(x) \vee f(y) = 0)$ for all $x, y \in G$.*

Proof.   Suppose that $x \wedge y = 0$ implies $f(x) \wedge f(y) = 0$ for all $x, y \in G$. Let $a, b \in G$ and $a \wedge b = c$. Then $(a - c) \wedge (b - c) = 0$ implies

$$f(a - c) \wedge f(b - c) = (f(a) - f(c)) \wedge (f(b) - f(c)) = 0,$$

so $f(a) \wedge f(b) = f(c)$. We also have that

$$
\begin{aligned}
f(a \vee b) &= f(-(-a \wedge -b)) \\
&= -(f(-a \wedge -b)) \\
&= -(-f(a) \wedge -f(b)) \\
&= f(a) \vee f(b).
\end{aligned}
$$

$\square$

A *totally ordered field* is a field whose additive group is a totally ordered group and product of two positive elements is still positive. For instance, the field $\mathbb{Q}$ of all rational numbers and the field $\mathbb{R}$ of all real numbers are both totally ordered fields with respect to usual order between real numbers. Let F be a totally ordered field and $a \in F$. Then either $a \geq 0$ or $a < 0$, so $a^2 \geq 0$ in either case. Thus the identity element 1 is positive since $1 = 1^2$. A consequence of this simple fact is that the field $\mathbb{C}$ of all complex numbers cannot be made into a totally ordered field since $i^2 = -1$, where $i = \sqrt{-1}$ is the imaginary unit.

Let F be a totally ordered field and V be a left (right) vector space over F. V is called a *vector lattice* over F if V is an ℓ-group and for all $\alpha \in F^+$ and $v \in V^+$, $\alpha v \in V^+$ ($v\alpha \in V^+$). We note that the addition on V is commutative. In case that $F = \mathbb{R}$, a vector lattice is usually called a *Rieze space*. A *convex vector sublattice* W of V is a subspace of V and a convex ℓ-subgroup of V. An element $\alpha \in F^+$ is called an *f-element* on V if $v \wedge u = 0 \Rightarrow \alpha v \wedge u = 0$ for all $v, u \in V$. More generally, for a unital totally ordered ring T and a left (right) module M over T, M is called an *ℓ-module* if its additive group is an ℓ-group and for any $\alpha \in T^+, x \in M^+$, $\alpha x \in M^+$ ($x\alpha \in M^+$). An ℓ-module is called an *f-module* if each element in T^+ is an f-element on M.

Theorem 1.13. *Let V be a vector lattice over a totally ordered field F.*

(1) Each positive element of F is an f-element on V, that is, V is an f-module over F. Thus any polar is a convex vector sublattice.

(2) Suppose that $v_1, \cdots, v_k \in V$ are disjoint. Then for any $\alpha_1, \cdots, \alpha_k \in F$, $\alpha_1 v_1 + \cdots + \alpha_k v_k \geq 0$ if and only if each $\alpha_i \geq 0$.

(3) Any disjoint subset of V is linear independent over F.

Proof. (1) For any $\beta \in F$ with $\beta > 0$, since F is totally ordered and $1 > 0$, $\beta^{-1} > 0$. Suppose that $0 < \alpha \in F$. If $v \wedge u = 0$, for $v, u \in V$, then

$$0 \leq (\alpha + 1)^{-1}(\alpha v \wedge u) \leq (\alpha + 1)^{-1}((\alpha + 1)v \wedge (\alpha + 1)u) \leq v \wedge u = 0,$$

and hence $(\alpha + 1)^{-1}(\alpha v \wedge u) = 0$ and $\alpha v \wedge u = 0$. Therefore each positive element of F is an f-element on V. For $X \subseteq V$, we already know that $X^{\perp}$ is a convex ℓ-subgroup. $X^{\perp}$ is also a subspace of V over F since $\forall \alpha \in F, v \in V$, $|\alpha v| = |\alpha||v|$ (Exercise 17) and V is an f-module over F.

(2) If each $\alpha_j \geq 0$, then $\alpha_1 v_1 + \cdots + \alpha_k v_k \geq 0$. Conversely suppose that $\alpha_1 v_1 + \cdots + \alpha_k v_k \geq 0$ and suppose that $\alpha_1 < 0, \ldots, \alpha_n < 0$, and $\alpha_{n+1} \geq 0, \ldots, \alpha_k \geq 0$, where $1 \leq n < k$. Then $-\alpha_1 v_1 - \ldots - \alpha_n v_n \leq \alpha_{n+1} v_{n+1} + \ldots + \alpha_k v_k$, and hence by Theorem 1.5(7), we have

$$\begin{aligned}
-\alpha_1 v_1 &= (-\alpha_1 v_1) \wedge (-\alpha_1 v_1 - \ldots - \alpha_n v_n) \\
&\leq (-\alpha_1 v_1) \wedge (\alpha_{n+1} v_{n+1} + \ldots + \alpha_k v_k) \\
&\leq (-\alpha_1 v_1 \wedge \alpha_{n+1} v_{n+1}) + \ldots + (-\alpha_1 v_1 \wedge \alpha_k v_k) \\
&= 0
\end{aligned}$$

by (1) since $v_1 \wedge v_{n+1} = \ldots = v_1 \wedge v_k = 0$. Thus $-\alpha_1 v_1 = 0$, so $\alpha_1 = 0$, which is a contradiction. Thus each $\alpha_j \geq 0$.

(3) This is a direct consequence of (2). $\qquad \square$

1.2.2 *Structure theorems of ℓ-groups and vector lattices*

In this section, we prove some algebraic structure theorems for ℓ-groups and vector lattices that contain basic elements. This theory was initially developed by P. Conrad and it plays important roles in study of ℓ-groups.

Let G be an ℓ-group. An element $0 < a \in G$ is called a *basic* element if for any $c, d \in G^+$, $c, d \leq a$ implies c and d are comparable, that is, either $c \geq d$ or $c \leq d$. A nonzero polar is called a *minimal polar* if it does not contain any nonzero polar.

Theorem 1.14. *Let G be an ℓ-group.*

(1) For $0 < a \in G$, a is basic if and only if $a^{\perp\perp}$ is totally ordered.

(2) Let a, b be basic elements. Then either $a \wedge b = 0$ or $a^{\perp\perp} = b^{\perp\perp}$, and $a^{\perp\perp} = b^{\perp\perp}$ if and only if a and b are comparable.

(3) For $0 < a \in G$, a is a basic element if and only if for any $0 < b \leq a$, $b^{\perp\perp} = a^{\perp\perp}$.

(4) For $0 < a \in G$, a is a basic element if and only if $a^{\perp\perp}$ is a minimal polar.

Proof. We first note that for any $x \in G$, $x^{\perp\perp}$ is a convex ℓ-subgroup by Theorem 1.8(2).

(1) Suppose that a is basic. Let $x, y \in a^{\perp\perp}$ and $x \wedge y = 0$. Then $(a \wedge x) \wedge (a \wedge y) = 0$ implies that $a \wedge x = 0$ or $a \wedge y = 0$ since $a \wedge x$ and $a \wedge y$ are comparable. Thus $x \in a^{\perp}$ or $y \in a^{\perp}$, so $x = x \wedge x = 0$ or $y = y \wedge y = 0$. Hence $a^{\perp\perp}$ is totally ordered (Exercise 12). Conversely, suppose that $a^{\perp\perp}$ is totally ordered and let $0 \leq x, y \leq a$. Then $x, y \in a^{\perp\perp}$ implies that x and y are comparable, so a is basic.

(2) Suppose that $a \wedge b \neq 0$. Let $0 \leq x \in a^{\perp\perp}$. Take $0 < y \in b^{\perp}$. Then $y \wedge b = 0$ implies that $(x \wedge y) \wedge (a \wedge b) = 0$. Since $x \wedge y$ and $a \wedge b$ are both in $a^{\perp\perp}$ that is totally ordered by (1), we must have $x \wedge y = 0$ or $a \wedge b = 0$. Since $a \wedge b \neq 0$, $x \wedge y = 0$, so $x \in b^{\perp\perp}$ and $a^{\perp\perp} \subseteq b^{\perp\perp}$. Similarly $b^{\perp\perp} \subseteq a^{\perp\perp}$. Therefore, $a^{\perp\perp} = b^{\perp\perp}$.

If $a^{\perp\perp} = b^{\perp\perp}$, then $a, b \in a^{\perp\perp}$ implies that a and b are comparable by (1). Conversely, if a and b are comparable, then $a \wedge b \neq 0$, so $a^{\perp\perp} = b^{\perp\perp}$.

(3) If a is basic and $0 < b \leq a$, then b is also basic, so $b^{\perp\perp} = a^{\perp\perp}$. Conversely, suppose that the condition is true and $0 < x, y \leq a$. Let $z = x - x \wedge y$ and $w = y - x \wedge y$. Then $z \wedge w = 0$ and $z, w \in a^{\perp\perp}$. Suppose that $z \neq 0$. Then $0 < z \leq a$ implies that $z^{\perp\perp} = a^{\perp\perp}$, so $w \in z^{\perp\perp}$. On the other hand, $z \wedge w = 0$ implies that $w \in z^{\perp}$, and hence $w \wedge w = 0$, so $w = 0$. Thus $y \leq x$. Similarly, if $w \neq 0$, then $x \leq y$. Therefore a is basic.

(4) Suppose that $a^{\perp\perp}$ is a minimal polar and $0 < b \leq a$. Then $0 \neq b^{\perp\perp} \subseteq a^{\perp\perp}$, so $b^{\perp\perp} = a^{\perp\perp}$. Therefore a is basic by (3). Now suppose that a is basic and $\{0\} \neq X^{\perp} \subseteq a^{\perp\perp}$ for some $X \subseteq G$. Take $0 < x \in X^{\perp}$. Then $x \in a^{\perp\perp}$ implies that x is also basic. Thus $x^{\perp\perp} = a^{\perp\perp}$ by (2). Hence $a^{\perp\perp} = x^{\perp\perp} \subseteq X^{\perp\perp\perp} = X^{\perp}$, so $X^{\perp} = a^{\perp\perp}$. Therefore $a^{\perp\perp}$ is a minimal polar. $\square$

Corollary 1.1. *Let G be an ℓ-group and $a \in G$. If a is a basic element, then $a^{\perp\perp}$ is a maximal convex totally ordered subgroup in the sense that for any convex totally ordered subgroup M of G if $a^{\perp\perp} \subseteq M$, then $M = a^{\perp\perp}$.*

Proof. Suppose that M is a convex totally ordered subgroup containing $a^{\perp\perp}$ and $0 < g \in M$. Since M is convex and totally ordered, g is basic and hence $a, g \in M$ implies that $a^{\perp\perp} = g^{\perp\perp}$ by Theorem 1.14(2). Thus $g \in a^{\perp\perp}, \forall g \in M^{+}$, so $M = a^{\perp\perp}$. $\square$

Let G be an ℓ-group. A subset S of G is called a *basis* if

(i) each element in S is basic, and
(ii) S is a maximal disjoint set of G.

Equivalently a subset S of G is a basis if S is a disjoint set of basic elements with $S^\perp = \{0\}$ (Exercise 13). In this book, terminology basis means the basis defined above. For the basis of a vector space we always call it a vector space basis.

Theorem 1.15. *Let G be an ℓ-group.*

(1) G has a basis if and only if G satisfies

> *(∗) each $0 < g \in G$ is greater than or equal to at least one basic element.*

(2) If G satisfies the following condition (C), then G has a basis.

> *(C) Each $0 < g \in G$ is greater than at most a finite number of disjoint elements.*

Proof. (1) If $G = \{0\}$, then the result is trivially true. Let $G \neq \{0\}$. Suppose that S is a basis for G. Then $S \neq \emptyset$. For $0 < g \in G \setminus S$, there is an $a \in S$ such that $a \wedge g > 0$ since $S^\perp = \{0\}$. Then $a \wedge g$ is basic since a is basic, and $a \wedge g \leq g$. Conversely suppose that G satisfies (∗). Let

$$\mathcal{M} = \{A \mid A \text{ is a disjoint set of basic elements of } G\}.$$

Clearly $\mathcal{M} \neq \emptyset$ since if a is a basic element of G, then $\{a\} \in \mathcal{M}$. $\mathcal{M}$ is a partially ordered set with respect to set inclusion. Let $\{A_i \mid i \in I\}$ be a chain in $\mathcal{M}$, then it is easy to check that $\cup_{i \in I} A_i \in \mathcal{M}$. So by Zorn's Lemma, $\mathcal{M}$ has a maximal element, say S. We show that S is a basis. To this end, we just need to show that $S^\perp = \{0\}$. Suppose that $0 < g \in S^\perp$ and $g \geq b$ for some basic element b. Then $b \in S^\perp$, so $S \subsetneq S \cup \{b\}$ and $S \cup \{b\}$ is disjoint, which contradicts with the fact that S is maximal in $\mathcal{M}$. Hence $S^\perp = \{0\}$ and S is a basis of G.

(2) We show that (∗) in (1) is satisfied, so G has a basis. For $0 < g \in G$, consider $T = \{x \in G \mid 0 < x \leq g\}$. If T contains no disjoint elements, then T is totally ordered (Exercise 14), so g is basic. Suppose that T contains n disjoint elements $x_1, \cdots, x_n$ and any $n+1$ elements in T are not disjoint for some positive integer n. We claim that each x_i is a basic element for $i = 1, \cdots, n$. Suppose that $0 \leq y, z \leq x_i$ and y, z are not comparable. Let $y \wedge z = w$. Then $(y-w) \wedge (z-w) = 0$ with $(y-w) > 0$ and $(z-w) > 0$, and hence the set $\{(y-w), (z-w), x_1, \cdots, x_{i-1}, x_{i+1}, \cdots, x_n\} \subseteq T$ is disjoint since $(y-w), (z-w) \leq x_i$, which is a contradiction. Thus y, z must be comparable, so x_i is basic. Therefore each $0 < g \in G$ is greater than or equal to a basic element. $\square$

For finite-dimensional vector lattices, condition (C) in Theorem 1.15 is satisfied.

Corollary 1.2. *Let V be a vector lattice over a totally ordered field F. If V is finite-dimensional over F as a vector space, then V satisfies (C) in Theorem 1.15, and hence V has a basis.*

Proof. By Theorem 1.13(3), any disjoint subset of V is linearly independent over F, then that V is finite-dimensional over F implies that V contains at most a finite number of disjoint elements, so condition (C) in Theorem 1.15 is satisfied. □

A vector lattice V over a totally ordered field F is called *Archimedean over F* if for $a, b \in V^+$, $\alpha a \leq b$ for all $\alpha \in F^+$ implies that $a = 0$. Certainly if V is Archimedean, then V is Archimedean over F (Exercise 15). However if F is not a totally ordered Archimedean field, then the fact that V is Archimedean over F may not imply that V is Archimedean. For instance, any totally ordered field that is not Archimedean is an Archimedean vector lattice over itself.

Theorem 1.16. *Let G be an ℓ-group.*

(1) If A and B are convex totally ordered subgroups of G, then $A \subseteq B$ or $A \supseteq B$ or $A \cap B = \{0\}$.

(2) If A and B are maximal convex totally ordered subgroups of G, then either $A = B$ or $A \cap B = \{0\}$.

Proof. (1) Suppose $A \nsubseteq B$ and $A \nsupseteq B$. Then there exist $0 < a \in A \setminus B$ and $0 < b \in B \setminus A$. Since A and B are convex, $a \wedge b \in A \cap B$. Let $0 \leq c \in A \cap B$. Since $c, a \in A$, c and a are comparable, so it follows from $a \notin B$ that $c \leq a$. Similarly, $c \leq b$. Thus $c \leq a \wedge b$ for any $0 \leq c \in A \cap B$. Take $c = 2(a \wedge b) \in A \cap B$. Then $2(a \wedge b) \leq (a \wedge b)$ implies that $a \wedge b = 0$, so for any $0 \leq c \in A \cap B$, $c = 0$. Therefore $A \cap B = \{0\}$.

(2) If A and B are maximal convex totally ordered subgroups, then $A \subseteq B$ or $A \supseteq B$ implies that $A = B$. Thus by (1), we have either $A = B$ or $A \cap B = \{0\}$. □

We note that Theorem 1.16 is true for convex totally ordered subspaces of a vector lattice over a totally ordered field.

The following Theorem 1.17 is the structure theorem of vector lattices we need when we consider the structure of a class of ℓ-algebras in chapter 2, and the result is actually true for ℓ-groups.

For a vector lattice V over a totally ordered field F, let $\{V_i \mid i \in I\}$ be a family of convex vector sublattices of V over F. Define

$$\sum_{i \in I} V_i = \{v_1 + \cdots + v_k \mid v_j \in V_{k_j}\}.$$

We leave it as an exercise to verify that $\sum_{i \in I} V_i$ is the convex vector sublattice of V over F generated by $\{V_i \mid i \in I\}$ (Exercise 18). The sum $\sum_{i \in I} V_i$ is called a *direct sum*, denoted by $\oplus_{i \in I} V_i$, if $V_i \cap V_j = \{0\}, \forall i \neq j$.

We have used the same symbol $\oplus_{i \in I}$ to denote the direct sum of convex ℓ-subgroups of an ℓ-group before. Later we will also use it to denote the direct sum of ℓ-ideals of an ℓ-ring. The reader should be able to tell the meaning of the symbol from context without confusion.

Theorem 1.17. *Let V be a vector lattice over a totally ordered field F. If V satisfies condition (C) in Theorem 1.15 and no maximal convex totally ordered subspace of V is bounded above, then V is a direct sum of maximal convex totally ordered subspaces over F.*

Proof. By Theorem 1.15(2), V has a basis S. For each $s \in S$, $s^{\perp\perp}$ is a maximal convex totally ordered subspace of V by Theorem 1.14(1) and Corollary 1.1. We show that V is a direct sum of $s^{\perp\perp}$, $s \in S$. Since for $s, t \in S$, if $s \neq t$, then $s^{\perp\perp} \cap t^{\perp\perp} = \{0\}$ by Theorems 1.14(2) and 1.16(2), we just need to show that V is a sum of $s^{\perp\perp}$, $s \in S$.

Let $0 < a \in V$. By condition (C), we may assume that there are k disjoint basic elements $v_1, \cdots, v_k$ less than or equal to a for some positive integer k, and a is not greater than or equal to $k+1$ disjoint basic elements. Since $S^{\perp} = \{0\}$, for each $i = 1, \cdots, k$, there exists an $s_i \in S$ such that $v_i \wedge s_i \neq 0$. We show that $a \in s_1^{\perp\perp} + \cdots + s_k^{\perp\perp}$. For $i = 1, \cdots, k$, each $s_i^{\perp\perp}$ is a maximal convex totally ordered subspace of V, and hence there exists $0 < x \in s_1^{\perp\perp}$ such that $x \not\leq a$ since $s_1^{\perp\perp}$ is not bounded above. Let $a \wedge x = a_1$. Then $(a - a_1) \wedge (x - a_1) = 0$ and $0 < x - a_1 \in s_1^{\perp\perp}$, so $((a - a_1) \wedge s_1) \wedge (x - a_1) = 0$ implies $(a - a_1) \wedge s_1 = 0$ since $(a - a_1) \wedge s_1 \in s_1^{\perp\perp}$ that is totally ordered. Let $a - a_1 = a_1'$. Then $a = a_1 + a_1'$ with $a_1 \in s_1^{\perp\perp}$ and $a_1' \wedge s_1 = 0$. Again there exists $0 < y \in s_2^{\perp\perp}$ such that $y \not\leq a_1'$. Suppose that $a_1' \wedge y = a_2$. Then $(a_1' - a_2) \wedge (y - a_2) = 0$ with $y - a_2 > 0$, so similarly $(a_1' - a_2) \wedge s_2 = 0$. Let $a_1' - a_2 = a_2'$. We have $a_1' = a_2 + a_2'$ with $a_2 \in s_2^{\perp\perp}$ and $a_2' \wedge s_2 = 0$. Hence $a = a_1 + a_2 + a_2'$ with $a_2' \wedge s_1 = 0$ and $a_2' \wedge s_2 = 0$ since $a_2' \leq a_1'$. Continuing this progress, we have $a = a_1 + a_2 + \cdots + a_k + a_k'$ with $a_k' \wedge s_1 = \cdots = a_k' \wedge s_k = 0$. If $a_k' > 0$, then there exists an element $t \in S$ such that $a_k' \wedge t \neq 0$, so $a_k' \wedge t \leq a$ is a basic element. Thus $(a_k' \wedge t) \wedge v_j \neq 0$

for some $1 \leq j \leq k$. By Thoerem 1.14, $(a'_k \wedge t)^{\perp\perp} = v_j^{\perp\perp} = s_j^{\perp\perp}$, and hence $a'_k \wedge t$ and s_j are comparable, which is a contradiction since $a'_k \wedge s_j = 0$. Hence $a'_k = 0$ and $a = a_1 + a_2 + \cdots + a_k \in s_1^{\perp\perp} + \cdots + s_k^{\perp\perp}$. This completes the proof. $\qquad\square$

1.3 Lattice-ordered rings and algebras

In this section, we introduce lattice-ordered rings, provide examples, and prove basic properties of them. All rings are associative, and a ring may not have the identity element with respect to its multiplication.

1.3.1 *Definitions, examples, and basic properties*

A *partially ordered ring* is a ring R whose additive group is a partially ordered group and for any $a, b \in R$, if $a \geq 0$ and $b \geq 0$ then $ab \geq 0$. The positive cone of a partially ordered ring R is the positive cone of its additive partially ordered group: $R^+ = \{r \in R \mid r \geq 0\}$. The following result is the ring analogue of Theorems 1.2 and 1.3. We leave the proof as an exercise (Exercise 19).

Theorem 1.18. *Let R be a partially ordered ring with positive cone $P = R^+$. Then*

(1) $P + P \subseteq P$,
(2) $PP \subseteq P$,
(3) $P \cap -P = \{0\}$.

Conversely, if R is a ring and P is a subset that satisfies the above three conditions, then the relation defined by for all $x, y \in R$, $x \leq y$ if $y - x \in P$ makes R into a partially ordered ring with positive cone P.

A partially ordered ring R is called a *lattice-ordered ring* (*ℓ-ring*), or a *totally ordered ring* (*o-ring*) if the partial order on R is a lattice order, or a total order. Certainly an *o*-ring is an *ℓ*-ring. A ring is called *unital* if it has the multiplicative identity element, denoted by 1, and an *ℓ*-ring is called *ℓ-unital* if it is unital and $1 > 0$. We will see later that a unital *ℓ*-ring may not be *ℓ*-unital. A *lattice-ordered field* (*ℓ-field*) or a *totally ordered field* (*o-field*) is a field, and an *ℓ*-ring or an *o*-ring. Similarly a *lattice-ordered division ring* or a *totally ordered division ring* is a division ring, and an *ℓ*-ring or *o*-ring. Let F be a totally ordered field. A *lattice-ordered algebra* (*ℓ-algebra*) A over F is an algebra and an ℓ-ring such that for all $\alpha \in F$, $a \in A$, $\alpha \geq 0$ and $a \geq 0$ implies $\alpha a \geq 0$. So with respect to the addition and scalar multiplication, A is a vector lattice over F. An *ℓ-ideal* of an ℓ-ring is an ideal and a convex ℓ-subgroup. By Theorem 1.8(1), an ideal I of an ℓ-ring R is an ℓ-ideal of R if and only if $|r| \leq |x|$, for any $x \in I$ and $r \in R$, implies $r \in I$. Similarly define *left ℓ-ideal* and *right ℓ-ideal* for an ℓ-ring. It is clear that an ℓ-ring itself and $\{0\}$ are (left, right) ℓ-ideals. An ℓ-ring R is called *ℓ-simple* if it contains no other ℓ-ideals except R, $\{0\}$, and $R^2 \neq \{0\}$. An (left, right) ℓ-ideal of an ℓ-algebra A is an (left, right) ℓ-ideal of the ℓ-ring A and also a subspace of A over F. Clearly the intersection of any family of (left, right) ℓ-ideals is an (left, right) ℓ-ideal. Let X be a subset of an ℓ-ring R, $\langle X \rangle$ denotes the intersection of all ℓ-ideals of R containing X and $\langle X \rangle$ is called the ℓ-ideal *generated by* X. If $X = \{x\}$, then $\langle x \rangle$ is used for $\langle X \rangle$.

For ℓ-rings R and S, an *ℓ-homomorphism* from R to S is a ring homomorphism and a lattice homomorphism from R to S. For an ℓ-homomorphism $\varphi : R \to S$ of two ℓ-rings, define the *kernel of* φ as $\mathrm{Ker}(\varphi) = \{r \in R \mid \varphi(r) = 0\}$. Then $\mathrm{Ker}(\varphi)$ is an ℓ-ideal of R. An *ℓ-isomorphism* between two ℓ-rings is a one-to-one and onto ℓ-homomorphism, and two ℓ-rings R and S are called *ℓ-isomorphic*, denoted by $R \cong S$, if there exists an ℓ-isomorphism between them. Let I be an ℓ-ideal of an ℓ-ring R. Then R/I becomes an ℓ-ring and the elements in R/I are denoted by $a + I$, $a \in R$ (Exercise 20). The projection $\pi : R \to R/I$ is an ℓ-homomorphism between two ℓ-rings. An ℓ-ring R is called *Archimedean* if its additive ℓ-group is Archimedean, and an ℓ-algebra A over a totally ordered field F is called *Archimedean over* F if A is Archimedean over F as a vector lattice over F.

For a family of ℓ-rings $\{R_i \mid i \in I\}$, the cartesian product $\Pi_{i \in I} R_i = \{\{a_i\} \mid a_i \in R_i\}$, where $\{a_i\}$ denotes a function from I to $\cup R_i$ that maps each i to a_i, becomes an ℓ-ring with respect to the addition:

$$\{a_i\} + \{b_i\} = \{a_i + b_i\},$$

the multiplication:

$$\{a_i\}\{b_i\} = \{a_i b_i\},$$

and the order:

$$\{a_i\} \geq 0 \text{ if each } a_i \geq 0 \text{ in } R_i.$$

Then $\Pi_{i \in I} R_i$, together with those operations, is called the *direct product* of the family $\{R_i \mid i \in I\}$. The *direct sum* of $\{R_i \mid i \in I\}$ is $\oplus_{i \in I} R_i = \{\{a_i\} \in \Pi_{i \in I} R_i \mid$ only finitely many $a_i \neq 0\}$.

The following result gives us simple methods to construct lattice orders on rings to make them into ℓ-rings and to construct new lattice orders from existing lattice orders. For a unital ring R, an element u is called a *unit* if u has an inverse with respect to the multiplication of R.

Theorem 1.19.

(1) Let A be an algebra over a totally ordered field F and let B be a vector space basis of A over F. If for all $a, b \in B$, ab is a linear combination of elements in B with positive scalars in F, then A can be made into an ℓ-algebra by defining an element of A is positive if each scalar in its unique linear combination of distinct elements in B is positive.

(2) Suppose that R is a unital ℓ-ring (ℓ-algebra) with positive cone P and $u > 0$ is a unit. Then uP is the positive cone of a lattice order on R to make it into an ℓ-ring (ℓ-algebra).

Proof. (1) A linear combination of vectors over F is called a *positive linear combination* if each scalar in the combination belongs to F^{+}. Let P consist of all positive linear combinations of vectors in B. Then three conditions in Theorem 1.18 are satisfied, and $F^{+}P \subseteq P$. For $a \in A$, a can be uniquely written as $a = \alpha_1 v_1 + \cdots + \alpha_k v_k$ for distinct $v_i \in B$, and scalars $\alpha_i \in F$. Then it is straightforward to verify that

$$a \wedge 0 = (\alpha_1 \wedge 0)v_1 + \cdots + (\alpha_k \wedge 0)v_k$$

and

$$a \vee 0 = (\alpha_1 \vee 0)v_1 + \cdots + (\alpha_k \vee 0)v_k$$

(Exercise 21). Thus the order is a lattice order and A is an ℓ-algebra over F.

(2) Obviously uP is closed under the addition of R, and since $u \in P$, uP is also closed under the multiplication of R. Finally that $uP \subseteq P$ implies $(uP) \cap -(uP) = \{0\}$. Thus R is a partially ordered ring with the positive cone uP. To see that uP is the positive cone of a lattice order, we consider the mapping $f : R \to R$ defined by for all $a \in R$, $f(a) = ua$. Since u is a unit, f is a group isomorphism of the additive group of R. For any $a, b \in R$,

$$a \wedge_{(uP)} b = f(f^{-1}(a) \wedge_P f^{-1}(b)) = u(u^{-1}a \wedge_P u^{-1}b),$$

where $a \wedge_{(uP)} b$ is the greatest lower bound of a and b with respect to uP, and $u^{-1}a \wedge_P u^{-1}b$ is the greatest lower bound of $u^{-1}a$ and $u^{-1}b$ with respect to P (Exercise 22). Similarly,

$$a \vee_{(uP)} b = f(f^{-1}(a) \vee_P f^{-1}(b)) = u(u^{-1}a \vee_P u^{-1}b).$$

Finally it is clear that if P is the positive cone of an ℓ-algebra, then uP is also closed under positive scalar multiplication. □

A vector space basis B of an algebra is called a *multiplicative basis* if for any $a, b \in B$, $ab \in B$ or $ab = 0$. By Theorem 1.19(1), if an algebra A has a multiplicative basis, then A can be made into an ℓ-algebra in which B is a basis, that is, B is a disjoint set of basic elements with $B^{\perp} = \{0\}$. For instance, standard matrix units e_{ij}, $1 \leq i, j \leq n$ is a multiplicative basis for matrix algebra $M_n(F)$ over a totally ordered field F.

More generally, for an algebra A over a totally ordered field F, a vector space basis B is called a *multiplicative basis over F^+* if for any $a, b \in B$, $ab = \alpha c$ for some $\alpha \in F^+$ and $c \in B$. Similarly by Theorem 1.19(1) again, if A has a multiplicative basis over F^+, A can be made into an ℓ-algebra over F with B as a basis. For instance, in the field $A = \mathbb{Q}[\sqrt{2}]$, $B = \{1, \sqrt{2}\}$ is a multiplicative basis over $\mathbb{Q}^+$.

An important application of Theorem 1.19(2) is constructing lattice orders on an ℓ-unital ℓ-ring such that 1 is not positive. For an ℓ-unital ℓ-ring R, take a positive unit u such that u^{-1} is not positive. Then R is an ℓ-ring with the positive cone uR^+ and since $u^{-1} \notin R^+$, $1 \notin uR^+$, so R is not ℓ-unital with respect to this lattice order.

Now we present some examples of ℓ-rings.

Example 1.3.

(1) Suppose that R is an ℓ-ring and $M_n(R)$ is the $n \times n$ matrix ring over R with $n \geq 2$. Define a matrix $(a_{ij}) \geq 0$ if each $a_{ij} \geq 0$ in R. Clearly three conditions in Theorem 1.18 are satisfied and the product of a positive scalar and a positive matrix is still positive. It is easily verified that for any two matrices (a_{ij}) and (b_{ij}),

$$(a_{ij}) \wedge (b_{ij}) = (a_{ij} \wedge b_{ij}) \quad \text{and} \quad (a_{ij}) \vee (b_{ij}) = (a_{ij} \vee b_{ij}).$$

Hence $M_n(R)$ is an ℓ-ring with positive cone $M_n(R^+)$. This lattice order on $M_n(R)$ is called the *entrywise order*. Clearly if R is ℓ-unital, then identity matrix is positive with respect to the entrywise order. For a totally ordered field F, $M_n(F)$ is an ℓ-algebra over F with respect to the entrywise order.

Let e_{ij} be the standard matrix units in matrix rings, namely, the $(i, j)^{th}$ entry in e_{ij} is 1 and other entries in e_{ij} are zero. As we mentioned before, $\{e_{ij} \mid i, j = 1, \cdots, n\}$ is multiplicative and hence it is a basis of $M_n(F)$ over F with respect to the entrywise order.

Let f be the matrix $f = e_{11} + e_{12} + e_{21} + e_{33} + \cdots + e_{nn}$. Then $f \in M_n(F^+)$ and $f^{-1} = e_{12} + e_{21} - e_{22} + e_{33} + \cdots + e_{nn} \notin M_n(F^+)$. Thus by Theorem 1.19(2), $f M_n(F^+)$ is the positive cone of an ℓ-algebra $M_n(F)$ over F in which $1 \not> 0$.

(2) Suppose that G is a group (semigroup) and F is a totally ordered field. Let $F[G] = \{\sum \alpha_i g_i \mid \alpha_i \in F, g_i \in G\}$ be the group (semigroup) algebra over F. In this case the operation on G is written as multiplication. Define $\sum \alpha_i g_i \geq 0$ if each $\alpha_i \geq 0$, that is, the positive cone is $F^+[G]$. Then $F[G]$ is an ℓ-algebra over F, and the lattice order is called the *coordinatewise order*. Clearly $1G = \{1g \mid g \in G\}$, where 1 is the identity element of F, is a basis and also a vector space basis of $F[G]$ over F.

A difference between examples (1) and (2) is that the identity matrix in $M_n(F)$ is not a basic element but the identity element in $F[G]$ is basic.

(3) Let F be a totally ordered field and $R = F[x]$ be the polynomial ring over F. Except the lattice order on R defined in (2), we consider some other lattice orders on R. Let $p(x) = a_n x^n + \cdots + a_k x^k \in R$ with $a_i \in F$, $0 \leq k \leq n$, and $a_k, a_n \neq 0$. If we define $p(x) > 0$ by $a_n > 0$, then R is a totally ordered algebra and the ordering is called *lexicographic ordering*. If we define $p(x) > 0$ by $a_k > 0$, then R is also a totally ordered algebra and the ordering is called *antilexicographic ordering*. Both total orders are not Archimedean over F (Exercise 23).

Let's construct more lattice orders on $R = F[x]$. Fix a positive integer n, define the positive cone P_n on R as follows. For a polynomial $p(x) = a_k x^k + \cdots + a_0$ of degree k. If $k \leq n$, define $p(x) \geq 0$ if $a_k > 0$ and $a_0 \geq 0$, and if $k > n$, then define $p(x) \geq 0$ if $a_k > 0$. Then three conditions in Theorem 1.18 are satisfied and $F^+ P_n \subseteq P_n$, so R is a partially ordered algebra over F. Moreover, for a polynomial $p(x) = a_k x^k + a_{k-1} x^{k-1} + \cdots + a_1 x + a_0$ of degree k,

$$
p(x) \vee 0 = \begin{cases}
p(x), & \text{if } k > n, a_k > 0, \\
0, & \text{if } k > n, a_k < 0, \\
p(x), & \text{if } k \leq n, a_k > 0, a_0 \geq 0, \\
0, & \text{if } k \leq n, a_k < 0, a_0 \leq 0, \\
p(x) - a_0, & \text{if } k \leq n, a_k > 0, a_0 < 0, \\
a_0, & \text{if } k \leq n, a_k < 0, a_0 > 0.
\end{cases}
$$

We leave the verification of these facts to the reader (Exercise 24).

Then R becomes an ℓ-algebra over F that has *squares positive* in the sense that for each $r \in R$, $r^2 \geq 0$.

1.3.2 Some special ℓ-rings

Let R be an ℓ-ring. An element $a \in R^+$ is called a *d-element* if

$$\text{for all } x, y \in R, x \wedge y = 0 \Rightarrow ax \wedge ay = xa \wedge ya = 0,$$

and a is called an *f-element* if

$$\text{for all } x, y \in R, x \wedge y = 0 \Rightarrow ax \wedge y = xa \wedge y = 0.$$

Each f-element is clearly a d-element. An ℓ-ring R is called a d-ring (f-ring) if each element in R^+ is a d-element (f-element). Define

$$d(R) = \{a \in R^+ \mid a \text{ is a } d\text{-element}\}$$

and

$$f(R) = \{a \in R \mid |a| \text{ is an } f\text{-element}\}.$$

We may also define left and right d-element. An element $a \in R^+$ is called a *left d-element* (*right d-element*) if

$$\text{for all } x, y \in R, x \wedge y = 0 \Rightarrow ax \wedge ay = 0 \ (xa \wedge ya = 0).$$

Example 5.1 shows that generally a left d-element may not be a right d-element. Left and right f-element may be defined similarly.

Theorem 1.20. *Let R be an ℓ-ring.*

(1) An element $a \in R^+$ is a d-element if and only if for all $x, y \in R$, $a(x \wedge y) = ax \wedge ay$ and $(x \wedge y)a = xa \wedge ya$.

(2) Suppose that $a \in R^+$ is invertible, then a is a d-element if and only if $a^{-1} \in R^+$.

(3) For all $x, y \in R$, $|xy| \leq |x||y|$, and the equality holds if and only if R is a d-ring.

(4) The $d(R)$ is a convex subset of R that is closed under the multiplication of R, but generally $d(R)$ is not closed under the addition of R.

(5) $f(R)$ is a convex ℓ-subring of R and an f-ring.

(6) R is an f-ring if and only if for each $a \in R$, $a^\perp$ is an ℓ-ideal of R.

(7) If R is a d-ring (f-ring) and I is an ℓ-ideal of R, then R/I is a d-ring (f-ring).

Proof. (1) Suppose that a is a d-element. Let $x, y \in R$. Then $(x - (x \wedge y)) \wedge (y - (x \wedge y)) = 0$ implies

$$a(x - (x \wedge y)) \wedge a(y - (x \wedge y)) = (x - (x \wedge y))a \wedge (y - (x \wedge y))a = 0,$$

so $ax \wedge ay = a(x \wedge y)$ and $xa \wedge ya = (x \wedge y)a$ by Theorem 1.5(1). The converse is trivial.

(2) Let 1 be the identity element of R. Suppose that a is a d-element. We have $a((-1) \vee 0) = (-a) \vee 0 = 0$ implies that $(-1) \vee 0 = 0$, so $1 = 1^+ - 1^- = 1^+ \geq 0$. Then $a(a^{-1} \wedge 0) = 1 \wedge 0 = 0$ implies that $a^{-1} \wedge 0 = 0$, that is, $a^{-1} \geq 0$. Conversely, suppose a and a^{-1} are both positive. If $x \wedge y = 0$ for $x, y \in R$, then

$$0 \leq a^{-1}(ax \wedge ay) \leq (a^{-1}ax \wedge a^{-1}ay) = x \wedge y = 0,$$

so $ax \wedge ay = 0$. Similarly $xa \wedge ya = 0$. Thus a is a d-element.

(3)
$$\begin{aligned}
|xy| &= |(x^+ - x^-)(y^+ - y^-)| \\
&= |x^+y^+ - x^-y^+ - x^+y^- + x^-y^-| \\
&\leq x^+y^+ + x^-y^+ + x^+y^- + x^-y^- \\
&= |x||y|.
\end{aligned}$$

Suppose that R is a d-ring. For $x, y \in R$, since

$$\begin{aligned}
0 &\leq (x^+y^+ + x^-y^-) \wedge (x^-y^+ + x^+y^-) \\
&\leq (x^+y^+ \wedge x^-y^+) + (x^+y^+ \wedge x^+y^-) + (x^-y^- \wedge x^-y^+) + (x^-y^- \wedge x^+y^-) \\
&= (x^+ \wedge x^-)y^+ + x^+(y^+ \wedge y^-) + x^-(y^- \wedge y^+) + (x^- \wedge x^+)y^- \\
&= 0,
\end{aligned}$$

we have

$$\begin{aligned}
|xy| &= |(x^+y^+ + x^-y^-) - (x^-y^+ + x^+y^-)| \\
&= (x^+y^+ + x^-y^-) + (x^-y^+ + x^+y^-) \\
&= |x||y|,
\end{aligned}$$

by Theorem 1.7(1) and (3).

Conversely suppose that $|xy| = |x||y|$ for all $x, y \in R$. If $z \wedge w = 0$ for some $z, w \in R$, then $|z - w| = z + w$ by Theorem 1.7(1) and (3), so for any $a \in R^+$, $|a(z - w)| = |a||z - w| = a(z + w) = az + aw$ implies $az \wedge aw = 0$ (Exercise 25). Similarly $z \wedge w = 0$ implies $za \wedge wa = 0$. Thus R is a d-ring.

(4) Suppose that $a, b \in d(R)$ and $c \in R$ with $a \leq c \leq b$. If $x \wedge y = 0$ for $x, y \in R$, then $0 \leq ax \wedge ay \leq cx \wedge cy \leq ax \wedge ay = 0$ implies $cx \wedge cy = 0$.

Similarly $xc \wedge yc = 0$. Thus $c \in d(R)$. It is clear that $d(R)$ is closed under the multiplication in R by the definition of d-element. In the Example 1.3(1), each standard matrix unit e_{ij} is a d-element, but the sum of two d-elements may not be a d-element. For example, $e_{12} + e_{11}$ is not a d-element since $e_{12} \wedge e_{22} = 0$, but

$$(e_{12} + e_{11})e_{12} \wedge (e_{12} + e_{11})e_{22} = e_{12} \wedge e_{12} \neq 0.$$

(5) Let $a, b \in f(R)$ and $x \wedge y = 0$ for $x, y \in R$. Then by Theorems 1.7(5) and 1.5(7),

$$0 \leq |a - b|x \wedge y \leq (|a| + |b|)x \wedge y \leq (|a|x \wedge y) + (|b|x \wedge y) = 0,$$

so $|a - b|x \wedge y = 0$. Similarly, $x|a - b| \wedge y = 0$. Thus $|a - b|$ is an f-element and hence $a - b \in f(R)$. We also have

$$0 \leq |ab|x \wedge y \leq (|a||b|)x \wedge y = 0,$$

so $|ab|x \wedge y = 0$. Similarly, $x|ab| \wedge y = 0$. Thus $|ab|$ is an f-element and hence $ab \in f(R)$. Finally if $|x| \leq |a|$ for some $a \in f(R)$, $x \in R$, then clearly $|x|$ is an f-element, so $x \in f(R)$. Hence $f(R)$ is a convex ℓ-subring of R and an f-ring.

(6) Suppose that R is an f-ring and $a \in R$. We already know that $a^{\perp}$ is a convex ℓ-subgroup of the additive ℓ-group of R. Let $b \in a^{\perp}$ and $r \in R$, then $|b| \wedge |a| = 0$ implies $|r||b| \wedge |a| = |b||r| \wedge |a| = 0$, and hence $|rb| \wedge |a| = |rb| \wedge |a| = 0$ by (3). Thus $rb, br \in a^{\perp}$ and $a^{\perp}$ is an ℓ-ideal of R. Conversely suppose that for each $a \in R$, $a^{\perp}$ is an ℓ-ideal of R. Let $x \wedge y = 0$ for $x, y \in R$ and $r \in R^{+}$. Then $x \in y^{\perp}$ implies $rx, xr \in y^{\perp}$, so $rx \wedge y = xr \wedge y = 0$, namely r is an f-element of R for each $r \in R^{+}$.

(7) Let R be a d-ring and I be an ℓ-ideal of R. Suppose that $(x + I) \wedge (y + I) = 0$ and $z + I \geq 0$. We may assume that $z \geq 0$. Then $x \wedge y = w \in I$ implies that $(x - w) \wedge (y - w) = 0$, and hence

$$z(x - w) \wedge z(y - w) = 0 \text{ and } (x - w)z \wedge (y - w)z = 0.$$

Thus

$$(z + I)(x + I) \wedge (z + I)(y + I) = 0, (x + I)(z + I) \wedge (y + I)(z + I) = 0$$

in R/I, that is, R/I is a d-ring. Similarly to show that if R is an f-ring, then R/I is an f-ring. $\square$

Some fundamental properties of f-rings are summarized in the following results. An ℓ-ring R is said to be a *subdirect product* of the family of ℓ-rings $\{R_i \mid i \in I\}$ if R is an ℓ-subring of the direct product $\Pi_{i \in I} R_i$ such that

$\pi_k(R) = R_k$ for every $k \in I$, where $\pi_k : \Pi_{i \in I} R_i \to R_k$ is the canonical ℓ-epimorphism, that is, for $\{a_i\} \in \Pi_{i \in I} R_i$, $\pi_k(\{a_i\}) = a_k$. An ℓ-ring R is called *subdirectly irreducible* if R contains a smallest nonzero ℓ-ideal, that is, the intersection of all nonzero ℓ-ideals is a nonzero ℓ-ideal. For instance, ℓ-simple ℓ-rings are subdirectly irreducible.

Lemma 1.1. *Let R be an ℓ-ring and $a \in R$.*

(1) $\langle a \rangle = \{x \in R \mid |x| \leq n|a| + r|a| + |a|s + t|a|u, n \in \mathbb{Z}^+, r, s, t, u \in R^+\}$.

(2) If R is commutative, then $\langle a \rangle = \{x \in R \mid |x| \leq n|a| + r|a|, n \in \mathbb{Z}^+, r \in R^+\}$.

(3) If R is ℓ-unital, then $\langle a \rangle = \{x \in R \mid |x| \leq t|a|u, t, u \in R^+\}$.

(4) Suppose that R is an f-ring. If $x \wedge y = 0$ for any $x, y \in R$, then $\langle x \rangle \cap \langle y \rangle = \{0\}$.

Proof. (1) Let

$$I = \{x \in R \mid |x| \leq n|a| + r|a| + |a|s + t|a|u, n \in \mathbb{Z}^+, r, s, t, u \in R^+\}.$$

Suppose that $x, y \in I$. Then

$$|x| \leq n|a| + r|a| + |a|s + t|a|u \quad \text{and} \quad |y| \leq n_1|a| + r_1|a| + |a|s_1 + t_1|a|u_1,$$

where $n, n_1 \in \mathbb{Z}^+$, $r, r_1, s, s_1, t, t_1, u, u_1 \in R^+$. Thus

$$|x - y| \leq |x| + |y| \leq (n+n_1)|a| + (r+r_1)|a| + |a|(s+s_1) + (t+t_1)|a|(u+u_1),$$

so $x - y \in I$. Hence I is a subgroup of R. It is clear that for any $x \in I$ and $r \in R$, $rx, xr \in I$. It follows that I is an ideal. If $|r| \leq |x|$ for some $r \in R$ and $x \in I$, then clearly $r \in I$ by the definition of I. Hence I is an ℓ-ideal. Since $a \in I$ and every ℓ-ideal containing a contains I, we have $\langle a \rangle = I$.

(2) and (3) are direct consequences of (1).

(4) Let $0 \leq a \in \langle x \rangle \cap \langle y \rangle$. Then

$$a \leq nx + rx + xs + uxv \quad \text{and} \quad a \leq n_1 y + r_1 y + y s_1 + u_1 y v_1$$

for some $r, r_1, s, s_1, u, u_1, v, v_1 \in R^+$ and positive integers n, n_1. Since R is an f-ring, $x \wedge y = 0$ implies

$$(rx + xs + uxv + nx) \wedge (r_1 y + y s_1 + u_1 y v_1 + n_1 y) = 0,$$

by Theorem 1.5(7), so $a = 0$. Thus $\langle x \rangle \cap \langle y \rangle = \{0\}$. $\qquad \square$

Theorem 1.21.

(1) An ℓ-ring R is ℓ-isomorphic to a subdirect product of a family of ℓ-rings $\{R_i \mid i \in I\}$ if and only if there is a family of ℓ-ideals $\{J_i \mid i \in I\}$ such that $R \cong R_k/J_k$ for each $k \in I$ and the intersection of $\{J_i \mid i \in I\}$ is zero.

(2) A subdirectly irreducible f-ring is totally ordered.

(3) An ℓ-ring is an f-ring if and only if it is a subdirect product of totally ordered rings.

Proof. (1) We may assume that R is a subdirect product of a family of ℓ-rings $\{R_i \mid i \in I\}$. Define $J_i = \mathrm{Ker}(\pi_i) \cap R$. Then J_i is an ℓ-ideal of R and $R/J_i \cong R_i$ for each $i \in I$. Suppose that $\{a_i\} \in \cap_{i \in I} J_i$. Then $\pi_k(\{a_i\}) = a_k = 0$ for all $k \in I$, so $\{a_i\} = 0$, that is, $\cap_{i \in I} J_i = \{0\}$. Conversely suppose there is a family of ℓ-ideals $\{J_i \mid i \in I\}$ of R such that $\cap_{i \in I} J_i = \{0\}$ and $R/J_i \cong R_i$. Then the ℓ-ring $\{\{a + J_i\}_{i \in I} \mid a \in R\}$ is a subdirect product of the family of ℓ-rings $\{R/J_i \mid i \in I\}$ and $R \cong \{\{a + J_i\}_{i \in I} \mid a \in R\}$ (Exercise 27).

(2) Let R be a subdirectly irreducible f-ring. Suppose that $x \wedge y = 0$, for $x, y \in R$. By Lemma 1.1 $\langle x \rangle \cap \langle y \rangle = \{0\}$, and hence either $\langle x \rangle = \{0\}$ or $\langle y \rangle = \{0\}$, so either $x = 0$ or $y = 0$. Hence for any $x \in R$, either $x^- = 0$ or $x^+ = 0$ since $x^+ \wedge x^- = 0$, that is, R is totally ordered.

(3) Let R be an f-ring. For each element $a \in R$, $a \neq 0$, define

$$\mathcal{M}_a = \{I \mid I \text{ is an } \ell\text{-ideal and } a \notin I\}.$$

Then $\mathcal{M}_a \neq \emptyset$ since $\{0\}$ is in $\mathcal{M}_a$. $\mathcal{M}_a$ is a partially ordered set by set inclusion. For a subset $\{I_k\}$ of $\mathcal{M}_a$ that is totally ordered, the union $\cup_k I_k$ is an ℓ-ideal of R with $a \notin \cup_k I_k$. Thus by Zorn's Lemma, $\mathcal{M}_a$ has a maximal element, denoted by I_a. The quotient ℓ-ring R/I_a is subdirectly irreducible with the smallest nonzero ℓ-ideal $\langle a + I_a \rangle$, so by Theorem 1.20(7) and (2), R/I_a is totally ordered. Consider

$$x \in J = \bigcap_{0 \neq a \in R} I_a.$$

If $x \neq 0$, then $J \subseteq I_x$ implies $x \in I_x$, which is a contradiction. Thus $J = \{0\}$, and hence by (1), R is a subdirect product of totally ordered rings $\{R/I_a \mid a \in R, a \neq 0\}$. The converse is trivial (Exercise 29). $\square$

An important method of proving properties of f-rings is first to consider totally ordered rings and then use the fact that an f-ring is a subdirect product of totally ordered rings.

Theorem 1.22. *Let R be an f-ring.*

(1) If $a \wedge b = 0$ for $a, b \in R$, then $ab = 0$. Thus R has squares positive.

(2) If R is Archimedean, then R is commutative.

(3) If R is unital, then each idempotent element of R is in the center of R.

(4) If R is unital and $a^n = 1$ for some $a \in R^+$ and some positive integer n, then $a = 1$.

Proof. (1) If $a \wedge b = 0$, then $ab \wedge b = 0$, and $ab \wedge ab = 0$, so $ab = 0$. For any $x \in R$,

$$x^2 = (x^+ - x^-)^2 = (x^+)^2 - x^+ x^- - x^- x^+ + (x^-)^2 = (x^+)^2 + (x^-)^2 \geq 0,$$

since $x^+ \wedge x^- = 0$ implies $x^+ x^- = x^- x^+ = 0$.

(2) We show that given $a, b \geq 0$, for any positive integer n, $n|ab - ba| \leq a^2 + b^2$. We first assume that R is totally ordered with $a > b$. Since R is Archimedean, there exists an integer k such that $ka \leq nb < (k+1)a$. Let $nb = ka + r$ with $0 \leq r < a$. We have

$$n|ab - ba| = |a(ka + r) - (ka + r)a| = |ar - ra| \leq a^2 \leq a^2 + b^2.$$

If R is an f-ring, then R is ℓ-isomorphic to a subdirect product of totally ordered rings, so by Theorem 1.21(1), there exist ℓ-ideals I_k such that each R/I_k is a totally ordered ring and intersection of I_k is equal to zero. Given $0 \leq a, b \in R$ and a positive integer n, by previous argument we have

$$n|(a + I_k)(b + I_k) - (b + I_k)(a + I_k)| \leq (a + I_k)^2 + (b + I_k)^2$$

in R/I_k for each k. Then

$$n|ab - ba| + I_k = (n|ab - ba| + I_k) \wedge (a^2 + b^2 + I_k)$$

in R/I_k for each k, so

$$n|ab - ba| - (n|ab - ba| \wedge (a^2 + b^2)) \in I_k$$

for each k. Thus $n|ab - ba| - (n|ab - ba| \wedge (a^2 + b^2)) = 0$, and hence $n|ab - ba| \leq a^2 + b^2$ in R for any positive integer n. It follows that $ab - ba = 0$ since R is Archimedean, so $ab = ba$ for $a, b \geq 0$. Since each element in an ℓ-ring is a difference of two positive elements, R is commutative.

(3) We notice that a unital f-ring must be ℓ-unital by (1). First suppose that R is totally ordered and $e \in R$ is an idempotent element. Since e is idempotent, $1 - e$ is also idempotent. By (1) each idempotent element is positive, so $e, 1 - e \geq 0$. If $e \leq 1 - e$, then $e = e^2 \leq (1 - e)e = 0$, so $e = 0$. If $1 - e \leq e$, then $1 - e = 0$, so $e = 1$. Therefore we have proved that in a unital totally ordered ring, there exist only two idempotent elements, that is, 1 and 0.

Suppose now that R is an f-ring. Then there are ℓ-ideals $\{I_k\}$ of R such that $\cap_k I_k = \{0\}$ and each R/I_k is a totally ordered ring. Let e be an idempotent of R, by the above argument, since R/I_k is a unital totally ordered ring, $e + I_k = 0 + I_k$ or $1 + I_k$ in R/I_k, so for any $a \in R$,

$$(e + I_k)(a + I_k) = (a + I_k)(e + I_k)$$

in R/I_k, that is, $(ea - ae) \in I_k$ for each k. Hence $ea - ae = 0$, and $ea = ae$ for each $a \in R$. Therefore e is in the center of R.

(4) As we have done before, we first assume that R is totally ordered. If $1 < a$, then $1 < a \leq a^2 \leq \cdots \leq a^n = 1$, which is a contradiction. Similarly, $a \not< 1$. Thus $a = 1$. So the result is true in a unital totally ordered ring. For an f-ring R, there are ℓ-ideals $\{J_k\}$ of R such that $\cap_k J_k = \{0\}$ and each R/J_k is a totally ordered ring. Let $a \in R^+$ with $a^n = 1$. Then for each k, $(a + J_k)^n = 1 + J_k$ in R/J_k and $a + J_k \in (R/J_k)^+$, so $a + J_k = 1 + J_k$ in R/J_k for each k. Thus $a - 1 \in J_k$ for each k, so $a - 1 = 0$, $a = 1$. □

An ℓ-ring R is called an *almost f-ring* if for all $a, b \in R$, $a \wedge b = 0 \Rightarrow ab = 0$, or equivalently $x^+ x^- = 0$ for all $x \in R$. By Theorem 1.22(1), each f-ring is an almost f-ring and each almost f-ring has squares positive.

The following are two immediate consequences of Theorem 1.22. (1) Any $n \times n$ matrix ring over any unital ring cannot be made into an f-ring if $n \geq 2$ since it contains idempotent elements that are not in the center. (2) Any nontrivial finite group algebra $F[G]$ over a totally ordered field F cannot be made into an f-ring such that $(G \setminus \{e\}) \cap F[G]^+ \neq \emptyset$ since for any element g in G there exists a positive integer n such that $g^n = e$, where e is the identity element of group G. In particular $F[G]$ cannot be made into a totally ordered ring with the exception when G is a trivial group. However, a finite group algebra $F[G]$ may be made into an f-ring with $(G \setminus \{e\}) \cap F[G]^+ = \emptyset$ as shown in the following example.

Example 1.4. Consider $R = \mathbb{Q}[G]$ with $G = \{e, a\}$ and $a^2 = e$. Define $u = \frac{1}{2}(e + a)$ and $v = \frac{1}{2}(e - a)$. Then $u^2 = u$, $v^2 = v$, $uv = 0$, and $\{u, v\}$ is linearly independent over $\mathbb{Q}$. Thus $P = \mathbb{Q}^+ u + \mathbb{Q}^+ v$ is the positive cone of a lattice order. Clearly u and v are both f-elements, so R is an f-ring. We note that $1 = u + v$ and $a = u - v \not> 0$.

A group ring of an infinite group may be made into a totally ordered ring. The simplest example will be the group ring $F[G]$ of an infinite cyclic group $G = \{g^n \mid n \in \mathbb{Z}\}$. Define an element $\sum_{i=-m}^{n} \alpha_i g^i \geq 0$ if $\alpha_n > 0$. Then $F[G]$ is a totally ordered ring with

$$\cdots < g^{-2} < g^{-1} < 1 < g < g^2 < \cdots .$$

By Theorem 1.22, an Archimedean f-ring is commutative. But an f-algebra over a totally ordered field F that is Archimedean over F may not be commutative. For instance, any totally ordered division algebra that is Archimedean over its center is such an example. However if a totally ordered division ring is algebraic over its center, then it is commutative by Albert's Theorem. We refer the reader to [Steinberg (2010)] for the proof of Albert's Theorem.

Theorem 1.23 (Albert's Theorem). *Let D be a totally ordered division ring. If $a \in D$ is algebraic over the center of D, then a is in the center.*

In the following we consider some properties of ℓ-ideals of an ℓ-ring. Suppose that R is an ℓ-ring and $I_1, \cdots, I_n$ be ℓ-ideals of R. Define

$$I_1 + \cdots + I_n = \{a \in R \mid a = a_1 + \cdots + a_n, a_i \in I_i\}.$$

Theorem 1.24. *Let R be an ℓ-ring and $I_1, \cdots, I_n, I$ be ℓ-ideals of R.*

(1) $I_1 + \cdots + I_n$ is an ℓ-ideal of R which is the ℓ-ideal generated by $\{I_1, \cdots, I_n\}$.

(2) $(I_1 + \cdots + I_n) \cap I = (I_1 \cap I) + \cdots + (I_n \cap I)$.

(3) Each ℓ-ideal of R/I is of the form J/I, where J is an ℓ-ideal of R containing I, and the mapping $J \to J/I$ is a one-to-one correspondence between the set of all ℓ-ideals of R which contain I and the set of all ℓ-ideals of R/I.

Proof. (1) It is clear that $I_1 + \cdots + I_n$ is an ideal of R. Suppose that $|x| \leq |a_1 + \cdots + a_n|$ for some $x \in R$ and $a_i \in I_i$. Then $|x| \leq |a_1| + \cdots + |a_n|$ implies that $|x| = x_1 + \cdots + x_n$ with $0 \leq x_i \leq |a_i|$, and hence $|x| \in I$ since each $x_i \in I_i$. Then similarly $0 \leq x^+, x^- \leq |x|$ implies that $x^+, x^- \in I$. Thus $x = x^+ - x^- \in I$ and I is an ℓ-ideal.

(2) This follows from Theorem 1.9. It can be proved directly as follows. Clearly

$$(I_1 + \cdots + I_n) \cap I \supseteq (I_1 \cap I) + \cdots + (I_n \cap I).$$

Take $0 \leq a \in (I_1 + \cdots + I_n) \cap I$. Then $a = a_1 + \cdots + a_n$, where each $a_i \in I_i$. Then $a = |a| \leq |a_1| + \cdots + |a_n|$ implies that $a = x_1 + \cdots + x_n$, where $0 \leq x_i \leq |a_i|$. Since $x_i \leq a \in I$, each $x_i \in I$. Hence each $x_i \in I_i \cap I$ and $a \in (I_1 \cap I) + \cdots + (I_n \cap I)$. Since each element in $(I_1 + \cdots + I_n) \cap I$ is a difference of two positive elements, we have $(I_1 + \cdots + I_n) \cap I \subseteq (I_1 \cap I) + \cdots + (I_n \cap I)$. Therefore $(I_1 + \cdots + I_n) \cap I = (I_1 \cap I) + \cdots + (I_n \cap I)$.

(3) The proof of these facts is the same to the similar results in general ring theory, so we omit the proof. □

There are some other properties on ℓ-ideals that are similar to the properties on ideals in general ring theory. For example, if I, J are ℓ-ideals of an ℓ-ring R, then

$$I/(I \cap J) \cong (I + J)/J \quad \text{and} \quad (R/I)/(J/I) \cong R/J \text{ if } I \subseteq J.$$

We leave the verification to the reader.

1.3.3 ℓ-radical and ℓ-prime ℓ-ideals

Suppose that R is an ℓ-ring and I, J are ℓ-ideals of R. The ring theoretical product IJ is not an ℓ-ideal of R in general. We use $\langle IJ \rangle$ to denote the ℓ-ideal generated by IJ. An ℓ-ideal I is called *nilpotent* if $I^n = \{0\}$ for some positive integer n, and if $I^n = \{0\}$ and $I^k \neq \{0\}$ for any positive integer $k < n$, then n is called *nilpotent of index*. If I, J are both nilpotent ℓ-ideals, then $I + J$ is also a nilpotent ℓ-ideal by $(I + J)/I \cong J/I \cap J$ (Exercise 30).

Definition 1.2. The ℓ-radical of an ℓ-ring R is the set

$$\ell\text{-}N(R) = \{a \in R \mid x_0|a|x_1|a| \cdots x_{n-1}|a|x_n = 0 \text{ for some } n = n(a) \text{ and}$$
$$\text{for all } x_0, \cdots, x_n \in R\}.$$

Theorem 1.25. *Suppose that R is an ℓ-ring.*

(1) $\ell\text{-}N(R)$ is an ℓ-ideal, which is the union of all of the nilpotent ℓ-ideals of R. Each element in $\ell\text{-}N(R)$ is nilpotent.

(2) If R is commutative, then $\ell\text{-}N(R) = \{a \in R \mid |a| \text{ is nilpotent}\}$.

(3) If R is an ℓ-ring which satisfies the ascending or descending chain condition on ℓ-ideals, then $\ell\text{-}N(R)$ is nilpotent.

Proof. (1) If I is an nilpotent ℓ-ideal, then evidently each element in I is contained in $\ell\text{-}N(R)$. Conversely suppose $a \in R$ and there exists a positive integer n such that $x_0|a|x_1|a| \cdots x_{n-1}|a|x_n = 0$ for all $x_0, \cdots, x_n \in R$. Then

$$(|a|R)^{n+1} = (R|a|)^{n+1} = (R|a|R)^n = (|a| + |a|R + R|a| + R|a|R)^{2n+1} = 0$$

implies the ℓ-ideal generated by a is nilpotent. Thus $\ell\text{-}N(R)$ is the union of all of the nilpotent ℓ-ideals. $\ell\text{-}N(R)$ is closed under the addition of R since the sum of two nilpotent ℓ-ideals is still nilpotent. Clearly $a^{2n(a)+1} = 0$, for each element a in $\ell\text{-}N(R)$.

(2) Let $x \in R$ and $|x|$ is nilpotent. Then by Lemma 1.1,

$$\langle x \rangle = \{u \in R \mid |u| \leq n|x| + r|x|, \text{ for some } n \geq 1 \text{ and } r \in R^+\}.$$

Since R is commutative and $|x|$ is nilpotent, $\langle x \rangle$ is nilpotent, so $x \in \ell\text{-}N(R)$ (Exercise 31).

(3) If R satisfies the ascending chain condition on ℓ-ideals, then it contains a maximal nilpotent ℓ-ideal M. For any nilpotent ℓ-ideal I, $M + I$ is nilpotent and $M \subseteq M + I$ implies $M = M + I$, so $I \subseteq M$. Thus $\ell\text{-}N(R) = M$ is nilpotent.

Suppose that R satisfies the descending chain condition on ℓ-ideals. We denote $\ell\text{-}N(R)$ just by N. For an ℓ-ideal H of R, define $H^{(2)} = \langle H^2 \rangle$, $H^{(3)} = \langle HH^{(2)} \rangle$, and $H^{(n)} = \langle HH^{(n-1)} \rangle$ for any $n \geq 2$. Then $N^{(n)}$ are ℓ-ideals and $N \supseteq N^{(2)} \supseteq \cdots \supseteq N^{(n)} \supseteq \cdots$, so by descending chain condition on ℓ-ideals, we have $N^{(k)} = N^{(k+1)} = N^{(k+2)} = \cdots$ for some positive integer k. Let $M = N^{(k)}$. Then $M = M^{(2)} = M^{(3)} = \cdots$. Assume that $M \neq \{0\}$. Then the set

$$\mathcal{N} = \{I \in R \mid I \text{ is an } \ell\text{-ideal of } R, I \subseteq M, MIM \neq \{0\}\}$$

is not empty since $M^{(3)} = M$, so there exists a minimal element K in $\mathcal{N}$. Take $0 < a \in K$ with $MaM \neq \{0\}$ and define

$$J = \{c \in R \mid |c| \leq uav, u, v \in M^+\}.$$

Then J is an ℓ-ideal of R with $\{0\} \neq J \subseteq K$ and $MJM \neq \{0\}$ (Exercise 32). So $J \in \mathcal{N}$, and hence $J = K$. Thus $a \leq uav$ for some $u, v \in M^+$. Therefore $a \leq uav \leq u^2av^2 \leq \cdots \leq u^nav^n = 0$ for some positive integer n since $M \subseteq \ell\text{-}N(R)$ and each element in $\ell\text{-}N(R)$ is nilpotent, so $a = 0$, which is a contradiction. Hence we must have $M = \{0\}$, so $N^{(k)} = \{0\}$ implies that $(\ell\text{-}N(R))^k = \{0\}$. $\qquad\square$

Let R be an ℓ-ring. An ℓ-ideal I is called *proper* if $I \neq R$. An ℓ-ideal P is called an *ℓ-prime ℓ-ideal* of R if P is proper and for any two ℓ-ideals I, J of R, $IJ \subseteq P$ implies $I \subseteq P$ or $J \subseteq P$. For ℓ-ideals I, J, it is clear that $IJ \subseteq P$ if and only if $\langle IJ \rangle \subseteq P$, so the definition of ℓ-prime ℓ-ideal is independent of the choice of $IJ \subseteq P$ or $\langle IJ \rangle \subseteq P$.

An ℓ-ring R is called *ℓ-prime* if $\{0\}$ is an ℓ-prime ℓ-ideal. It is clear that a proper ℓ-ideal I of R is ℓ-prime if and only if R/I is an ℓ-prime ℓ-ring. A ring R is called a *domain* if $a, b \in R$, $a \neq 0$ and $b \neq 0$ implies $ab \neq 0$, and an ℓ-ring R is called an *ℓ-domain* if for any $a, b \in R$, $a > 0$ and $b > 0$ implies $ab > 0$. Certainly if an ℓ-ring is a domain then it is an ℓ-domain, but an ℓ-domain may not be a domain as shown by the following example. However, an f-ring is a domain if and only if it is an ℓ-domain by Theorem 1.20(3).

Example 1.5. Let $S = \{a, b\}$ be the semigroup with the multiplication $ab = ba = a^2 = b^2 = a$, and $\mathbb{R}[S]$ be the semigroup ℓ-algebra with real coefficients defined in example 1.3(2). Then $\mathbb{R}[G]$ is an ℓ-domain (Exercise 33). Since $(a - b)^2 = 0$, $\mathbb{R}[G]$ is not a domain. We notice that $\mathbb{R}[G]$ is an Archimedean and commutative ℓ-ring in which the square of each element is positive since $(\alpha a + \beta b)^2 = (\alpha + \beta)^2 a \geq 0$.

A nonempty subset M of an ℓ-ring R is called an *m-system* if $M \subseteq R^+$ and for any $a, b \in M$ there is an $x \in R^+$ such that $axb \in M$. A nonempty subset S of R is called *multiplicative closed* if for any $a, b \in S$, $ab \in S$. It is clear that if $S \subseteq R^+$ is a multiplicative closed subset of R, then S is an m-system.

Theorem 1.26. *Let R be an ℓ-ring.*

(1) Suppose that P is an ℓ-prime ℓ-ideal and I is an ℓ-ideal of R. If $I^n \subseteq P$ for some positive integer n, then $I \subseteq P$.

(2) A proper ℓ-ideal P of R is ℓ-prime if and only if $a, b \in R^+$ and $aR^+b \subseteq P \Rightarrow a \in P$ or $b \in P$. In particular, if R is commutative, then a proper ℓ-ideal is ℓ-prime if and only if $a, b \in R^+, ab \in P \Rightarrow a \in P$ or $b \in P$.

(3) A proper ℓ-ideal of R is ℓ-prime if and only if $R^+ \setminus P$ is an m-system.

(4) Suppose that M is an m-system of R and I is an ℓ-ideal of R with $I \cap M = \emptyset$. Then I is contained in an ℓ-prime ℓ-ideal P with $P \cap M = \emptyset$.

Proof. (1) Since $I^n \subseteq P$, $I \langle I^{n-1} \rangle \subseteq P$ (Exercise 34), and hence $I \subseteq P$ or $I^{n-1} \subseteq P$. If $I^{n-1} \subseteq P$, by continuing the above procedure, we will eventually have $I \subseteq P$.

(2) Suppose that P is ℓ-prime and $aR^+b \subseteq P$, for some $a, b \in R^+$. Then $\langle R^+ aR^+ \rangle \langle R^+ bR^+ \rangle \subseteq P$, so $\langle R^+ aR^+ \rangle \subseteq P$ or $\langle R^+ bR^+ \rangle \subseteq P$. If $\langle R^+ aR^+ \rangle \subseteq P$, then $\langle a \rangle^3 \subseteq P$, and hence $\langle a \rangle \subseteq P$ by (1). Hence $a \in P$. Similarly if $\langle R^+ bR^+ \rangle \subseteq P$, then $b \in P$. Conversely suppose that the given condition is true, and suppose that I, J are ℓ-ideals of R with $IJ \subseteq P$ and $I \not\subseteq P$. Then there is $0 \leq a \in I \setminus P$. For any $0 \leq b \in J$, $aR^+b \subseteq IJ \subseteq P$ implies $b \in P$. Thus $J \subseteq P$, so P is ℓ-prime.

Suppose that R is commutative. Let P be an ℓ-prime ℓ-ideal of R and $ab \in P$ for some $a, b \in R^+$. Then $aR^+b = R^+(ab) \subseteq P$ implies that $a \in P$ or $b \in P$. Conversely let P be a proper ℓ-ideal of R and for any $a, b \in R^+$, $ab \in P$ implies that $a \in P$ or $b \in P$. Let I, J be ℓ-ideals of R with $IJ \subseteq P$. If $I \not\subseteq P$, then there exists $0 \leq a \in I \setminus P$, so for any $0 \leq b \in J$, $ab \in P$ implies that $b \in P$. Thus $J \subseteq P$ and P is ℓ-prime.

(3) follows immediately from (2).

(4) Let

$$\mathcal{N} = \{J \mid J \text{ is an } \ell\text{-ideal}, I \subseteq J, \text{ and } J \cap M = \emptyset\}.$$

Then $I \in \mathcal{N}$. If $\{J_i\}$ is a chain in $\mathcal{N}$, then $\cup J_i$ is an ℓ-ideal and $(\cup J_i) \cap M = \emptyset$. By Zorn's Lemma, $\mathcal{N}$ has a maximal element P. We show that P is ℓ-prime. Let $a, b \in R^+$, $aR^+b \subseteq P$, and $a, b \notin P$. Then $\langle P, a \rangle \cap M \neq \emptyset$. Let $z_1 \in \langle P, a \rangle \cap M$. Then

$$z_1 \leq n_1 a + r_1 a + a s_1 + u_1 a v_1 + p_1, \ n_1 \geq 0, r_1, s_1, u_1, v_1 \in R^+, p_1 \in P^+.$$

Similarly there exists $z_2 \in \langle P, b \rangle \cap M$. Then

$$z_2 \leq n_2 b + r_2 b + b s_2 + u_2 b v_2 + p_2, \ n_2 \geq 0, r_2, s_2, u_2, v_2 \in R^+, p_2 \in P^+.$$

Since M is an m-system, there is $x \in R^+$ such that $z_1 x z_2 \in M$. On the other hand,

$$z_1 x z_2 \leq (n_1 a + r_1 a + a s_1 + u_1 a v_1 + p_1) x (n_2 b + r_2 b + b s_2 + u_2 b v_2 + p_2)$$

implies $z_1 x z_2 \in P$ since $aR^+b \subseteq P$, which contradicts with $P \cap M = \emptyset$. Thus $aR^+b \subseteq P$ implies $a \in P$ or $b \in P$, so P is ℓ-prime by (2). $\qquad \square$

Theorem 1.27. *Let R be an f-ring.*

(1) *For each $k \geq 1$, $N_k = \{a \in R \mid a^k = 0\}$ is a nilpotent ℓ-ideal of R. Thus $\ell\text{-}N(R) = \{a \in R \mid a \text{ is nilpotent}\}$.*

(2) *If R is ℓ-prime, then R is a totally ordered domain.*

(3) *A proper ℓ-ideal P of R is ℓ-prime if and only if for any $a, b \in R$, $ab \in P$ implies that $a \in P$ or $b \in P$.*

(4) *A proper ℓ-ideal P of R is ℓ-prime if and only if for any $a, b \in R$, $a \wedge b \in P$ implies that $a \in P$ or $b \in P$ and for any $c \in R$, $c^2 \in P$ implies that $c \in P$.*

Proof. (1) We first assume that R is totally ordered. Let $a, b \in N_k$. Then $|a - b| \leq |a| + |b| \leq 2|a|$ or $2|b|$, and hence $|a - b|^k = 0$, $(a - b)^k = 0$, that is, $(a - b) \in N_k$. If $|x| \leq |a|$ for some $a \in N_k$ and $x \in R$, then $|x^k| = |x|^k \leq |a|^k = |a^k| = 0$, so $x^k = 0$ and $x \in N_k$. Take $0 \leq a \in N_k$ and $0 \leq x \in R$. Without loss of generality, suppose $ax \leq xa$. Then

$$0 \leq (ax)^k \leq (xa)^k = x(ax)^{k-1}a \leq x(xa)^{k-1}a \leq \cdots \leq x^k a^k = 0,$$

and similarly

$$0 \leq (xa)^k = x(ax)^{k-1}a \leq x(xa)^{k-1}a = x^2(ax)^{k-2}a^2 \leq \cdots \leq x^k a^k = 0,$$

so $(ax)^k = (xa)^k = 0$. Thus N_k is an ℓ-ideal, and it is clear that $(N_k)^k = 0$.

Since an f-ring is a subdirect product of totally ordered rings, N_k is also an nilpotent ℓ-ideal of it. We leave the verification of this fact as an exercise (Exercise 35).

(2) Suppose that T is an ℓ-prime f-ring and $x \in T$. Since $x^+x^- = 0$ and T contains no nonzero nilpotent element, $(x^-R^+x^+)^2 = \{0\}$ implies $x^-R^+x^+ = \{0\}$. Then T is ℓ-prime implies that $x^- = 0$ or $x^+ = 0$, that is, R is totally ordered. Let $ab = 0$ for some $a, b \in R$. Then $|a||b| = 0$, so $|a|^2 = 0$ or $|b|^2 = 0$ since $|a| \leq |b|$ or $|b| \leq |a|$. Therefore $a = 0$ or $b = 0$ and R is a domain.

(3) Suppose that P is an ℓ-prime ℓ-ideal of R. Then R/P is an ℓ-prime f-ring, and hence by (2) $ab \in P$ implies that $a \in P$ or $b \in P$. The converse is clearly true.

(4) If P is ℓ-prime, then R/P is totally ordered by (2). Since $a \wedge b = 0$ in R implies that $(a + P) \wedge (b + P) = 0$ in R/P, $a + P = 0$ or $b + P = 0$, so $a \in P$ or $b \in P$. Conversely suppose that $a \wedge b \in P$ implies that $a \in P$ or $b \in P$ and $c^2 \in P$ implies that $c \in P$, for $a, b, c \in R$. Assume that $xy \in P$ for some $x, y \in R$. Then since $(|x| \wedge |y|)^2 \leq |x||y| = |xy| \in P$, $(|x| \wedge |y|)^2 \in P$, so $|x| \wedge |y| \in P$ and hence $|x| \in P$ or $|y| \in P$. Hence $x \in P$ or $y \in P$, that is, P is ℓ-prime. □

For an ℓ-ring R, its *p-radical*, denoted by $\ell\text{-}P(R)$, is the intersection of all of the ℓ-prime ℓ-ideals of R. A ring is called *reduced* if it contains no nonzero nilpotent element, and an ℓ-ring is called ℓ-*reduced* if it contains no nonzero positive nilpotent element.

Theorem 1.28. *Let R be an ℓ-ring.*

(1) $\ell\text{-}N(R) \subseteq \ell\text{-}P(R)$ *and each element of* $\ell\text{-}P(R)$ *is nilpotent. If R is commutative or an f-ring, then* $\ell\text{-}N(R) = \ell\text{-}N(P)$.
(2) *The p-radical of* $R/\ell\text{-}P(R)$ *is zero.*
(3) $\ell\text{-}N(R) = \{0\}$ *if and only if* $\ell\text{-}P(R) = \{0\}$.
(4) *Suppose that* $\ell\text{-}N(R) = \{0\}$. *If R is a d-ring or an almost f-ring, then R is a reduced f-ring.*

Proof. (1) Since every nilpotent ℓ-ideal is contained in each ℓ-prime ℓ-ideal by Theorem 1.26, $\ell\text{-}N(R) \subseteq \ell\text{-}P(R)$. Suppose that $a \in R$ is not nilpotent. Then $|a|$ is not nilpotent and $\{|a|^n \mid n \geq 1\}$ is an m-system not containing zero, so Theorem 1.26(4) implies that there exists an ℓ-prime ℓ-ideal I such that $\{|a|^n \mid n \geq 1\} \cap I = \emptyset$, and hence $a \notin I$, so $a \notin \ell\text{-}P(R)$. Thus each element in $\ell\text{-}P(R)$ is nilpotent.

If R is commutative or an f-ring, then ℓ-$N(R) = \{x \in R \mid |x|$ is nilpotent$\}$, so ℓ-$P(R) \subseteq \ell$-$N(R)$.

(2) Each ℓ-ideal of R/I can be expressed as J/I, where J is an ℓ-ideal of R containing I. Also J/I is ℓ-prime in R/I if and only if J is ℓ-prime in R (Exercise 36). Hence ℓ-$P(R/\ell$-$P(R)) = \{0\}$.

(3) Suppose that ℓ-$N(R) = \{0\}$. If ℓ-$P(R) \neq \{0\}$, take $0 < a_0 \in \ell$-$P(R)$. Then $\langle a_0 \rangle^n \neq \{0\}$ for any positive integer n, so $\langle R^+ a_0 R^+ \rangle^2 \neq \{0\}$ since $\langle a_0 \rangle^3 \subseteq \langle R^+ a_0 R^+ \rangle$, and hence there is $b_0 \in R^+$ such that $a_1 = a_0 b_0 a_0 \neq 0$. Similarly, there is $b_1 \in R^+$ such that $a_2 = a_1 b_1 a_1 \neq 0$. Continuing inductively, we obtain $a_n = a_{n-1} b_{n-1} a_{n-1} \neq 0$ for all $n \geq 1$. It follows that $\{a_i \mid i \geq 0\}$ is an m-system not containing 0 (Exercise 37), so by Theorem 1.26(4) there is an ℓ-prime ℓ-ideal P such that $P \cap \{a_i \mid i \geq 0\} = \emptyset$. Thus $a_0 \notin P$, which is a contradiction, and hence ℓ-$P(R) = \{0\}$.

(4) Suppose first that R is a d-ring. Since ℓ-$N(R) = \{0\}$, by (3) R is a subdirect product of ℓ-prime ℓ-rings which are d-rings (Exercise 38). We show that an ℓ-prime d-ring D is a totally ordered domain. Let $a \in D^+$ with $aD^+ = \{0\}$ or $D^+ a = \{0\}$. Then $aD^+ a = \{0\}$, so D is ℓ-prime implies $a = 0$. Let $x \wedge y = 0$ for $x, y \in D$ and $c, d \in D^+$. Then

$$0 \leq d(cx \wedge y) = (dcx \wedge dy) \leq (dc + d)x \wedge (dc + d)y = 0$$

implies $d(cx \wedge y)$ for all $d \in D^+$, and hence $cx \wedge y = 0$. Similarly, $xc \wedge y = 0$. Hence D is an f-ring. Thus by Theorem 1.27(4), D is totally ordered and a domain. Therefore, R is a reduced f-ring.

Now suppose that R is an ℓ-prime almost f-ring. We first show that if $a \in R^+$ and $a^2 = 0$, then $a = 0$. Let $z \in R^+$. We claim that $aza = 0$. Suppose that $x = az - za$. If $x^+ = 0$, then $az \leq za$ implies $aza \leq za^2 = 0$, so $aza = 0$. Similarly $x^- = 0$ implies $aza = 0$. In the following we assume that $x^+ \neq 0$ and $x^- \neq 0$. Then $x^+ x^- = x^- x^+ = 0$ implies for any $y, w \in R^+$, $(x^- yx^+)^2 = (x^+ wx^-)^2 = 0$. By Theorem 1.22(1), for each element $u \in R$, $u^2 \geq 0$, so $(x^- yx^+ - a)^2 \geq 0$ implies that

$$0 \leq ax^- yx^+ + x^- yx^+ a \leq (x^- yx^+)^2 + a^2 = 0.$$

Thus $ax^- yx^+ = 0$ for all $y \in R^+$. It follows that $ax^- = 0$ since $x^+ \neq 0$ and R is ℓ-prime. Similarly, $(x^+ wx^-)^2 = 0$ and $a^2 = 0$ for all $w \in R^+$ implies $ax^+ = 0$. Thus $ax = ax^+ - ax^- = 0$, so $a(az - za) = 0$, and hence $aza = 0$.

Therefore in any case we have $aza = 0$ for any $z \in R^+$, that is, $aR^+ a = \{0\}$. It follows that $a = 0$ since R is ℓ-prime. Hence R contains no nonzero positive nilpotent element, that is, R is ℓ-reduced. Now let $a, b \in R^+$ with $ab = 0$. Then for any $z \in R^+$, $(bza)^2 = 0$, so $bza = 0$, that is, $bR^+ a = \{0\}$.

Thus $a = 0$ or $b = 0$, and R is an ℓ-domain. Therefore $x^+ = 0$ or $x^- = 0$ for any $x \in R$ since $x^+ x^- = 0$. Hence R is totally ordered and a domain.

If R is an almost f-ring with ℓ-$N(R) = \{0\}$, then R is a subdirect product of ℓ-prime almost f-rings, and hence it is a subdirect product of totally ordered domains. Therefore R is a reduced f-ring. $\square$

By Theorem 1.28(4), a reduced almost f-ring is an f-ring. Interestingly a reduced partially ordered ring satisfying a similar relation to almost f-rings is also an f-ring.

Theorem 1.29. *For a reduced partially ordered ring R, if for any $a \in R$, there exist $a_1, a_2 \in R^+$ such that $a = a_1 - a_2$ and $a_1 a_2 = a_2 a_1 = 0$, then R is an f-ring.*

Proof. We fist show that zero is the greatest lower bound of a_1, a_2. Suppose that $c \le a_1, a_2$ and $c = c_1 - c_2$ with $c_1, c_2 \in R^+$ and $c_1 c_2 = c_2 c_1 = 0$. Then $0 \le c_1^2 = c_1(c_1 + c_2) = c_1 c \le c_1 a_1$ and $0 \le c_1^2 \le a_2 c_1$ implies that $0 \le c_1^4 \le c_1 a_1 a_2 c_1 = 0$, and hence $c_1^4 = 0$ and $c_1 = 0$ since R is reduced. Thus $c = -c_2 \le 0$.

Next we show that $a_1 = a \vee 0$. Clearly $a_1 \ge a, 0$. Suppose that $b \ge a, 0$ for some $b \in R$. Then

$$a_1 - b \le a_1, a_2 \Rightarrow a_1 - b \le 0,$$

so $a_1 \le b$. Thus $a_1 = a \vee 0$. It is straightforward to check that for any $a, b \in R$, $a \vee b = [(a - b) \vee 0] + b$, and hence the partial order is a lattice order.

Now it is easy to check that R is an almost f-ring, so it is an f-ring. $\square$

An ℓ-prime ℓ-ideal P of an ℓ-ring R is called *minimal* if any ℓ-ideal of R properly contained in P is not an ℓ-prime ℓ-ideal of R. For instance, in an ℓ-domain, $\{0\}$ is the unique minimal ℓ-prime ℓ-ideal.

Theorem 1.30. *Let R be an ℓ-ring.*

(1) Each ℓ-prime ℓ-ideal of R contains a minimal ℓ-prime ℓ-ideal.

(2) An ℓ-prime ℓ-ideal P is minimal if and only if any m-system properly containing $R^+ \setminus P$ contains 0.

(3) If R is ℓ-reduced, then an ℓ-prime ℓ-ideal P is minimal if and only if for each $x \in P$ with $x \ge 0$, there exists $y \notin P$ with $y \ge 0$ such that $xy = 0$.

(4) If R is ℓ-reduced, then for each minimal ℓ-prime ℓ-ideal P, R/P is an ℓ-domain. Thus an ℓ-ring is ℓ-reduced if and only if it is a subdirect product of ℓ-domains.

Proof.   (1) Let P be an ℓ-prime ℓ-ideal. Consider

$$\mathcal{M} = \{N \mid N \subseteq P \text{ and } N \text{ is an } \ell\text{-prime } \ell\text{-ideal}\}.$$

Then $P \in \mathcal{M}$. Partially order $\mathcal{M}$ by set inclusion. For a chain $\{P_i \mid i \in I\} \subseteq \mathcal{M}$, By Theorem 1.26(3), $J = \cap_{i \in I} P_i$ is an ℓ-prime ℓ-ideal since $R^+ \setminus J = \cup_{i \in I} (R^+ \setminus P_i)$ is an m-system. By Zorn's Lemma (or Exercise 4), $\mathcal{M}$ has a minimal element, which is a minimal ℓ-prime ℓ-ideal contained in P.

(2) Suppose that P is a minimal ℓ-prime ℓ-ideal and $(R^+ \setminus P) \subsetneqq M$ for some m-system M. If $0 \notin M$, then by Theorem 1.26(4) there exists an ℓ-prime ℓ-ideal I such that $M \cap I = \emptyset$. It follows that $I \subseteq P$, and hence $I = P$. Then $M \subseteq R^+ \setminus P$, which is a contradiction. Thus $0 \in M$. Conversely, suppose that I is an ℓ-prime ℓ-ideal and $I \subseteq P$. Then $(R^+ \setminus P) \subseteq (R^+ \setminus I)$, and if this inclusion is proper, then $0 \in R^+ \setminus I$, which is a contradiction. Hence we must have $R^+ \setminus P = R^+ \setminus I$, and hence $I = P$. Therefore P is a minimal ℓ-prime ℓ-ideal.

(3) "$\Leftarrow$" Let P be an ℓ-prime ℓ-ideal. By (1), there is a minimal ℓ-prime ℓ-ideal Q such that $Q \subseteq P$. If $Q \neq P$, then take $0 < x \in P \setminus Q$. By the assumption, we can find $y \notin P$ and $y \geq 0$ such that $xy = 0$. Then $(yR^+x)^2 = 0$ and R is ℓ-reduced implies $yR^+x = 0$, and hence $y \in Q \subseteq P$, which is a contradiction. Therefore we must have $Q = P$, so P is minimal.

"$\Rightarrow$" Let P be a minimal ℓ-prime ℓ-ideal of R and $0 \leq a \in P$. Define

$$S = \{a_1 a a_2 a \cdots a_n a a_{n+1} \mid n \geq 1,\ a_i \in R^+ \setminus P\} \cup (R^+ \setminus P).$$

Then $(R^+ \setminus P) \subsetneqq S$ since, for instance, $a_1 a a_2 \in P \cap S$, and S is an m-system, so by (2), $0 \in S$. Thus $a_1 a a_2 a \cdots a_n a a_{n+1} = 0$ for some $n \geq 1$.

We observe that if $uv = 0$ for $u, v \in R^+$, then $(vu)^2 = 0$ implies $vu = 0$ since R is ℓ-reduced, and hence $(uxv)^2 = 0$ for any $x \in R^+$. Thus $uxv = 0$. This observation tells us that if $uv = 0$, then we may insert any $x \geq 0$ between them to get $uxv = 0$. We use this basic fact to show that if $x_1 \cdots x_i x_{i+1} \cdots x_k = 0$, for some $k \geq 2$ and each $x_j \in R^+$, then $x_1 \cdots x_{i+1} x_i \cdots x_k = 0$. In fact, by inserting the terms $x_{i+1}, (x_{i+2} \cdots x_k)(x_1 \cdots x_{i-1}), x_i$ into $x_1 \cdots x_i x_{i+1} \cdots x_k = 0$, we get

$$x_1 \cdots x_{i-1}(x_{i+1}) x_i (x_{i+2} \cdots x_k)(x_1 \cdots x_{i-1}) x_{i+1}(x_i) x_{i+2} \cdots x_k = 0,$$

so R is ℓ-reduced and $[(x_1 \cdots x_{i-1}(x_{i+1}) x_i (x_{i+2} \cdots x_k))]^2 = 0$ imply $x_1 \cdots x_{i-1} x_{i+1} x_i \cdots x_k = 0$. This analysis shows that in a zero product of positive elements, we may interchange the order of two elements and the product is still zero. Using this idea, from $a_1 a a_2 a \cdots a_n a a_{n+1} = 0$, we have $a_1 \cdots a_{n+1}a^n = 0$. Since $a_1, a_2 \in R^+ \setminus P$, $a_1x_1a_2 \in R^+ \setminus P$ for some $x_1 \in R^+$, and hence $a_1x_1a_2x_2a_3 \in R^+ \setminus P$ for some $x_2 \in R^+$. Continuing this process, we have

$$a_1x_1a_2x_2a_3x_3 \cdots a_nx_na_{n+1} \in R^+ \setminus P$$

for some $x_1, \cdots, x_n \in R^+$. Now $a_1aa_2a \cdots a_naa_{n+1} = 0$ implies

$$a_1a_2 \cdots a_na_{n+1}a^n = 0 \implies a_1x_1a_2x_2a_3x_3 \cdots a_nx_na_{n+1}a^n = 0.$$

Let $y = a_1x_1a_2x_2 \cdots a_nx_na_{n+1}$. Then $0 \le y \notin P$ and $ya^n = 0$, so $(ay)^n = 0$. Therefore $ay = 0$.

(4) Suppose that P is a minimal ℓ-prime ℓ-ideal of R and suppose that $a + P \in R/P$ with $a \in R^+ \setminus P$. We assume that $a^2 \in P$. Then there exists $0 \le y \notin P$ such that $a^2y = 0$, and hence $(aya)^2 = (aya)(aya) = 0$, so $aya = 0$. It follows that $(ay)^2 = 0$, so $ay = 0$. Then $(yR^+a)^2 = 0$, and hence $yR^+a = 0$. Now P is ℓ-prime implies $a \in P$ or $y \in P$, which is a contradiction. Therefore $a^2 \notin P$ and hence R/P is ℓ-reduced. Since R/P is ℓ-prime and ℓ-reduced, R/P is an ℓ-domain (Exercise 39).

If R is ℓ-reduced, then the intersection of all minimal ℓ-prime ℓ-ideals of R is zero by Theorem 1.28(3), and hence R is isomorphic to a subdirect product of ℓ-domains by previous argument. It is clear that the subdirect product of ℓ-domains is ℓ-reduced. □

In the following we consider the ℓ-radical of an ℓ-algebra over a totally ordered field F. The main result is to show that if the ℓ-radical is zero, then a finite-dimensional ℓ-algebra is Archimedean over F. Clearly, for an ℓ-algebra A over a totally ordered field F, $\ell\text{-}N(A)$ is closed under the scalar multiplication, that is, $\ell\text{-}N(A)$ is an ℓ-ideal of ℓ-algebra A. Let V be a vector lattice over a totally ordered field F. An element $a \in V^+$ is called a *strong unit of V over F* if for every $x \in V$, there is an $\alpha_x \in F$ such that $x \le \alpha_x a$.

Theorem 1.31.

(1) Every finite-dimensional vector lattice V has a strong unit.

(2) Let A be an ℓ-algebra over a totally ordered field F with strong unit. The set

$$i(A) = \{a \in A \mid \alpha|a| \le u \text{ for every strong unit } u \text{ and every } \alpha \in F^+\}$$

is an ℓ-ideal of A, called i-ideal, and $i(A)$ contains no strong unit of A over F. A is Archimedean over F if and only if $i(A) = \{0\}$.

(3) If A is finite-dimensional, then ℓ-ideal $i(A)$ is nilpotent. Thus if ℓ-$N(A) = \{0\}$, then A is Archimedean over F.

Proof. (1) Let $v_1, \cdots, v_n$ be a vector space basis of V over F for some positive integer n. Then $u = |v_1| + \cdots + |v_n|$ is a strong unit. In fact, for any $v \in V$, $v = \alpha_1 v_1 + \cdots + \alpha_n v_n$, $\alpha_i \in F$, implies

$$v \leq |v| \leq |\alpha_1||v_1| + \cdots + |\alpha_n||v_n| \leq (|\alpha_1| + \cdots + |\alpha_n|)u.$$

(2) Let $x, y \in i(A)$. For $\alpha \in F^+$ and a strong unit u,

$$2\alpha|x - y| \leq 2\alpha|x| + 2\alpha|y| \leq u + u = 2u$$

implies $\alpha|x - y| \leq u$, so $x - y \in i(A)$. Clearly for any $a \in i(A)$ and $\alpha \in F$, $\alpha a \in i(A)$, and $|y| \leq |x|$ with $y \in A$ and $x \in i(A)$ implies that $y \in i(A)$. Thus $i(A)$ is a convex vector sublattice of A. Suppose $x \in i(A)$ and $a \in A$. For a strong unit u, $|a|u \leq \beta u$ for some $0 < \beta \in F$. Hence for any $\alpha \in F^+$, $\alpha\beta|ax| \leq \alpha\beta|a||x| \leq |a|u \leq \beta u$, so $\alpha|ax| \leq u$. Thus $ax \in i(A)$. Similarly $xa \in i(A)$. Thus $i(A)$ is an ℓ-ideal of A. Finally for a strong unit u of A, $2u \not\leq u$ implies $u \notin i(A)$.

If A is Archimedean over F, then clearly $i(A) = \{0\}$. Suppose $i(A) = \{0\}$ and $\alpha x \leq y$ for some $x, y \in A^+$ and all $\alpha \in F^+$. For any strong unit u, there is $\beta \in F^+$ such that $y \leq \beta u$, so $\alpha x \leq u$ for all $\alpha \in F^+$. Thus $x \in i(A)$, and hence $x = 0$. Therefore A is Archimedean over F.

(3) Let $I = i(A)$. As we have done before, define $I^{(2)} = \langle I^2 \rangle$, $I^{(3)} = \langle II^{(2)} \rangle, \cdots, I^{(n)} = \langle II^{(n-1)} \rangle$ for any $n \geq 2$. Clearly $I^n \subseteq I^{(n)}$ for any $n \geq 2$. We show that if $I^{(k)} \neq 0$, then $I^{(k+1)}$ is properly contained in $I^{(k)}$ for $k \geq 2$. Since $I^{(k)}$ is finite-dimensional as a vector lattice over F, by (1) $I^{(k)}$ will have a strong unit u_k. Let u be a strong unit of A. If $a \in I^{(k+1)}$, then $|a| \leq \sum |x_i y_i|$, where $x_i \in I$ and $y_i \in I^{(k)}$. Then for some $\beta, \gamma \in F^+$ we have $|x_i| \leq \beta u_1$ and $|y_i| \leq \gamma u_k$, so $|x_i y_i| \leq |x_i||y_i| \leq \beta\gamma u_1 u_k$. Since $I^{(k)}$ is an ℓ-ideal of A, $uu_k \in I^{(k)}$, so $uu_k \leq \delta u_k$ for some $\delta \in F^+$. Hence for all $\alpha \in F^+$,

$$\beta\gamma\delta\alpha|x_i y_i| \leq \beta\gamma\delta\alpha|x_i||y_i| \leq \alpha\beta^2\gamma^2\delta u_1 u_k \leq \beta\gamma uu_k \leq \beta\gamma\delta u_k,$$

so $\alpha|x_i y_i| \leq u_k$ for all $\alpha \in F^+$. Let v_k be an arbitrary strong unit of $I^{(k)}$. Then $u_k \leq \lambda v_k$ for some $0 < \lambda \in F^+$. Thus for all $\alpha \in F^+$, $\lambda\alpha|x_i y_i| \leq u_k \leq \lambda v_k$, and hence $\alpha|x_i y_i| \leq v_k$. Thus $|x_i y_i| \in i\left(I^{(k)}\right)$, so $\sum |x_i y_i| \in i\left(I^{(k)}\right)$. Therefore $a \in i\left(I^{(k)}\right)$ and $I^{(k+1)} \subseteq i\left(I^{(k)}\right) \subsetneq I^{(k)}$ by (2).

Now since A is finite-dimensional over F, there must be a positive integer k such that $I^{(k)} = \{0\}$, so $I^k = 0$. It follows that $I = i(A) \subseteq \ell$-$N(A)$, and hence if ℓ-$N(A) = \{0\}$, then $i(A) = \{0\}$, so by (2) A is Archimedean over F. □

The following result is a direct consequence of Theorems 1.31 and 1.17.

Corollary 1.3. *Suppose that A is a finite-dimensional ℓ-algebra over a totally ordered field F. If A is Archimedean over F, then as a vector lattice over F, A is a finite direct sum of maximal convex totally ordered subspaces of A over F. In particular, if ℓ-N(A) = {0}, then A is a finite direct sum of maximal convex totally ordered subspaces of A over F.*

Proof. If A is Archimedean over F, then A has no maximal convex totally ordered subspace that is bounded above. Since that A is finite-dimensional implies that condition (C) in Theorem 1.15 is satisfied, Theorem 1.17 applies. □

The following result gives the further relation between f-elements and d-elements in an ℓ-unital ℓ-domain.

Theorem 1.32. *Let R be an ℓ-unital ℓ-domain.*

(1) If a is a d-element, then either a is an f-element or $a \wedge 1 = 0$.

(2) If a is a d-element, then either the set $\{a^n \mid n \geq 0\}$, where $a^0 = 1$, is disjoint or a^k is an f-element for some $k \geq 1$.

(3) If for $0 < a \in R$, a^k is an f-element, then a is a d-element and a basic element.

Proof. We first notice that since R is an ℓ-domain, $f(R)$ is a totally ordered domain.

(1) Suppose that $a \wedge 1 = b > 0$. Then b is an f-element since $b \leq 1$. Let $x, y \in R$ such that $x \wedge y = 0$. Then $0 \leq ax \wedge by \leq ax \wedge ay = 0$ since $b \leq a$ and a is a d-element. Then $ax \wedge by = 0$ implies $bax \wedge by = 0$, so $b(ax \wedge y) = 0$. Hence $ax \wedge y = 0$ since R is an ℓ-domain and $b > 0$. Similarly, $xa \wedge y = 0$. Therefore a is an f-element.

(2) Suppose that for any $n \geq 1$, a^n is not an f-element. Then by (1), $a^n \wedge 1 = 0$ for any $n \geq 1$. Thus for any positive integer i, j with $1 \leq i < j$, $a^j \wedge a^i = a^i(a^{j-i} \wedge 1) = 0$, so the set $\{a^n \mid n \geq 0\}$ is disjoint.

(3) If $x \wedge y = 0$, then $a^{k-1}(ax \wedge ay) \leq a^k x \wedge a^k y = 0$, so $ax \wedge ay = 0$. Similarly $xa \wedge ya = 0$. Thus a is a d-element. Let $0 \leq b, c \leq a$. Then $0 \leq a^{k-1}b, a^{k-1}c \leq a^k \in f(R)$ which is totally ordered, so $a^{k-1}b$ and $a^{k-1}c$ are comparable. Thus b, c are comparable, that is, a is a basic element. □

As an application of Theorem 1.32, we determine all the lattice orders on polynomial ring $F[x]$, where F is a totally ordered field, such that x is a d-element.

Corollary 1.4. *Let $R = F[x]$ be an ℓ-algebra over F in which x is a d-element. Then either $R^+ = F^+[x]$ or $f[R] = F[x^k]$ for some $k \geq 1$, and $R = f(R) + f(R)x + \cdots + f(R)x^{k-1}$ with $R^+ = f(R)^+ + f(R)^+x + \cdots + f(R)^+x^{k-1}$.*

Proof. Since x is a d-element, $(-1 \vee 0)x = -x \vee 0 = 0$ implies $1^- = 0$, so $1 > 0$. By Theorem 1.32(2), either $\{x^n \mid n \geq 0\}$ is disjoint or there exists positive integer k such that $x^k \in f(R)$. In the first case, it is clear that $R^+ = F^+[x]$. In the second case, suppose that k is the smallest positive integer such that $x^k \in f(R)$. Then by Theorem 1.32(1), $\{1, x, \cdots, x^{k-1}\}$ is a disjoint set. Let $E = F[x^k]$. Then $R = E + Ex + \cdots + Ex^{k-1}$ and $R^+ = E^+ + E^+x + \cdots + E^+x^{k-1}$ since E^+ consists of f-elements. Then $E = f(R)$. $\qquad\square$

Exercises

(1) Let $(A, \leq)$ be a partially ordered set with the partial order $\leq$. Using Zorn's Lemma to show that $\leq$ can be extended to a total order on A, that is, there exists a total order on A which is an extension of $\leq$.

(2) Let A be a nonempty set. Define $\leq$ on A by $\forall a, b \in A$, $a \leq b$ if $a = b$. Show that $\leq$ is a partial order on A and if A has more than one element, then it is not a lattice order.

(3) Prove that the power set P_A of a set A is a complete distributive lattice under the partial order of set inclusion defined in Example 1.1.

(4) Let $(A, \leq)$ be a nonempty partially ordered set. Prove, by Zorn's Lemma, that if each subset of A that is a chain has a lower bound in A, then A contains a minimal element.

(5) Let G be a partially ordered group. Suppose that $g \vee 0$ exists for any $g \in G$. Prove that G is an ℓ-group and for any $f, g \in G$,

$$f \vee g = [(f - g) \vee 0] + g \text{ and } f \wedge g = g - [(-f + g) \vee 0].$$

(6) Verify Example 1.2(2) and (3).

(7) Let G be an ℓ-group and $a \wedge b = 0$ for $a, b \in G$. Prove that $na \wedge mb = 0$ for any positive integers n and m.

(8) Let G be an ℓ-group and $X \subseteq G$. Prove that $X \subseteq X^{\perp\perp}$ and $X^{\perp\perp\perp} = X^\perp$.

(9) Let G be an ℓ-group and G_1, G_2 be distinct convex ℓ-subgroups of G. Prove that if for any $a \in G_1^+, b \in G_2^+$, $a + b = b + a$, then for any $x \in G_1, y \in G_2$, $x + y = y + x$.

(10) For an ℓ-group G and a normal convex ℓ-subgroup N of G, prove that if $x + N = x_1 + N$ and $y + N = y_1 + N$, then

$$(x \vee y) + N = (x_1 \vee y_1) + N \text{ and } (x \wedge y) + N = (x_1 \wedge y_1) + N.$$

(11) Prove that the projection $\pi : G \to G/N$, where G is an ℓ-group and N is a normal convex ℓ-subgroup of G, preserves sup and inf.

(12) Let G be an ℓ-group. Prove that if $\forall x, y \in G$, $x \wedge y = 0 \Rightarrow x = 0$ or $y = 0$, then G is totally ordered.

(13) Let G be an ℓ-group and S be a subset of G. Then S is a basis of G if and only if S is a disjoint set of basic elements and $S^{\perp} = \{0\}$.

(14) Let G be an ℓ-group and $0 < g \in G$. Define $T = \{x \in G \mid 0 < x \leq g\}$. Suppose that for any $x, y \in T$, $x \wedge y \neq 0$. Prove that any two elements in T are comparable.

(15) Let V be a vector lattice over a totally ordered field F. Prove that if V is Archimedean, then V is Archimedean over F.

(16) Let V be a vector lattice over a totally ordered Archimedean field F. Prove that if V is Archimedean over F, then V is Archimedean.

(17) Let V be a vector lattice over a totally ordered field F. Prove that $\forall \alpha \in F, v \in V$, $|\alpha v| = |\alpha||v|$.

(18) Let V be a vector lattice over a totally ordered field F and $\{V_i \mid i \in I\}$ be a collection of convex vector sublattices of V. Prove

$$\sum_{i \in I} V_i = \{v \in V \mid v = v_1 + \cdots + v_k, v_j \in V_{k_j}\}$$

is a convex vector sublattice of V and $\sum_{i \in I} V_i$ is the convex vector sublattice generated by the family $\{V_i \mid i \in I\}$.

(19) Prove Theorem 1.18.

(20) Let R be an ℓ-ring and I be an ℓ-ideal. Prove that R/I is an ℓ-ring with respect to the partial order $a + I \leq b + I$ if $a \leq b + c$ for some $c \in I$.

(21) Verify $a \wedge 0$ and $a \vee 0$ in Theorem 1.19(1).

(22) Suppose that R is an ℓ-ring with the positive cone P and $u \in P$ is a unit. Prove that uP is the positive cone of an ℓ-ring R.

(23) Prove both total orders defined in Example 1.3(3) are not Archimedean over F.

(24) For the polynomial algebra $R = F[x]$ over a totally ordered field F, fix a positive integer $n \geq 2$. Define $p(x) = a_k x^k + \cdots + a_1 x + a_0 \geq 0$ if $k > n$ and $a_k > 0$, or if $k \leq n$ and $a_k > 0$, $a_0 \geq 0$. Prove that R is an ℓ-ring with squares positive.

(25) Let R be an ℓ-ring and $x, y \in R^+$. Prove that $|x - y| = x + y$ if and only if $x \wedge y = 0$.

(26) Let $\varphi : R \to S$ be an ℓ-homomorphism of the two ℓ-rings R and S. Show that $\mathrm{Ker}(\varphi)$ is an ℓ-ideal of R.

(27) Let R be an ℓ-ring and $\{J_i \mid i \in I\}$ be a family of ℓ-ideals of R with $\cap_{i \in I} J_i = \{0\}$. Prove that $\{\{a + J_i\}_{i \in I} \mid a \in R\}$ is an ℓ-subring of the direct product $\Pi_{i \in I} R/J_i$, and $R \cong \{\{a + J_i\}_{i \in I} \mid a \in R\}$.

(28) Let $\pi : R \to R/I$ be the projection. Prove that if N is an ℓ-ideal of R/I, then there exists an ℓ-ideal $J \supseteq I$ such that $\pi(J) = N$. Thus each ℓ-ideal of R/I can be written as J/I for some ℓ-ideal $J \supseteq I$ in R.

(29) Prove that if an ℓ-ring R is a subdirect product of totally ordered rings, then R is an f-ring.

(30) Let R be an ℓ-ring and I, J nilpotent ℓ-ideals of R. Show that $I + J$ is also nilpotent.

(31) For a commutative ℓ-ring R and a nilpotent element x, prove $\langle x \rangle$ is a nilpotent ℓ-ideal of R.

(32) Prove that the J defined in Theorem 1.25(3) is an ℓ-ideal.

(33) Verify that the semigroup ℓ-ring in Example 1.5 is an ℓ-domain.

(34) Let I and P be ℓ-ideals of an ℓ-ring R. Prove if $I^n \subseteq P$ for some positive integer n, then $I\langle I^{n-1} \rangle \subseteq P$.

(35) For an f-ring R, prove that $N_k = \{a \in R \mid a^k = 0\}$ is an ℓ-ideal of R.

(36) For an ℓ-ring R and a proper ℓ-ideal I, if J is an ℓ-ideal containing I, then J/I is ℓ-prime in R/I if and only if J is ℓ-prime in R.

(37) Verify that $\{a_i \mid i \geq 0\}$ in Theorem 1.28(3) is an m-system.

(38) Let R be a d-ring or an almost f-ring. Prove that for any ℓ-ideal I of R, R/I is also a d-ring or an almost f-ring.

(39) Prove that an ℓ-prime and ℓ-reduced ℓ-ring is an ℓ-domain, and an ℓ-reduced o-ring is a domain.

(40) Consider the polynomial ring $R = \mathbb{R}[x]$. Prove that if R is an ℓ-ring with squares positive, $x \in R^+$, and $1 \wedge x^n = 0$, for a fixed positive integer n, then the lattice order on R is P_n defined in Example 1.3(3).

(41) Consider the field $\mathbb{Q}[\sqrt{2}] = \{\alpha + \beta\sqrt{2} \mid \alpha, \beta \in \mathbb{Q}\}$. Prove that the positive cone of an ℓ-field $\mathbb{Q}[\sqrt{2}]$ in which $1 \not> 0$ is equal to uP, where P is the positive cone of an ℓ-field $\mathbb{Q}[\sqrt{2}]$ with $1 > 0$ and $u \in P$ is invertible with $u^{-1} \notin P$.

(42) Describe all the lattice orders on group algebra $\mathbb{R}[G]$, where G is a group of order 2.

(43) Consider ring $R = \mathbb{R} \times \mathbb{R}$. Define the positive cone on R by $P = \{(a, b) \mid b > 0\} \cup \{(0, 0)\}$. Prove that (R, P) is a partially ordered ring,

but not an ℓ-ring.

(44) Consider $n \times n$ matrix algebra $M_n(\mathbb{R})$ ($n \geq 2$). Define the positive cone

$$P = \{(a_{ij}) \mid a_{nj} = 0, j = 1, \cdots, n - 1 \text{ and } a_{nn} > 0\} \cup \{(0,0)\}.$$

Prove that $(M_n(\mathbb{R}), P)$ is a partially ordered ring, however it is not an ℓ-ring.

(45) Consider polynomial ring $\mathbb{R}[x]$. Define the positive cone

$$P = \{f(x) \mid \text{ each coefficient of } f(x) \text{ is strictly positive}\} \cup \{0\}.$$

Prove that $(\mathbb{R}[x], P)$ is a partially ordered ring, but not an ℓ-ring.

(46) Prove that a unital d-ring must be an f-ring.

(47) Let $R = x\mathbb{R}[x]$ be the ring of polynomials with zero constant over $\mathbb{R}$. Order R lexicographically by defining $a_n x^n + \cdots + a_1 x > 0$ if $a_n > 0$. Then R is a totally ordered ring. Define $A = \{(x, a, y, z) \mid a \in \mathbb{R}, x, y, z \in R\}$ with the coordinatewise addition and following multiplication

$$(x, a, y, z)(x', a', y', z') =$$
$$(2xx' + ax' + a'x, aa', x(y' + z') + x'(y + z) + a'y + ay',$$
$$x(y' + z') + x'(y + z) + a'z + az').$$

Then A becomes a ring with identity $(0, 1, 0, 0)$. Define the positive cone as $(x, a, y, z) \geq 0$ if

$$x > 0, \text{ or } x = 0 \text{ and } a > 0, \text{ or } x = a = 0, \text{ and } y \geq 0 \text{ and } z \geq 0.$$

Prove that A is a commutative ℓ-ring in which the identity element is a weak unit in the sense that $1 \wedge a = 0$ implies that $a = 0$ for any $a \in A$, however A is not an f-ring.

(48) Prove that an ℓ-ring R is an almost f-ring if and only if for any $a \in R$, $|a|^2 = a^2$.

(49) Prove that an ℓ-ring is an f-ring if and only if for any $a, b \in R^+$, $\langle a \wedge b \rangle = \langle a \rangle \cap \langle b \rangle$.

(50) Let R be an ℓ-ring and I be an ℓ-ideal of R. I is called ℓ-*semiprime* if for any ℓ-ideal H of R, $H^k \subseteq I$ for some positive integer implies that $H \subseteq I$. Prove that an ℓ-ideal I is ℓ-semiprime if and only if for any $a \in R^+$, $aR^+a \subseteq I$ implies that $a \in I$.

Chapter 2

Lattice-ordered algebras with a d-basis

In this chapter we present the structure theory of unital finite-dimensional Archimedean ℓ-algebras over a totally ordered field with a d-basis. The structure theory on this class of ℓ-algebras is similar to Wedderburn's structure theory of finite-dimensional algebras in general ring theory.

2.1 Examples and basic properties

G. Birkhoff and R. S. Pierce started a systematic study of ℓ-rings in their paper "Lattice-ordered Rings" published in 1956. Based on their study of various examples of ℓ-rings, they observed that since "in general, lattice-ordered algebras can be quite pathological", general structure theorems are very difficult to find. Therefore, they suggested studying special classes of ℓ-rings. One class in particular has been studied intensively is that of f-rings, whose general structure is much better understood today.

However, M. Henriksen pointed out that the class of f-rings excludes many important examples of ℓ-rings and ℓ-algebras [Henriksen (1995)]. For instance, neither matrix and triangular matrix ℓ-algebras with the entrywise order nor group ℓ-algebras and polynomial ℓ-rings with the coordinatewise order are f-rings. Henriksen's observations prompted researchers to look beyond f-rings, for new classes of ℓ-rings and ℓ-algebras that contain these important examples and, at the same time, maintain good structure theory. In particular, Henriksen suggested the following problem as a place to start (Problem 4, [Henriksen (1995)]):

> Develop a structure theory for a class of lattice-ordered rings that include semigroup algebras over $\mathbb{R}$. If S is a multiplicative semigroup, $s_1, s_2, ..., s_n \in S$ and $a_1, a_2, ..., a_n \in \mathbb{R}$, let $\sum a_i s_i \geq 0$

if $a_i \geq 0$ for $1 \leq i \leq n$. Do this at least for a class of semigroups large enough to include $\{1, x, ..., x^n, ...\}$ and the semigroup of unit matrices $\{E_{ij}\}$ (where E_{ij} has a 1 in row i and column j, and zeros elsewhere for $1 \leq i \leq n$ and $1 \leq j \leq n$).

For general ℓ-rings there is no good structure theory because the defining condition $(a, b \geq 0 \Rightarrow ab \geq 0)$ that relates the order and the multiplication is pretty loose. The challenge is then to find appropriate stronger conditions. One way of keeping some of the advantages of f-rings and d-rings while at the same time broadening the class of ℓ-rings and ℓ-algebras under consideration is the following thoughts. We know that a d-ring is an ℓ-ring whose positive cone consists entirely of d-elements. We may broaden the condition by requiring only that the positive cone be *generated* by d-elements. This modification motivates the following definition.

Definition 2.1. Let R be an ℓ-ring. A subset S of R is called a *d-basis* if S is a basis of the additive ℓ-group of R, defined in Chapter 1, and each element in S is a d-element of R.

In this chapter we will study algebraic structure of unital finite-dimensional Archimedean ℓ-algebra over a totally ordered field with a d-basis. This class of ℓ-rings contains rich examples. Before we provide some examples, we prove that the identity element in such ℓ-algebras must be positive. Throughout this chapter F always denotes a totally ordered field and all ℓ-rings and ℓ-algebras are nontrivial. Recall the condition (C) for an ℓ-group G from Theorem 1.15.

(C) Each $0 < g \in G$ is greater than at most a finite number of disjoint elements.

Theorem 2.1. *Let A be a unital Archimedean ℓ-algebra over F with a d-basis and satisfy condition (C).*

(1) Each basic element of A is a d-element.
(2) The identity element $1 > 0$.

Proof. Let S be a d-basis of A. By Theorem 1.17, A is the direct sum of $s^{\perp\perp}$, $s \in S$, considered as a vector lattice over F.

(1) Let x be a basic element. Then $x \in s_j^{\perp\perp}$ for some $s_j \in S$. Since A is Archimedean over F, there exists $\alpha \in F^+$ such that $\alpha s_j \nleq x$, so $x \leq \alpha s_j$ since $s_j^{\perp\perp}$ is totally ordered. Thus x is a d-element, so each basic element of A is a d-element.

(2) Suppose that $1 = x_1 + \ldots + x_k$, where $x_1 \in s_{i_1}^{\perp\perp}, \ldots, x_k \in s_{i_k}^{\perp\perp}$ and $s_{i_1}, \cdots, s_{i_k}$ are distinct basic elements. If $x_j > 0$, since x_j is basic, x_j is a d-element, then $(1^-)x_j = (-1 \vee 0)x_j = -x_j \vee 0 = 0$, and if $x_j < 0$, then $(1^-)(-x_j) = 0$ by the above argument. Thus in both cases, $(1^-)x_j = 0$, for $j = 1, \ldots, k$, so

$$1^- = (1^-)1 = (1^-)(x_1 + \ldots + x_k) = (1^-)x_1 + \ldots + (1^-)x_k = 0.$$

Thus $1 = 1^+ - 1^- = 1^+ > 0$. $\square$

For a unital finite-dimensional Archimedean ℓ-algebra A over F with a d-basis, since a disjoint subset of A must be linearly independent by Theorem 1.13, a d-basis must be finite. We also notice that a d-basis of an ℓ-algebra may not be a vector space basis since it may not span the whole space.

Now we provide some examples of ℓ-rings and ℓ-algebras that have a d-basis.

Example 2.1.

(1) Any totally ordered ring has a d-basis with one element, and any f-ring or d-ring has a d-basis if and only if their additive ℓ-group has a basis.

(2) The matrix ℓ-algebra $M_n(F)$ with the entrywise order has a d-basis $\{e_{ij} \mid 1 \le i, j \le n\}$, where e_{ij} are standard matrix units. Similarly, let $T_n(F)$ be the $n \times n$ *upper triangular* matrix ℓ-algebra over F with the entrywise order. Then $T_n(F)$ also has a d-basis consisting of standard matrix units $\{e_{ij} : 1 \le i \le j \le n\}$.

(3) Let $F[G]$ be the group ℓ-algebra of a group G with the coordinatewise order. Then $1G = \{1g \mid g \in G\}$, where 1 is the identity element of F, is a d-basis. Moreover, let S be a semigroup satisfying *cancellation law*, namely, for any $r, s, t \in S$, $rs = rt$ or $sr = tr$ implies $s = t$. Then, using the coordinatewise order, the semigroup algebra $F[S]$ becomes an ℓ-algebra over F with $1S$ as a d-basis. Especially, polynomial rings over F in one or more variables are ℓ-algebras with a d-basis with respect to the coordinatewise order.

It is easily seen that the d-bases in the previous examples are also vector space bases over F. This is not always the case, as example (4) illustrates.

(4) Let $F[[x]] = \{\sum_{i \ge 0} \alpha_i x^i \mid \alpha_i \in F\}$ be the *ring of formal power series* over F. Then it is an ℓ-algebra over F with respect to the coordinatewise order. The set $\{x^n \mid n \ge 0\}$ is a d-basis, but not a vector

space basis over F since the set does not span $F[[x]]$ as a vector space over F. Similarly consider the field $F((x))$ of all *formal Laurent series* $f(x) = \sum_{-\infty}^{\infty} \alpha_i x^i$, where among the coefficients $\alpha_i \in F$ with $i < 0$, only finitely many can be nonzero. Again, with respect to the coordinatewise order, $F((x))$ becomes an ℓ-field with the d-basis $\{x^n : n \in \mathbb{Z}\}$ which is not a vector space basis over F.

(5) Let $K = \mathbb{Q}(b)$ be the finite extension field of $\mathbb{Q}$, where $0 < b \in \mathbb{R}$ satisfies an irreducible polynomial $x^n - \alpha$ over $\mathbb{Q}$ with $0 < \alpha \in \mathbb{Q}$. Then $K = \{\alpha_0 + \alpha_1 b + ... + \alpha_{n-1}b^{n-1} \mid \alpha_i \in \mathbb{Q}\}$ with respect to the coordinatewise order, is an ℓ-field since $b^n = \alpha > 0$. Since b is a d-element by Theorem 1.32(3), $\{1, b, \cdots, b^{n-1}\}$ is a d-basis of K as well as a vector space basis over $\mathbb{Q}$.

Next we list all 2-dimensional and 3-dimensional unital ℓ-algebras with a d-basis which is also a vector space basis. For simplicity, F is assumed to be a totally ordered subfield of $\mathbb{R}$.

Example 2.2. Let A be a unital ℓ-algebra over F with a d-basis D containing two elements that is also a vector space basis of A over F.

(1) If 1 is not basic, then A is a 2-dimensional f-algebra. Therefore $A \cong F \oplus F$.
(2) If 1 is basic, then we may assume that $1 \in D$. Let 1 and $0 < a \in A$ form a d-basis for A. Then $a^2 = \alpha 1 + \beta a$ for some $\alpha, \beta \in F^+$. Since $1 \wedge a = 0$, $a \wedge a^2 = 0$. We must have $\beta = 0$. Thus $a^2 = \alpha 1$ with $\alpha \geq 0$.

  (a) If $\alpha = 0$, then $A = 1F \oplus aF$ as a vector lattice over F with $a^2 = 0$.

    Now suppose $\alpha > 0$.

  (b) If $\sqrt{\alpha} \in F$, let $b = (\sqrt{\alpha})^{-1}a$. Then $b^2 = 1$ and $A = 1F \oplus bF \cong F(G)$, where G is a cyclic group of order 2.
  (c) If $\sqrt{\alpha} \notin F$, let $\sqrt{\alpha} = b \in \mathbb{R}$. Then $A = 1F \oplus aF \cong F(b)$, where $F(b)$ is the quadratic extension field of F with the coordinatewise order defined in Example 2.1(5).

Example 2.3. Let A be a unital ℓ-algebra over F with a d-basis D containing three elements that is also a vector space basis. Then A is isomorphic to one of the following ℓ-algebras over F. The verification of this fact is left to the reader (Exercise 1).

(1) $F \oplus F \oplus F$, a direct sum of three copies of F, so it is an f-algebra.
(2) $T_2(F)$, where $T_2(F)$ is the 2×2 upper triangular matrix ℓ-algebra.

(3) $Fe \oplus Ff \oplus Fa$, as a vector lattice with $1 = e + f$, $fa = af = a$ and $a^2 = 0$.

(4) $F \oplus F[G]$, where G is a cyclic group of order 2.

(5) $F \oplus F(b)$, where $0 < b \in \mathbb{R} \setminus F$, $b^2 \in F$, and $F(b)$ is the ℓ-field in Example 2.1(5).

(6) $F1 \oplus Fa \oplus Fb$, as a vector lattice where $a^2 = b^2 = ab = ba = 0$.

(7) $F1 \oplus Fa \oplus Fa^2$, as a vector lattice with $a^3 = 0$.

(8) $F[G]$, where G is a cyclic group of order 3.

(9) $F(b)$, where $0 < b \in \mathbb{R} \setminus F$ and $b^3 \in F$, $F(b)$ is the ℓ-field in Example 2.1(5).

In all of above examples, each d-basis, joint with 0, forms a semigroup with 0, that is, the product of two basic elements is either zero or again a basic element. However this observation is not true in general, as shown in the following example.

Example 2.4. Let A be the 4-dimensional vector space over F with the vector space basis $\{1, a, b, c\}$. With the coordinatewise order, A is a vector lattice over F. The multiplication table of the basis is defined as follows.

	1	a	b	c
1	1	a	b	c
a	a	$b+c$	0	0
b	b	0	0	0
c	c	0	0	0

Then A is an ℓ-algebra over F with $\{1, a, b, c\}$ as a d-basis, and $a^2 = b + c$ is not basic (Exercise 2). We note that $M = Fa + Fb + Fc$ is the unique maximal ℓ-ideal of A, so A is not ℓ-simple. If A is ℓ-simple, then, by Lemma 2.3, the product of two basic elements is either zero or a basic element.

We next present some properties of a unital Archimedean ℓ-algebra A over F with a d-basis that satisfies condition (C). By Theorem 1.17 as a vector lattice, A is a direct sum of maximal convex totally ordered subspaces of A over F. Since $1 > 0$, $1 = c_1 + \cdots + c_n$, where $\{c_1, \cdots, c_n\}$ is a disjoint set of basic elements for some positive integer n. Since $c_i \leq 1$ for $i = 1, \cdots, n$, each c_i is an f-element, so $c_i \wedge c_j = 0$ implies $c_i c_j = c_i c_j \wedge c_i c_j = 0$ for $i \neq j$. Then for each $i = 1, \cdots, n$, $c_i = 1c_i = (c_1 + \cdots + c_n)c_i = c_i^2$. That is, each c_i is idempotent.

Theorem 2.2. *Let A be a unital Archimedean ℓ-algebra over F with a d-basis and let A satisfy condition (C). Suppose $1 = c_1 + \cdots + c_n$ where $n \geq 1$*

and $\{c_1, \cdots, c_n\}$ *is a disjoint set of basic elements.*

(1) *For each basic element* $a \in A$, *there exists* c_i *such that* $c_i a = a$ *and* $c_k a = 0$ *for* $k \neq i$. *Similarly, there exists* c_j *such that* $a c_j = a$ *and* $a c_k = 0$ *for* $k \neq j$.

(2) *For each basic element* $a \in A$,

    (i) a *is nilpotent; or*

    (ii) *there exists a positive integer* n_a *such that* $0 \neq a^{n_a} \in c_i^{\perp\perp}$ *for some* c_i; *or*

    (iii) *the set* $\{a^m \mid m \geq 1\}$ *is disjoint and* $a^m \wedge 1 = 0$ *for each* $m \geq 1$.

(3) *For each* $i = 1, \cdots, n$, $c_i^{\perp\perp}$ *is a convex totally ordered subalgebra and a domain with identity element* c_i, *and* $f(A) = c_1^{\perp\perp} + \cdots + c_n^{\perp\perp}$. *If* $x \in A$ *is a basic element and an idempotent* f-*element, then* $x = c_i$ *for some* $i = 1, \cdots, n$.

(4) *Let* I *be a right (left)* ℓ-*ideal of* A. *Then* $c_i I$ ($I c_i$) *is a right (left)* ℓ-*ideal of* A *and* $c_i I$ ($I c_i$) $\subseteq I$.

Proof. (1) Since $a = 1a = c_1 a + \cdots + c_n a$ and a is a basic element, $c_i a$ and $c_k a$ are comparable. On the other hand, since a is a d-element by Theorem 2.1(1), $c_i \wedge c_k = 0$ implies $c_i a \wedge c_k a = 0$ for $i \neq k$. Thus if $c_i a \neq 0$, then $c_k a = 0$ for any $k \neq i$, and hence $a = c_i a$. The other conclusion can be proved similarly.

(2) We do some analysis first. Suppose that $a^m \neq 0$ for some positive integer m. Then $a^m = \sum_{t=1}^{r} a_t$, where $\{a_1, \cdots, a_r\}$ is a disjoint set of basic elements. Suppose that $c_i a = a$ for some $i = 1, \cdots, n$. Then $c_i a^m = a^m$, so $c_i a_t = a_t$ for each $t = 1, \cdots, r$ since $a_t \leq a^m$. Take $c_j \neq c_i$. If $a_t \wedge c_j \neq 0$, then a_t and c_j are comparable by Theorem 1.14(2) since they are both basic elements. If $a_t \leq c_j$ then $a_t = c_i a_t \leq c_i c_j = 0$, which is a contradiction. If $c_j \leq a_t$, then $c_j = c_j^2 \leq c_j a_t = c_j c_i a = 0$, which is again a contradiction. Thus $a_t \wedge c_j = 0$ for any $t = 1, \cdots, r$ and $j \neq i$.

If for each $t = 1, \cdots, r$, we also have $a_t \wedge c_i = 0$, then $a_t \wedge 1 = 0$ for each t, and hence $a^m \wedge 1 = 0$. On the other hand, suppose that for some $s = 1, \cdots, r$, $a_s \wedge c_i \neq 0$, then $a_s \in c_i^{\perp\perp}$ by Theorem 1.14. We claim that $a^m = a_s$. We first notice that $a_s \in c_i^{\perp\perp}$ implies $a_s c_i = a_s$ by an argument similar to that in the previous paragraph. For any $t \neq s$, $a_t \wedge a_s = 0$ implies

$$0 \leq a_t a_s \wedge a_s a_t \leq a_t a^m \wedge a_s a^m = 0$$

since a is a d-element implies a^m is a d-element, so $a_t a_s \wedge a_s a_t = 0$. Then $a_s \in c_i^{\perp\perp}$ and A is Archimedean over F imply $c_i \leq \alpha a_s$ for some $0 < \alpha \in F$, and hence $a_t c_i \wedge c_i a_t = 0$. It follows from $c_i a_t = a_t$ that $a_t c_i \wedge a_t = 0$, so $a_t c_i \wedge a_t c_i = 0$ since c_i is an f-element. Thus $a_t c_i = 0$ for $t \neq s$, and hence $a^m = a^m c_i = (a_1 + \cdots + a_r) c_i = a_s c_i = a_s$.

So far we have proved that for a positive integer m, if $a^m \neq 0$, then either $a^m \wedge 1 = 0$ or $a^m \in c_i^{\perp\perp}$ for some c_i.

Hence if (i) and (ii) are not true, that is, if a is not nilpotent and $a^m \notin c_i^{\perp\perp}$ for any $m \geq 1$ and any c_i, then by above argument, $a^m \wedge 1 = 0$ for any positive integer, so a is a d-element implies for $r < s$, $a^s \wedge a^r = a^r(a^{s-r}) \wedge 1 = 0$, and hence the set $\{a^m \mid m \geq 1\}$ is disjoint and $a^m \wedge 1 = 0$ for all $m \geq 1$. That is, (iii) is true. We leave it as an exercise for the reader to show that any two statements of $(i), (ii), (iii)$ cannot be both true.

(3) We know that $c_i^{\perp\perp}$ is a convex totally ordered subspace for each $i = 1, \cdots, n$. Let $0 \leq x, y \in c_i^{\perp\perp}$. Since A is Archimedean over F and $c_i^{\perp\perp}$ is totally ordered, there exists $0 < \alpha \in F$ such that $x \leq \alpha c_i$. It follows that x is an f-element and hence $xy \in c_i^{\perp\perp}$. Therefore each $c_i^{\perp\perp}$ is a convex totally ordered subalgebra. Suppose $a^2 = 0$ for some $0 < a \in c_i^{\perp\perp}$. Again A is Archimedean implies $c_i \leq \beta a$ for some $0 < \beta \in F$, so $c_i = c_i^2 \leq \beta^2 a^2 = 0$, which is impossible. Thus $c_i^{\perp\perp}$ contains no nonzero nilpotent element and hence it is a domain (Exercise 39, Chapter 1).

Let $0 < x \in c_i^{\perp\perp}$, $i = 1, \ldots, n$. Since $c_i^{\perp\perp}$ is totally ordered and A is Archimedean, there exists $0 < \alpha \in F$ such that $0 < x \leq \alpha c_i$, so x is an f-element. Thus each $c_i^{\perp\perp} \subseteq f(A)$, and hence $c_1^{\perp\perp} + \ldots + c_n^{\perp\perp} \subseteq f(A)$. Let $0 < x \in f(A)$. Then $x = x1 = x c_1 + \ldots + x c_n$. Since x is an f-element, $x c_i \in c_i^{\perp\perp}$, $i = 1, \ldots, n$. Thus $x \in c_1^{\perp\perp} + \ldots + c_n^{\perp\perp}$, and hence $f(A) \subseteq c_1^{\perp\perp} + \ldots + c_n^{\perp\perp}$. Therefore $f(A) = c_1^{\perp\perp} + \ldots + c_n^{\perp\perp}$.

Let $x \in A$ be a basic element and an idempotent f-element. Then $x \in c_i^{\perp\perp}$ for some i. It follows from $x^2 = x$ that $x = c_i$ since $c_i^{\perp\perp}$ is a domain.

(4) Clearly $c_i I$ is a right ideal of A and a sublattice of A since c_i is an f-element. Let $a \in A^+$ and $b \in I^+$ with $a \leq c_i b$. We show that $a \in c_i I$. First we assume that b is a basic element. From (1), we have $c_i b = 0$ or $c_i b = b$, and hence $a \in I$ in either case. For any $j \neq i$, $c_j a \leq c_j c_i b = 0$ implies $c_j a = 0$, so $a = 1a = (c_1 + \cdots + c_n)a = c_i a \in c_i I$. In general case, let $b = b_1 + \cdots + b_k$, where $b_1, \cdots, b_k \in I$ are basic elements, and $a \leq c_i b_1 + \cdots + c_i b_k$. Thus $a = a_1 + \cdots + a_k$ with $0 \leq a_t \leq c_i b_t$ for some $a_1, \cdots, a_k \in A^+$ by Theorem 1.5. From the previous argument, each $a_t \in c_i I$, so $a \in c_i I$. Therefore $c_i I$ is a right ℓ-ideal of A. Finally $c_i I \subseteq I$ because of $c_i x = 0$ or x for each basic element x in I by (1). $\qquad\square$

Theorem 2.3. *Let A be a unital Archimedean ℓ-algebra over F with a d-basis and let A satisfy condition (C). For a convex ℓ-subalgebra H and an ℓ-ideal I, H and R/I are Archimedean ℓ-algebras with a d-basis satisfying condition (C).*

Proof.   Suppose that S is a d-basis of A. Then $H \cap S$ is a d-basis and H is Archimedean over F and satisfies condition (C) (Exercise 3).

Let I be an ℓ-ideal of A. For each $a \in A$, write $\bar{a} = a + I \in A/I$. Let S be a d-basis for A. We show that $V = \{\bar{s} \mid s \in S \setminus I\}$ is a d-basis for A/I. Let $0 \leq \bar{a}, \bar{b} \leq \bar{s} \in V$. Since $\bar{a}, \bar{b}$ are positive, we may assume that $a \geq 0$, and $b \geq 0$. Then we have $\bar{a} = \bar{a} \wedge \bar{s} = \overline{a \wedge s}$, so $a - (a \wedge s) = a_1 \in I$. Similarly, $b - (b \wedge s) = b_1 \in I$. Hence $a - a_1 = (a \wedge s) \leq s$ and $b - b_1 = (b \wedge s) \leq s$, so s is basic implies that $a - a_1$ and $b - b_1$ are comparable. If $a - a_1 \leq b - b_1$, then $\bar{a} = \overline{a - a_1} \leq \overline{b - b_1} = \bar{b}$. Similarly $b - b_1 \leq a - a_1$ implies that $\bar{b} \leq \bar{a}$. Thus $\bar{s}$ is basic in A/I. Now let $\bar{a} \wedge \bar{b} = 0$. Then $a \wedge b = c \in I$, and hence $(a - c) \wedge (b - c) = 0$. It follows that

$$s(a - c) \wedge s(b - c) = 0 \implies sa \wedge sb = sc \in I.$$

Thus $\bar{s}\bar{a} \wedge \bar{s}\bar{b} = 0$ in R/I. Similarly, $\bar{a}\bar{s} \wedge \bar{b}\bar{s} = 0$, that is, $\bar{s}$ is a d-element in A/I. Since A is a direct sum of $s^{\perp\perp}$, $s \in S$, A/I is a direct sum of $\bar{s_i}^{\perp\perp}$, where $\bar{s_i} \in V$, which implies that V is a d-basis of A/I and A/I satisfies condition (C).

Finally we show that A/I is Archimedean over F. To this end we just need to show that each $\bar{s_i}^{\perp\perp}$ is Archimedean over F for $\bar{s_i} \in V$ (Exercise 4). Let $0 < \bar{a}, \bar{b} \in \bar{s_i}^{\perp\perp}$ with $0 < a, b \in A$. Then $a = a_1 + \cdots + a_m$, where $a_1, \cdots, a_m$ are disjoint basic elements, and $b = b_1 + \cdots + b_\ell$, where $b_1, \cdots, b_\ell$ are disjoint basic elements. Since $a, b \notin I$, we may assume that $a_1, b_1 \notin I$. For any a_t, $1 < t \leq m$, $a_t \wedge a_1 = 0$ implies that $\overline{a_t} \wedge \overline{a_1} = 0$. It follows that $\overline{a_t} = 0$ since $\bar{a}$ is basic and $\bar{a} = \overline{a_1} + \cdots + \overline{a_m}$. Thus $\bar{a} = \overline{a_1}$ and a_1, s_i are comparable. Similarly $\bar{b} = \overline{b_1}$ and b_1, s_i are comparable. Now $0 < a_1, b_1 \in s_i^{\perp\perp}$ and $s_i^{\perp\perp}$ is totally ordered and Archimedean over F implies that there exist $0 < \alpha, \beta \in F$ such that $a_1 \leq \alpha b_1$ and $b_1 \leq \beta a_1$. Hence $\bar{a} = \overline{a_1} \leq \alpha\overline{b_1} = \alpha\bar{b}$ and $\bar{b} = \overline{b_1} \leq \beta\overline{a_1} = \beta\bar{a}$. Therefore $\bar{s_i}^{\perp\perp}$ is Archimedean over F. $\square$

A nonzero left (right) ℓ-ideal I is called *minimal* if for any nonzero left (right) ℓ-ideal J, $J \subseteq I$ implies that $J = I$.

Theorem 2.4. *Let A be a unital finite-dimensional Archimedean ℓ-algebra over F with a d-basis and $1 = c_1 + \cdots + c_n$, where $n \geq 1$ and $\{c_1, \cdots, c_n\}$ is a disjoint set of basic elements.*

(1) For $i = 1, \cdots, n$, $c_i^{\perp\perp}$ is a totally ordered field.

(2) If a, b are basic elements such that $ab \neq 0$ and one of them is not nilpotent, then ab is basic.

(3) Let I be a minimal right (left) ℓ-ideal of A. Then either $I^2 = \{0\}$ or $I = c_i A$ ($A c_i$) for some $i = 1, \cdots, n$.

Proof. (1) From Theorem 2.2(3) we know that $c_i^{\perp\perp}$ is a convex totally ordered subalgebra and a domain. If A is finite-dimensional over F, then $c_i^{\perp\perp}$ is also finite-dimensional over F, which implies $c_i^{\perp\perp}$ is a totally ordered division algebra over F. Now by Theorem 1.23, $c_i^{\perp\perp}$ is commutative and hence it is a totally ordered field.

(2) We first notice that since A is finite-dimensional and a disjoint set of A must be linearly independent over F by Theorem 1.13(3), the case (iii) in Theorem 2.2(2) cannot happen. Without loss of generality, we may assume that a is not nilpotent. Then, by Theorem 2.2(2), there exists a positive integer n_a such that $0 \neq a^{n_a} \in c_i^{\perp\perp}$ for some c_i, so $c_i a = a c_i = a$. Because of $ab \neq 0$, $c_i b = b$. Otherwise $c_j b = b$ for some $c_j \neq c_i$ and $ab = a c_i c_j b = 0$. Assume that $ab = a_1 + \cdots + a_r$, where $\{a_1, \cdots, a_r\}$ is a disjoint set of basic elements and $r \geq 1$. We claim that $r = 1$. Suppose $r > 1$. We have

$$a^{n_a} b = a^{n_a - 1} a_1 + \cdots + a^{n_a - 1} a_r,$$

and $a^{n_a} \in c_i^{\perp\perp}$ is an f-element. Thus $a^{n_a} b \in b^{\perp\perp}$ and $a^{n_a} b$ is a basic element, so $a^{n_a - 1} a_1$ and $a^{n_a - 1} a_2$ are comparable. On the other hand, $a_1 \wedge a_2 = 0$ and a is a d-element implies $a^{n_a - 1} a_1 \wedge a^{n_a - 1} a_2 = 0$. Therefore we must have $a^{n_a - 1} a_1 = 0$ or $a^{n_a - 1} a_2 = 0$. It follows that $a_1 = 0$ or $a_2 = 0$ (Exercise 5), which is a contradiction. Hence $r = 1$ and $ab = a_1$ is basic.

(3) We first notice that since A is finite-dimensional over F, each nonzero right ℓ-ideal contains a minimal right ℓ-ideal. Suppose that $I^2 \neq 0$. Then there exists a basic element $x \in I$ such that $xI \neq 0$. Define

$$J = \{a \in A \mid |a| \leq xr \text{ for some } 0 < r \in I\}.$$

Then J is a right ℓ-ideal (Exercise 6) and $J \subseteq I$. It follows from $xI \neq 0$ that $J \neq 0$. Thus by minimality of I, $J = I$, so $x \leq xr$ for some $0 < r \in I$. Let $r = r_1 + \ldots + r_m$, where $m \geq 1$ and $r_1, \ldots, r_m$ are disjoint basic elements. Since $r \in I$, each $r_j \in I$, and since x is a d-element, $xr_i \wedge xr_j = 0$ for $i \neq j$. Then $x \leq xr_1 + \ldots + xr_m$ and x is basic imply that $x \leq xr_j$ for some $j = 1, \ldots, m$ (Exercise 7). It follows from the fact that A is finite-dimensional over F that either r_j is nilpotent or $0 \neq r_j^{n_j} = w \in c_i^{\perp\perp}$ for some c_i, $n_j \geq 1$. If $r_j^u = 0$ for some positive integer u, then

$$x \leq xr_j \leq xr_j^2 \leq \ldots \leq xr_j^u = 0$$

implies that $x = 0$, which is a contradiction. Thus r_j is not nilpotent. Consequently $r_j^{n_j} = w \in I$ and $c_i \in I$ by $c_i \leq \alpha w$ for some $0 < \alpha \in F$. Hence $c_i A \subseteq I$, then $I = c_i A$ from the minimality of I and that $c_i A$ is a right ℓ-ideal implies $I = c_i A$.   □

Corollary 2.1. *Let A be a unital finite-dimensional ℓ-algebra over F with a d-basis and ℓ-$N(A) = \{0\}$. Suppose $1 = c_1 + \cdots + c_n$, where $n \geq 1$ and $\{c_1, \cdots, c_n\}$ is a disjoint set of basic elements. Then each $c_i A$ (Ac_i) is a minimal right (left) ℓ-ideal of A.*

Proof.   By Theorem 1.31(3), A is Archimedean over F. Let $I \subseteq c_i A$ be a minimal right ℓ-ideal of A. Since ℓ-$N(A) = \{0\}$, $I^2 \neq \{0\}$ (Exercise 8), and hence $I = c_j A$ for some c_j by Theorem 2.4(3). Then $c_j A \subseteq c_i A$ implies $i = j$, that is, $c_i A$ is a minimal right ℓ-ideal of A.   □

2.2   Structure theorems

In this section, we consider the structure of a unital finite-dimensional Archimedean ℓ-algebra over F with a d-basis.

2.2.1   *Twisted group ℓ-algebras*

Definition 2.2. Let G be a group. A function $t : G \times G \to F \setminus \{0\}$ is called a *positive twisting function* if t satisfies the following conditions,

(1) $t(g, h) > 0$, for all $g, h \in G$,
(2) $t(gh, f)t(g, h) = t(g, hf)t(h, f)$, for all $f, g, h \in G$,
(3) $t(g, e) = t(e, g) = 1$, where e is the identity element of G, for all $g \in G$.
    In the case that G is an abelian group, t is also commutative, that is,
(4) $t(g, h) = t(h, g)$, for all $g, h \in G$.

Define

$$F^t[G] = \{\sum_{i=1}^{n} \alpha_i g_i \mid \alpha_i \in F, g_i \in G\}.$$

With respect to the following operations, $F^t[G]$ becomes a vector lattice over F (Exercise 9). For $\sum_{i=1}^n \alpha_i g_i, \sum_{i=1}^n \beta_i g_i \in F^t[G]$, $\alpha \in F$,

$$\sum_{i=1}^n \alpha_i g_i + \sum_{i=1}^n \beta_i g_i = \sum_{i=1}^n (\alpha_i + \beta_i) g_i, \tag{2.1}$$

$$\alpha \sum_{i=1}^n \alpha_i g_i = \sum_{i=1}^n (\alpha \alpha_i) g_i, \tag{2.2}$$

$$\sum_{i=1}^n \alpha_i g_i \geq 0 \text{ if each } \alpha_i \in F^+. \tag{2.3}$$

Define multiplication in $F^t[G]$ by

$$(\sum_{i=1}^n \alpha_i g_i)(\sum_{j=1}^m \beta_j h_j) = \sum_{i=1}^n \sum_{j=1}^m (\alpha_i \beta_j) t(g_i, h_j)(g_i h_j),$$

where $(g_i h_j)$ is the product of g_i, h_j in the group G. The multiplication defined above is associative by Definition 2.2(2) and multiplication is distributive over the addition in $F^t[G]$ is clear by the definition. Thus $F^t[G]$ is an algebra over F. The condition (1) in Definition 2.2 implies the product of two positive elements is also positive, so $F^t[G]$ is an ℓ-algebra over F, called *twisted group ℓ-algebra* of G over F. In this book, $F^t[G]$ always denotes the ℓ-algebra with the coordinatewise order defined above. If G is abelian, then $F^t[G]$ is commutative by Definition 2.2(4). It is clear that $1G = \{1g \mid g \in G\}$ is a d-basis of the ℓ-algebra $F^t[G]$ over F. The identity element of $F^t[G]$ is $1e$, where 1 is the identity element of F. Sometimes we just identify $1G$ with G, so e is the identity element of $F^t[G]$ under this assumption.

Certainly if $t(g, h) = 1$ for all $g, h \in G$, then $F^t[G] = F[G]$ is the group ℓ-algebra. As an example, the ℓ-field $\mathbb{Q}[\sqrt{2}]$ with the coordinatewise order may be considered as a twisted group ℓ-algebra with $G = \{e, g\}$ being a cyclic group of order 2, and the twisting function t defined by

$$t(e, e) = t(e, g) = t(g, e) = 1 \text{ and } t(g, g) = 2.$$

We leave the verification of it as an exercise (Exercise 10).

Theorem 2.5. $F^t[G]$ *is an ℓ-simple ℓ-algebra over F.*

Proof. Let I be a nonzero ℓ-ideal of $F^t[G]$ and $0 < a = \sum_{i=1}^n \alpha_i g_i \in I$ with $\alpha_i \neq 0$. Then each $\alpha_i > 0$, so $0 < \alpha_1 g_1 \leq a$ implies $\alpha_1 g_1 \in I$. Suppose that $g \in G$ such that $g_1 g = e$ in G. Then $e = (\alpha_1 g_1)(\alpha_1^{-1} t(g_1, g)^{-1} g) \in I$. Therefore $I = F^t[G]$. $\qquad \square$

Some other properties of $F^t[G]$ include that the identity element is basic and it is an ℓ-domain (Exercise 11).

In the following section, we prove that a unital finite-dimensional ℓ-simple ℓ-algebra over F with a d-basis is ℓ-isomorphic to a matrix ℓ-algebra with the entrywise order over a twisted group ℓ-algebra of a finite group.

For more information on general twisted group rings, the reader is refereed to [Passman (2011)].

2.2.2 ℓ-simple case

Theorem 2.6. *Let A be a unital finite-dimensional ℓ-algebra over a totally ordered field F with a d-basis. If A is ℓ-simple, then A is ℓ-isomorphic to the matrix ℓ-algebra $M_n(K^t[G])$ with the entrywise order, where $n \geq 1$, K is a totally ordered field and a finite-dimensional ℓ-algebra over F, G is a finite group, t is a positive twisting function on G, and $K^t[G]$ is the twisted group ℓ-algebra of G over K.*

We prove the result by a series of steps. Since A is ℓ-simple, $\ell\text{-}N(A) = \{0\}$, so A is Archimedean over F by Theorem 1.31(3) and A is a finite direct sum of maximal convex totally ordered subspaces of A over F by Corollary 1.3. Let S be a d-basis of A over F. Then S is finite. Suppose that $1 = c_1 + \ldots + c_n$ with $n \geq 1$ and $c_1, \ldots, c_n$ are disjoint basic elements. For each $i = 1, \ldots, n$, define $K_i = c_i^{\perp\perp}$ and $H_i = c_i A c_i$. Then each K_i is a totally ordered field and finite-dimensional ℓ-algebra over F by Theorem 2.4(1), and each H_i is a convex ℓ-subalgebra of A over F (Exercise 12) with $K_i \subseteq H_i$, $i = 1, \cdots, n$.

Lemma 2.1. *For $i = 1, \cdots, n$, $H_i = c_i A c_i$ is ℓ-reduced.*

Proof.   Suppose that there exists $0 < x \in H_i = c_i A c_i$ with $x^k = 0$ for some positive integer k. Consider

$$I = \{a \in A \mid |a| \leq xr \text{ for some } r \in A^+\}.$$

Then I is the right ℓ-ideal generated by x. Since $x \in c_i A c_i \subseteq c_i A$, $I \subseteq c_i A$, and since $x \in I$, $I \neq 0$. Thus $I = c_i A$ because $c_i A$ is a minimal right ℓ-ideal by Corollary 2.1. So there exists $r \in A^+$ such that $c_i \leq xr$. It follows from $x \in c_i A c_i$ that $x c_i = x$, so $c_i \leq xr$ implies that $x = x c_i \leq x^2 r$. Then multiplying x from the left and r from the right of the inequality, we have

$$x \leq x^2 r \leq x^3 r^2 \leq \ldots \leq x^k r^{k-1} = 0,$$

which is a contradiction. Thus $H_i = c_i A c_i$ is ℓ-reduced. $\square$

Lemma 2.2. *For each $i = 1, \ldots, n$, K_i is contained in the center of H_i.*

Proof. Let $0 < z \in K_i$. To show that z is in the center of H_i, we just need to verify that $az = za$ for each basic element $a \in H_i$. By Lemma 2.1, a is not nilpotent, so by Theorem 2.2, there exists a positive integer $n_a \geq 1$ such that $a^{n_a} = w \in K_i$ with $w \neq 0$. If $az = 0$ or $za = 0$, then $wz = 0$ or $zw = 0$, which contradicts with the fact that K_i is a field. Thus $az \neq 0$ and $za \neq 0$. By Theorem 2.4(2), az and za are both basic elements. If $az \wedge za = 0$, then z is an f-element implies that $zaz \wedge zaz = 0$, and hence $zaz = 0$, which is a contradiction. Thus az and za are comparable. If $az < za$, then $a < zaz^{-1}$, where z^{-1} is the inverse of z in K_i. Hence

$$w = a^{n_a} \leq a^{n_a-1}(zaz^{-1}) \leq (zaz^{-1})^{n_a-1}(zaz^{-1}) = za^{n_a}z^{-1} = w.$$

From $w = a^{n_a-1}(zaz^{-1})$, we have $wa = aw = w(zaz^{-1})$, and hence $w(az) = w(za)$, so $az = za$, which contradicts with the fact that $az < za$. Similarly, $za < az$ is not possible. Thus $az = za$ for each basic element $a \in H_i$. Since each positive element in H_i is a sum of disjoint basic elements in H_i and each element in H_i is a difference of two positive elements in H_i, z commutes with each element of H_i, that is, z is in the center of H_i. Therefore K_i is contained in the center of H_i. □

For two basic elements $a, b \in H_i$, define $a \sim b$ if $a = zb$ for some $z \in K_i$. Then $\sim$ is an equivalence relation on H_i (Exercise 13). For a basic element $a \in H_i$, a' denotes its equivalence class and define

$$G_i = \{a' \mid a \in H_i \text{ is a basic element}\}$$

with the operation $a'b' = (ab)'$. Since H_i is ℓ-reduced, if $a, b \in H_i$ are basic elements, then ab is still a basic element by Theorem 2.4. It is clear that the operation is well-defined and associative with c_i' as the identity element (Exercise 14). For $a' \in G_i$, by Theorem 2.2(2), there exists a positive integer n_a such that $a^{n_a} \in K_i$. Thus $(a')^{n_a} = c_i'$. It follows that G_i is a group for $i = 1, \cdots, n$.

Let S_i be a d-basis for H_i. For a basic element $a \in H_i$, a is comparable with some $s \in S_i$. We claim that $a' = s'$. Since A is Archimedean over F, there exists $0 < \alpha \in F$ such that $a \leq \alpha s$. By Theorem 2.2, there exists a positive integer n_s such that $0 < s^{n_s} = u \in K_i$, and hence $as^{n_s-1} \leq \alpha s^{n_s} = \alpha u$ implies that $as^{n_s-1} = v \in K_i$. Thus $au = vs$. By Lemma 2.2, u is in the center of H_i, so we have

$$a = u^{-1}(ua) = u^{-1}(au) = u^{-1}(vs) = (u^{-1}v)s$$

and $u^{-1}v \in K_i$. Therefore $a \sim s$, that is, $a' = s'$. This shows that $G_i = \{s' \mid s \in S_i\}$. Hence G_i is a finite group with the order $|S_i|$.

Let $G = G_1 = \{s'_1, \ldots, s'_k\}$, where $\{s_1, \cdots, s_k\}$ is a d-basis of H_1, and let $K = K_1$. For s'_i and s'_j, let $s'_i s'_j = s'_u$. Then $s_i s_j = z_{ij} s_u$ for some $0 < z_{ij} \in K$. Define $t : G \times G \to K \setminus \{0\}$ by $t(s'_i, s'_j) = z_{ij}$. It is routine to verify that t is a positive twisting function (Exercise 15). Now we first form the twisted group ℓ-algebra $K^t[G]$, and then form matrix ℓ-algebra $M_n(K^t[G])$ with the entrywise order. Then $M_n(K^t[G])$ is an ℓ-algebra over F and we show that A and $M_n(K^t[G])$ are ℓ-isomorphic as ℓ-algebras over F.

Lemma 2.3. *For a basic element $a \in c_i A c_j$, $1 \leq i, j \leq n$, there exists a basic element $b \in c_j A c_i$ such that $ab = c_i$ and $ba = c_j$. As a consequence, the product of two basic elements is either zero or a basic element.*

Proof. Since A is ℓ-simple, $A = \langle a \rangle$, and hence there exist $r, s \in A^+$ such that $c_i \leq ras$. Suppose that $r = r_1 + \cdots + r_k$, where $r_1, \cdots, r_k$ are disjoint basic elements, and $s = s_1 + \cdots + s_\ell$, where $s_1, \cdots, s_\ell$ are disjoint basic elements. Then

$$c_i \leq \sum_{1 \leq u \leq k, 1 \leq v \leq \ell} r_u a s_v \Rightarrow c_i = \sum_{1 \leq u \leq k, 1 \leq v \leq \ell} b_{uv}$$

with $0 \leq b_{uv} \leq r_u a s_v$. Since c_i is basic, any two of $b_{uv}, 1 \leq u \leq k, 1 \leq v \leq \ell$, are comparable. We may assume $b_{u_1 v_1}$ is the largest one among b_{uv}, $1 \leq u \leq k$, $1 \leq v \leq \ell$, so $c_i \leq (k + \ell)b_{u_1 v_1} \leq (k + \ell)r_{u_1} a s_{v_1}$. Thus we have that $c_i \leq waz$ for some basic elements w and z. Then $c_i w = w$ and $z c_i = z$ by Theorem 2.2. Suppose that x is a basic element with $x \leq waz$ and $x \wedge c_i = 0$. Then $c_i x = x c_i = x$, and waz is a d-element implies that

$$x = x c_i \wedge c_i x \leq x(waz) \wedge c_i(waz) = 0,$$

which is a contradiction. Thus, since each positive element in A is a sum of disjoint basic elements, waz must be basic, and hence $c_i \leq (waz)$ implies that $waz = y \in K_i = c_i^{\perp\perp}$ by Theorem 1.14(2). From $waz \neq 0$ and $c_i a = a$, we have $wc_i = w$, so $w \in H_i$. By Lemma 2.1, w is not nilpotent, and hence there exists a positive integer n_w such that $w^{n_w} = q \in K_i$. Then

$$waz = y \Rightarrow q(az) = w^{n_w}(az) = w^{n_w - 1}y = yw^{n_w - 1},$$

since K_i is contained in the center of H_i, so $q(azw) = yw^{n_w} = yq = qy$. Thus $azw = y$, and $a(zwy^{-1}) = c_i$, where y^{-1} is the inverse of y in K_i. Let $b = zwy^{-1}$. Then $ab = c_i$ and b is a basic element by Theorem 2.4 since w is not nilpotent.

It is clear that $b \in c_j A c_i$. Then by a similar argument, there exists a basic element c such that $bc = c_j$. Thus $c \in c_i A c_j$, and

$$a = a c_j = a(bc) = (ab)c = c_i c = c.$$

Therefore $ab = c_i$ and $ba = c_j$.

Let x and y both be basic elements, and $xy \neq 0$. Suppose that $0 \leq u, v \leq xy$. Let $x \in c_i A c_j$. Then there exists a basic element $x_1 \in c_j A c_i$ such that $x x_1 = c_i$ and $x_1 x = c_j$, so $0 \leq x_1 u, x_1 v \leq x_1(xy) = y$. Thus $x_1 u$ and $x_1 v$ are comparable, and hence $u = x(x_1 u)$ and $v = x(x_1 v)$ are comparable. Therefore, xy is basic. $\qquad \square$

Since A is ℓ-simple, $A = \langle c_i \rangle$ for each $i = 1, \ldots, n$. Using the same argument as in the proof of Lemma 2.3, $c_1 \leq a_i c_i f_i$ for some basic element a_i and f_i. Since each $a_i \in c_1 A c_i$ is basic, by Lemma 2.3, there exist basic elements $b_i \in c_i A c_1$ such that

$$a_1 b_1 = c_1, \ b_1 a_1 = c_1$$
$$a_2 b_2 = c_1, \ b_2 a_2 = c_2$$
$$\vdots$$
$$a_n b_n = c_1, \ b_n a_n = c_n$$

Recall that $S = S_1$ is a d-basis for the convex ℓ-subalgebra $H_1 = c_1 A c_1$, and $K = K_1 = c_1^{\perp\perp}$.

Lemma 2.4. *For each basic element x of A, $x = b_i(zs)a_j$ for some $1 \leq i, j \leq n$, $z \in K$ and $s \in S$. Moreover if $x = b_u(wt)a_v$, where $w \in K$ and $t \in S$, then $u = i$, $v = j$, $w = z$ and $t = s$.*

Proof. We may assume that $x \in c_i A c_j$ for some i and j. Then $a_i x b_j \in c_1 A c_1$ is basic and $a_i x b_j \sim s$ for some $s \in S$, that is, $a_i x b_j = zs$ for some $0 < z \in K$. Hence $x = b_i(zs)a_j$.

If $x = b_u(wt)a_v$ for some $w \in K$ and $t \in S$. Then clearly $u = i$ and $v = j$, so $zs = wt$. If $s \neq t$, then $s \wedge t = 0$, and hence $zs \wedge wt = 0$ since z, w both are f-elements. This is a contradiction. Thus $s = t$, and we have $(z - w)s = 0$, so $z = w$. $\qquad \square$

Now we show that a unital finite-dimensional ℓ-simple ℓ-algebra with a d-basis is ℓ-isomorphic to $M_n(K^t[G])$. For $z \in K$ and $s' \in G$, $e_{ij}(zs') = (zs')e_{ij}$ denotes the matrix with ij^{th} entry equal to zs' and other entries equal to zero, $1 \leq i, j \leq n$.

Define the mapping $\varphi : A \to M_n(K^t[G])$ as follows. Define $\varphi(0) = 0$. For a basic element x of A, by Lemma 2.4, we may uniquely express x as

$x = b_i(zs)a_j$, where $z \in K$ and $s \in S$, $1 \leq i, j \leq n$. Define $\varphi(x) = e_{ij}(zs')$. For $0 < a \in A$, a can be uniquely expressed as a sum of disjoint basic elements, that is, $a = a_1 + \ldots + a_m$, where $a_1, \ldots, a_m$ are disjoint basic elements. Then define $\varphi(a) = \varphi(a_1) + \ldots + \varphi(a_m)$. Finally for each $0 \neq a \in A$, $\varphi(a) = \varphi(a^+) - \varphi(a^-)$.

If x and y in A are comparable basic elements. Then we must have $x = b_i(z_1 s)a_j$ and $y = b_i(z_2 s)a_j$ and z_1, z_2 are comparable. Then $x \pm y = b_i((z_1 \pm z_2)s)a_j$ implies that

$$\varphi(x \pm y) = e_{ij}((z_1 \pm z_2)s') = e_{ij}(z_1 s') \pm e_{ij}(z_2 s') = \varphi(x) \pm \varphi(y).$$

Thus it follows that φ preserves addition on A (Exercise 16).

Now consider the multiplication. Let $x = b_i(zs_1)a_j$ and $y = b_u(ws_2)a_v$ be two basic elements in A. If $u \neq j$, then $a_j b_u = 0$ implies that $xy = 0$, so $\varphi(xy) = 0$. On the other hand, $\varphi(x)\varphi(y) = e_{ij}(zs_1)e_{uv}(ws_2) = 0$. If $u = j$, then $xy = b_i((zw)s_1 s_2)a_v$, where $s_1 s_2 = t(s_1, s_2)(s_1 s_2)$ with $t(s_1, s_2) \in K$ and $(s_1 s_2) \in S$, and hence

$$\begin{aligned}
\varphi(xy) &= e_{iv}(zwt(s_1, s_2)(s_1 s_2)') \\
&= e_{iv}((zs_1')(ws_2')) \\
&= e_{ij}(zs_1')e_{uv}(ws_2') \\
&= \varphi(x)\varphi(y)
\end{aligned}$$

where the product $(zs_1')(ws_2')$ is in the twisted group ℓ-algebra $K^t[G]$. Thus it follows that φ preserves multiplication on A (Exercise 17). It is clear that φ is one-to-one and onto, and for any $a \in A$, $\alpha \in F$, $\varphi(\alpha a) = \alpha\varphi(a)$, and hence φ is an isomorphism between algebras A and $M_n(K^t[G])$ over F. Finally, for any $a \in A$, $\varphi(a) \geq 0$ if and only if $a \geq 0$ (Exercise 18), therefore φ is an ℓ-isomorphism between ℓ-algebras A and $M_n(K^t[G])$ over F. This completes the proof of Theorem 2.6.

In Corollaries 2.2 and 2.3 we consider some special cases of Theorem 2.6.

Corollary 2.2. *Let A be a unital finite-dimensional ℓ-simple ℓ-algebra over F. Then A is ℓ-isomorphic to the matrix ℓ-algebra $M_n(F^t[G])$ with the entrywise order, where $n \geq 1$, G is a finite group, t is a positive twisting function on G, and $F^t[G]$ is the twisted group ℓ-algebra of G over F, if and only if A contains a d-basis that is also a vector space basis of A over F.*

Proof. "$\Rightarrow$" It is clear that $\{e_{ij}(s) \mid 1 \leq i, j \leq n, s \in G\}$ is a d-basis and a vector space basis of A over F (Exercise 19).

"⇐" By Theorem 2.6, A is ℓ-isomorphic to the ℓ-algebra $M_n(K^t[G])$, where K is a totally ordered field and a finite-dimensional ℓ-algebra over F. Let S be a d-basis of A that is also a basis of A as a vector space over F, and let $0 < x, y \in K$. Since x, y are basic, there exist $s, t \in S$ such that $x = \alpha s$ and $y = \beta t$ for some $0 < \alpha, \beta \in F$. Then the fact that x and y are comparable implies that $s = t$. Therefore $K = Fs$, for some $s \in S$, is one-dimensional over F, and hence $M_n(K^t[G]) \cong M_n(F^t[G])$. □

Corollary 2.3. *Let A be a unital finite-dimensional ℓ-simple ℓ-algebra over F with a d-basis.*

(1) If 1 is basic or A is ℓ-reduced, then A is ℓ-isomorphic to the twisted group ℓ-algebra $K^t[G]$, where K is a totally ordered field and a finite-dimensional ℓ-algebra over F, G is a finite group, and t is a positive twisting function.

(2) If A is an f-algebra, then A is ℓ-isomorphic to a finite-dimensional totally ordered extension field of F.

(3) If A is commutative, then A is ℓ-isomorphic to $K^t[G]$ as in (1) with G being a finite commutative group.

Proof. From Theorem 2.6, A is ℓ-isomorphic to the ℓ-algebra $M_n(K^t[G])$.
(1) If 1 is basic or A is ℓ-reduced, then $n = 1$.
(2) If A is an f-algebra, then $n = 1$ and $G = \{e\}$.
(3) If A is commutative, then $n = 1$ and G is commutative. □

We next consider Theorem 2.6 when $F = \mathbb{R}$. First we state a well-known result that each unital finite-dimensional algebra can be considered as a subalgebra of a full matrix algebra.

Lemma 2.5. *Let B be a unital n-dimensional algebra over a field L. Then B can be considered as a subalgebra of $M_n(L)$ with the identity matrix as the identity element of B.*

Proof. Let $\{v_1, \cdots, v_n\}$ be a basis of B over L. For each $b \in B$, bv_i is a unique linear combination of $\{v_1, \cdots, v_n\}$, so there exists a unique $n \times n$ matrix $f_b \in M_n(L)$ such that

$$\begin{pmatrix} bv_1 \\ \vdots \\ bv_n \end{pmatrix} = f_b \begin{pmatrix} v_1 \\ \vdots \\ v_n \end{pmatrix}.$$

Define $\phi : B \to M_n(L)$ by $\phi(b) = f_b^T$, where f_b^T is the transpose of the matrix f_b. It is straightforward to check that ϕ is one-to-one and an algebra homomorphism (Exercise 20). Clearly ϕ maps the identity element of B to the identity matrix. $\qquad \square$

Corollary 2.4. *Let A be a unital finite-dimensional ℓ-simple ℓ-algebra over $\mathbb{R}$ with a d-basis, then A is ℓ-isomorphic to the ℓ-algebra $M_n(\mathbb{R}[H])$ where $n \geq 1$, H is a finite group, and $\mathbb{R}[H]$ is the group ℓ-algebra of H over $\mathbb{R}$.*

Proof. By Theorem 2.6, A is ℓ-isomorphic to the ℓ-algebra $M_n(K^t[G])$ over $\mathbb{R}$, where K is a totally ordered field and finite-dimensional ℓ-algebra over $\mathbb{R}$. If $\dim_{\mathbb{R}} K > 1$, then K is isomorphic to the field $\mathbb{C}$ of complex numbers, which is impossible since $\mathbb{C}$ cannot be a totally ordered field. Thus $\dim_{\mathbb{R}} K = 1$ and A is ℓ-isomorphic to the ℓ-algebra $M_n(\mathbb{R}^t[G])$.

We show that ℓ-algebra $\mathbb{R}^t[G]$ is actually a group ℓ-algebra $\mathbb{R}[H]$. Suppose that G contains k elements. Since G is a vector space basis of $\mathbb{R}^t[G]$ over $\mathbb{R}$, we may consider $\mathbb{R}^t[G]$ as a subalgebra of $M_k(\mathbb{R})$ containing the identity matrix by Lemma 2.5. For $a \in G$, since a is not nilpotent, by Theorem 2.2 there exists a positive integer n_a such that $a^{n_a} = \alpha_a e$ for some $0 < \alpha_a \in \mathbb{R}$, where e is the identity element of G. Then $\alpha_a = (\beta_a)^{n_a}$ for some $0 < \beta_a \in \mathbb{R}$. It follows that $(\beta_a^{-1} a)^{n_a} = e$. Let $\bar{a} = \beta_a^{-1} a$ and define $H = \{\bar{a} \mid a \in G\}$. We check that H is a group. For $g_i, g_j \in H$, $g_i g_j = \alpha g_k$ for some $0 < \alpha \in \mathbb{R}$ and $g_k \in H$, so $\det(g_i)\det(g_j) = \alpha^k \det(g_k)$, where $\det(g)$ denotes the determinant of a matrix g. For $g \in H$, since $g^m = e$ for some $m \geq 1$, $(\det(g))^m = 1$. It follows from $\det(g) \in \mathbb{R}$ that $\det(g) = \pm 1$. Hence $\det(g_i)\det(g_j) = \alpha^k \det(g_k)$ implies that $\alpha^k = 1$, so $\alpha = 1$ since $\alpha > 0$. Therefore $g_i g_j = g_k \in H$, so H is a finite group. Clearly H is also a d-basis for $\mathbb{R}^t[G]$, and hence $\mathbb{R}^t[G] = \mathbb{R}[H]$. Therefore the ℓ-algebra $M_k(\mathbb{R}^t[G])$ is equal to the ℓ-algebra $M_k(\mathbb{R}[H])$. $\qquad \square$

A totally ordered field F is called *real closed* if any proper algebraic extension field of F cannot be made into a totally ordered field. For instance, $\mathbb{R}$ is a real closed field. Corollary 2.4 is actually true for any real closed field.

Now we consider the uniqueness of the ℓ-isomorphism in Theorem 2.6.

Theorem 2.7. *Suppose that ℓ-algebras $M_{n_1}(K_1^{t_1}[G_1])$ and $M_{n_2}(K_2^{t_2}[G_2])$ are ℓ-isomorphic ℓ-algebras over F, where n_1, n_2 are positive integers, K_1, K_2 are totally ordered fields and finite-dimensional ℓ-algebras over F, G_1, G_2 are finite groups, and $t_1 : G_1 \times G_1 \to K_1 \setminus \{0\}$, $t_2 : G_2 \times G_2 \to K_2 \setminus \{0\}$ are positive twisting functions. Then $n_1 = n_2$ and ℓ-algebras $K_1^{t_1}[G_1]$ and $K_2^{t_2}[G_2]$ are ℓ-isomorphic. Moreover, K_1, K_2 are ℓ-isomorphic ℓ-algebras over F, and G_1, G_2 are isomorphic groups.*

Proof. For simplicity of notation, let $B = M_{n_1}(K_1^{t_1}[G_1])$ and $C = M_{n_2}(K_2^{t_2}[G_2])$. In B,

$$1_B = e_{11}(e_1) + \cdots + e_{n_1 n_1}(e_1),$$

where 1_B is the identity matrix in B, $e_{ii}(e_1)$ is the $n_1 \times n_1$ matrices with the ii^{th} entry equal to the identity element e_1 of G_1 and other entries equal to zero, $i = 1, \ldots, n_1$. By Theorem 2.2(3), B has at most n_1 basic elements that are also idempotent f-elements. Similarly, C has at most n_2 basic elements that are also idempotent f-elements. Therefore, that B and C are ℓ-isomorphic implies that $n_1 = n_2$.

Let $n = n_1 = n_2$ and $\varphi : B \to C$ be an ℓ-isomorphism between ℓ-algebras B and C. Since $1_B = e_{11}(e_1) + \cdots + e_{nn}(e_1)$, we have

$$
\begin{aligned}
1_C &= \varphi(1_B) \\
&= \varphi(e_{11}(e_1)) + \cdots + \varphi(e_{nn}(e_1)) \\
&= e_{11}(e_2) + \cdots + e_{nn}(e_2),
\end{aligned}
$$

where 1_C is the identity matrix of C and e_2 is the identity element of G_2. Then, since $\{e_{11}(e_2), \ldots, e_{nn}(e_2)\}$ and $\{\varphi(e_{11}(e_1)), \ldots, \varphi(e_{nn}(e_1))\}$ both are disjoint sets of basic elements that are also idempotent f-elements of C, we must have

$$\varphi(e_{11}(e_1)) = e_{i_1 i_1}(e_2), \ldots, \varphi(e_{nn}(e_1)) = e_{i_n i_n}(e_2),$$

where $\{i_1, \ldots, i_n\}$ is a permutation of $\{1, \ldots, n\}$. Let

$$E_1 = e_{11}(e_1) B e_{11}(e_1), \quad E_2 = e_{i_1 i_1}(e_2) C e_{i_1 i_1}(e_2).$$

It is clear that $\varphi|_{E_1} : E_1 \to E_2$ is an ℓ-isomorphism of the two ℓ-algebras (Exercise 21). Define $f : K_1^{t_1}[G_1] \to E_1$ by $f(x) = e_{11}(x)$ for all $x \in K_1^{t_1}[G_1]$. Then it is straightforward to verify that f is an ℓ-isomorphism of two ℓ-algebras (Exercise 22). Similarly $K_2^{t_2}[G_2]$ is ℓ-isomorphic to E_2. Therefore $K_1^{t_1}[G_1]$ and $K_2^{t_2}[G_2]$ are ℓ-isomorphic ℓ-algebras.

We also use φ to denote the ℓ-isomorphism from $K_1^{t_1}[G_1]$ to $K_2^{t_2}[G_2]$. By a direct calculation, we have $K_2 = e_2^{\perp\perp} = \varphi(e_1^{\perp\perp}) = \varphi(K_1)$ (Exercise 23), so $K_1 \cong K_2$. Moreover G_1 and G_2 have the same number of elements. Suppose that $G_1 = \{g_1, \cdots, g_k\}$ and $G_2 = \{h_1, \cdots, h_k\}$. Since $\varphi(g_j)$ is a basic element, $\varphi(g_j) = u_j h_{i_j}$ for unique $0 < u_j \in K_2$ and $h_{i_j} \in G_2$. Define $\theta : G_1 \to G_2$ by $\theta(g_j) = h_{i_j}$. For $g_r, g_s \in G_1$, suppose that $g_r g_s = g_t$ and $\varphi(g_r) = \alpha_r h_{i_r}$, $\varphi(g_s) = \alpha_s h_{i_s}$. Then $\varphi(g_r g_s) = \varphi(g_r)\varphi(g_s)$ implies that $\alpha_t h_{i_t} = (\alpha_r h_{i_r})(\alpha_s h_{i_s})$, and hence $h_{i_r} h_{i_s} = h_{i_t}$ in G_2. Hence $\theta(g_r g_s) = \theta(g_r)\theta(g_s)$, so θ is an isomorphism from G_1 to G_2. $\square$

2.2.3   General case

In this section we consider unital finite-dimensional Archimedean ℓ-algebras A with a d-basis over F that may not be ℓ-simple. We first consider the case that ℓ-$N(A) = \{0\}$. We notice that the results in Theorem 1.28 are true for ℓ-algebras.

Theorem 2.8. *Let A be a unital finite-dimensional ℓ-algebra over F with a d-basis. If ℓ-$N(A) = \{0\}$, then A is ℓ-isomorphic to a finite direct sum of unital finite-dimensional ℓ-simple ℓ-algebras over F with a d-basis. Thus A is ℓ-isomorphic to a direct sum of matrix ℓ-algebras with the entrywise order over twisted group ℓ-algebras of finite groups over F.*

Proof.   If ℓ-$N(A) = \{0\}$, then A is Archimedean over F, and hence A is a direct sum of maximal convex totally ordered subspaces of A over F. Also by Theorem 1.28, the intersection of ℓ-prime ℓ-ideals is zero. Since A is finite-dimensional, we may choose a finite number of ℓ-prime ℓ-ideals $P_1, \cdots, P_k$ such that $P_1 \cap \cdots \cap P_k = \{0\}$ (Exercise 24). We may also assume that the family $\{P_1, \cdots, P_k\}$ is minimal in the sense that no proper sub-family of it has intersection $\{0\}$.

   We show that each ℓ-prime ℓ-ideal P is a maximal ℓ-ideal. Suppose that $P \subseteq I$ and $P \neq I$ for some ℓ-ideal I of A. Define $J = \{a \in A \mid aI \subseteq P\}$. Clearly J is an ideal of A. Suppose that $|b| \leq |a|$ for some $a \in J$ and $b \in A$. Let $x \in I$ be a basic element. Since x is a d-element, $|bx| = |b|x \leq |a|x = |ax| \in P$ implies $bx \in P$. Then since each strictly positive element in I is a sum of disjoint basic elements in I, we have $bI \subseteq P$, that is, $b \in J$. Hence J is an ℓ-ideal of A. By the definition of J, $JI \subseteq P$, so $J \subseteq P$ since P is ℓ-prime and $I \not\subseteq P$.

   Suppose that $1 = c_1 + \cdots + c_n$, where $n \geq 1$ and $\{c_1, \cdots, c_n\}$ is a disjoint set of basic elements. If $c_i I = \{0\}$, then $c_i \in J \subseteq P \subseteq I$ implies that $c_i = c_i^2 = 0$, which is a contradiction. Thus for any c_i, $\{0\} \neq c_i I \subseteq c_i A$. From Theorem 2.2 and Corollary 2.1, $c_i I$ is a right ℓ-ideal and $c_i A$ is a minimal right ℓ-ideal, and hence $c_i I = c_i A \subseteq I$ for each $i = 1, \cdots, n$. Then $A = c_1 A + \cdots + c_n A$ implies $A \subseteq I$. Hence $I = A$ and P is a maximal ℓ-ideal of A.

   Since each P_i is a maximal ℓ-ideal of A, $P_i + (\cap_{j \neq i} P_j) = A$. Construct the direct sum $A/P_1 \oplus \cdots \oplus A/P_k$, each A/P_i is an ℓ-simple ℓ-algebra, and define the mapping $\varphi : a \to (a + P_1, \cdots, a + P_k)$. Clearly φ is one-to-one and an ℓ-homomorphism between two ℓ-algebras. For $a_i \in A$, $a_i = x_i + y_i$, where $x_i \in P_i, y_i \in \cap_{j \neq i} P_j$, $1 \leq i \leq k$. Let $a = y_1 + \cdots + y_k$. Then

$\varphi(a) = (a_1 + P_1, \cdots, a_k + P_k)$, that is, φ is also onto. Therefore φ is an ℓ-isomorphism between the two ℓ-algebras. □

We would like to present another proof of the result in Theorem 2.8 and further characterize those ℓ-simple components in the direct sum of Theorem 2.8.

Let R be an ℓ-ring and M be an ℓ-group that is also a right (left) R-module. Then M is called a right (left) ℓ-*module over* R if $xr \in M^+$ ($rx \in M^+$) whenever $x \in M^+, r \in R^+$. For ℓ-modules M and N, an ℓ-isomorphism φ is a module isomorphism from M to N such that for any $x, y \in M$, $\varphi(x \vee y) = \varphi(x) \vee \varphi(y)$ and $\varphi(x \wedge y) = \varphi(x) \wedge \varphi(y)$.

Let A be a unital finite-dimensional ℓ-algebra with a d-basis. Suppose that ℓ-$N(A) = 0$ and $1 = c_1 + \ldots + c_n$, where $n \geq 1$, and $c_1, \ldots, c_n$ are disjoint basic elements. By Corollary 2.1, $\{c_1 A, \ldots, c_n A\}$ consists of all the minimal right ℓ-ideals of A. For $i = 1, \ldots, n$, define A_i as the sum of all minimal right ℓ-ideals of A which are ℓ-isomorphic to $c_i A$ as right ℓ-modules over A.

Theorem 2.9. *Let A be a unital finite-dimensional ℓ-algebra with a d-basis. Suppose that ℓ-$N(A) = 0$ and $1 = c_1 + \ldots + c_n$, where $n \geq 1$, and $c_1, \ldots, c_n$ are disjoint basic elements.*

(1) For a minimal right ℓ-ideal I of A and a basic element x, if $xI \neq 0$, then xI is also a minimal right ℓ-ideal of A.

(2) For each $i = 1, \ldots, n$, A_i is an ℓ-ideal of A.

(3) For $1 \leq i, j \leq n$, if $c_i A$ and $c_j A$ are not ℓ-isomorphic as right ℓ-modules over A, then $A_i A_j = 0$.

(4) Each A_i is ℓ-simple and $A = A_1 \oplus \cdots \oplus A_k$ for some positive integer $k \leq n$.

Proof. (1) Since $A = c_1 A + \cdots + c_n A$ and xI is a right ℓ-ideal by Theorem 2.2,

$$xI = A \cap xI = (c_1 A \cap xI) + \cdots + (c_n A \cap xI)$$

by Theorem 1.9. Then since each $c_i A$ is a minimal right ℓ-ideal, we have either $c_i A \cap xI = \{0\}$ or $c_i A \cap xI = c_i A$. It follows that xI is a direct sum of some right ℓ-ideals in $\{c_1 A, \cdots, c_n A\}$. Since there exists a unique c_j such that $c_j x = x$, we must have that $xI = c_j A$, so xI is a minimal right ℓ-ideal of A.

(2) Since A_i is a right ℓ-ideal of A, it is sufficient to show that if I is a minimal right ℓ-ideal of A with $I \cong c_i A$ as right ℓ-modules of A, then

$xI \subseteq A_i$ for each basic element x of A. Suppose that $xI \neq 0$. Then by (1) xI is a minimal right ℓ-ideal of A. Define $\varphi : I \to xI$ by $\varphi(a) = xa$, $\forall a \in I$. Then φ is a homomorphism between right A-modules I and xI. Since x is a d-element, for any $a, b \in I$,

$$\varphi(a \wedge b) = x(a \wedge b) = (xa) \wedge (xb) = \varphi(a) \wedge \varphi(b).$$

Similarly $\varphi(a \vee b) = \varphi(a) \vee \varphi(b)$. Thus φ is an ℓ-homomorphism between ℓ-modules I and xI over A. Let H be the kernel of φ, that is, $H = \{a \in I \mid xa = 0\}$. Clearly H is a right ideal of A. Now let $b \in A$ and $a \in H$ with $|b| \leq |a|$. Then $b \in I$ and

$$|xb| = x|b| \leq x|a| = |xa| = 0,$$

so $xb = 0$ and hence $b \in H$. Therefore H is a right ℓ-ideal of A. It follows from the fact that I is a minimal right ℓ-ideal that either $H = I$ or $H = 0$. If $H = I$, then $xI = 0$, which is a contradiction. Hence $H = 0$, so φ is one-to-one. It is clear that φ is onto. Therefore $xI \cong I \cong c_i A$ as right ℓ-modules over A, so $xI \subseteq A_i$. Hence A_i is also a left ideal of A. This completes the proof of (2).

(3) Suppose that $I = c_i A$ and $J = c_j A$ are not ℓ-isomorphic. To show that $A_i A_j = 0$, it is enough to show that $IJ = 0$ (Exercise 25). If $IJ \neq 0$, then there exists a basic element $x \in I$ such that $xJ \neq 0$. By (1), xJ is a minimal right ℓ-ideal and by the proof of (2), $J \cong xJ$ as right ℓ-modules over A. On the other hand, $xJ \subseteq I$ since $x \in I$ and I is a right ℓ-ideal. Then $xJ = I$, so $J \cong xJ = I$, which is a contradiction. Hence $IJ = 0$.

(4) We may assume that $c_1 A, \cdots, c_k A$ are pairwise nonisomorphic ℓ-modules over A for some $1 \leq k \leq n$, and for each $j = 1, \cdots, n$, $c_j A$ is ℓ-isomorphic to one of $c_1 A, \cdots, c_k A$ as ℓ-modules over A. Then $A = A_1 \oplus \cdots \oplus A_k$ as the direct sum of ℓ-ideals $A_1, \cdots, A_k$ (Exercise 26).

Finally we show that each A_i, $i = 1, ..., k$, is an ℓ-simple ℓ-algebra. In fact, let $H \neq 0$ be an ℓ-ideal of A_i. Then H is an ℓ-ideal of A, so H contains a minimal right ℓ-ideal I of A, and hence $I \subseteq A_i$. Thus $I \cong c_i A$. Let J be a minimal right ℓ-ideal of A and $J \cong c_i A$. Then $J \cong I$. Suppose that $I = c_v A$ for some $v = 1, ..., n$ and $\varphi : I \to J$ be an ℓ-isomorphism of ℓ-modules over A. Then $c_v I = I$, so

$$J = \varphi(I) = \varphi(c_v I) = \varphi(c_v)I \subseteq \varphi(c_v)H \subseteq H,$$

since H is an ℓ-ideal of A. Hence $A_i \subseteq H$ by the definition of A_i, so $H = A_i$. Therefore A_i has no other ℓ-ideal except $\{0\}$ and A_i, that is, each A_i is ℓ-simple. $\square$

Finally consider a unital finite-dimensional Archimedean ℓ-algebra A over F with a d-basis. We show that $A = \ell\text{-}N(A) + H$, where H is a convex ℓ-subalgebra of A over F and $\ell\text{-}N(A) \cap H = \{0\}$. The proof of this result is based on the following characterization for basic elements that are not in $\ell\text{-}N(A)$.

Lemma 2.6. *Let A be a unital finite-dimensional Archimedean ℓ-algebra over F with a d-basis, and $1 = c_1 + \cdots + c_n$, where $n \geq 1$ and $\{c_1, \cdots, c_n\}$ is a disjoint set of basic elements. Suppose that $x \notin \ell\text{-}N(A)$ is a basic element. Then there exists a basic element y such that $xy = c_s$ and $yx = c_t$ for some c_s, c_t.*

Proof. We first assume that $\ell\text{-}N(A) = \{0\}$. By Theorem 2.8, A is ℓ-isomorphic to the ℓ-algebra $B = M_{n_1}(K_1^{t_1}[G_1]) \oplus \cdots \oplus M_{n_k}(K_k^{t_k}[G_k])$ for some positive integer k, where each $K_i^{t_i}[G_i]$ is a twisted group ℓ-algebra with the coordinatewise order and $M_{n_i}(K_i^{t_i}[G_i])$ is the matrix ℓ-algebra with entrywise order. For a basic element x in B, x is in some direct summand $M_{n_i}(K_i^{t_i}[G_i])$ with the form $e_{st}(\alpha g)$, where $0 < \alpha \in K_i$, $g \in G_i$, and $e_{st}(\alpha g)$ is the matrix with st^{th} entry equal to αg and other entries equal to zero. Take $y = e_{ts}(\alpha^{-1} t_i(g^{-1}, g)^{-1} g^{-1})$. Then we have $xy = e_{ss}(e_i)$ and $yx = e_{tt}(e_i)$ are both basic elements and idempotent f-elements in B, where e_i is the identity element of G_i. By Theorem 2.2(3), $e_{ss}(e_i) = c_s$ and $e_{tt}(e_i) = c_t$ for some $1 \leq s, t \leq n$.

For the general case, let $\overline{A} = A/\ell\text{-}N(A)$. We have $\ell\text{-}N(\overline{A}) = \{0\}$. For any element $a \in A$, denote $\overline{a} = a + \ell\text{-}N(A) \in \overline{A}$. Then $\overline{1} = \overline{c_1} + \ldots + \overline{c_n}$. For a basic element $x \notin \ell\text{-}N(A)$, $0 < \overline{x}$ is basic in $\overline{A}$ (Exercise 27), and hence by the previous paragraph there exists a basic element $\overline{y} \in \overline{A}$ such that $\overline{xy} = \overline{c_s}$ and $\overline{yx} = \overline{c_t}$. We may assume that y is basic in A.

From $xy - c_s \in \ell\text{-}N(A)$, we have that $xy - c_s$ is nilpotent. Suppose that $(xy - c_s)^\ell = 0$ and ℓ is an odd positive integer. Then $c_s(xy) = (xy)c_s = xy$ implies that $0 = (xy - c_s)^\ell = xd - c_s$ for some $d \in A$, and hence $c_s = |c_s| = |xd| = x|d|$ since x is a d-element. Let $|d| = d_1 + \cdots + d_t$, where $d_1, \cdots, d_t$ are disjoint basic elements. It follows from c_s is basic that $c_s = xd_m$ for some basic element d_m. Take $z = d_m$, then $xz = c_s$, and $z \notin \ell\text{-}N(A)$. Since $xc_t = x$, $xz \neq 0$ implies $c_t z = z$. By the same argument used above, there is a basic element w such that $zw = c_t$. Then we have

$$x = xc_t = x(zw) = (xz)w = c_s w = w$$

since $c_s w \neq 0$. Hence $xz = c_s$ and $zx = c_t$. $\qquad\square$

Theorem 2.10. *Let A be a unital finite-dimensional Archimedean ℓ-algebra over F with a d-basis. Then $A = \ell\text{-}N(A) + H$, where H is a convex ℓ-subalgebra of A and $\ell\text{-}N(A) \cap H = \{0\}$.*

Proof. As before assume that $1 = c_1 + \cdots + c_n$, where $n \geq 1$ and $\{c_1, \cdots, c_n\}$ is a disjoint set of basic elements. Define

$$H = \{a \in A \mid |a| \leq \sum_{i=1}^{k} a_i, \, a_1, \cdots, a_k \notin \ell\text{-}N(A)\},$$

and $\{a_1, \cdots, a_k\}$ is a disjoint set of basic elements. If two basic elements x and y are comparable, then $x^{\perp\perp} = y^{\perp\perp}$ implies that $x + y$ is also basic. Based on this fact, it is straightforward to check that H is a convex vector sublattice of A (Exercise 28). We show that $\ell\text{-}N(A) \cap H = \{0\}$. Suppose not, then $\ell\text{-}N(A) \cap H$ contains a basic element x. Since $x \in H$, $x \leq y$ for some basic element $y \notin \ell\text{-}N(A)$. By Lemma 2.6, there exists a basic element z such that $yz = c_i$ and $zy = c_j$ for some c_i, c_j, and hence $x \leq y$ implies $xz \leq c_i$. Let $xz = c$. We have $0 \neq c \in c_i^{\perp\perp}$ which is a field by Theorem 2.4, and $x = xc_j = xzy = cy$. Hence $c^{-1}x = c^{-1}cy = c_iy = y$, which implies that $y \in \ell\text{-}N(A)$ since $x \in \ell\text{-}N(A)$. This is a contradiction. Therefore $\ell\text{-}N(A) \cap H = \{0\}$.

Finally we show that H is closed under the multiplication of A. To this end, we show that for basic elements $x, y \notin \ell\text{-}N(A)$ with $xy \neq 0$, xy is also a basic element not in $\ell\text{-}N(A)$. By Lemma 2.6, there exists a basic element w such that $xw = c_s$ and $wx = c_t$ for some c_s, c_t, and hence $c_sx = x$ and $xc_t = x$. Since $xy \neq 0$, $c_ty = y$. Suppose that $xy = b_1 + \cdots + b_k$, where $k \geq 1$ and $b_1, \cdots, b_k$ are disjoint basic elements. We have

$$y = c_ty = (wx)y = wb_1 + \cdots + wb_k \text{ and } wb_i \wedge wb_j = 0, i \neq j,$$

since w is a d-element. It follows that $y = wb_u$ and $wb_v = 0$ for any $v \neq u$, so $c_sb_v = (xw)b_v = 0$. Since

$$c_s(xy) = c_sb_1 + \cdots + c_sb_k = xy = b_1 + \cdots + b_k,$$

$c_sb_v = b_v$ (Exercise 29), and hence $b_v = 0$ for any $v \neq u$. Hence $xy = b_u$ is a basic element. If $xy \in \ell\text{-}N(A)$, then $w(xy) = c_ty = y \in \ell\text{-}N(A)$, which is a contradiction. Hence $xy \notin \ell\text{-}N(A)$. Therefore H is a convex ℓ-subalgebra of A. $\square$

The following special cases are immediate consequence of Theorem 2.10. The verification is left to the reader (Exercise 35).

Theorem 2.11. *Let A be a unital finite-dimensional ℓ-algebra over F with a d-basis.*

(1) If A is Archimedean over F and commutative, then $A = \ell\text{-}N(A) + H$, where H is ℓ-isomorphic to a direct sum of twisted group ℓ-algebras of finite abelian groups.

(2) If A is ℓ-reduced, then A is ℓ-isomorphic to a direct sum of twisted group ℓ-algebras of finite groups.

Exercises

(1) Verify Example 2.3.

(2) Verify that the ℓ-algebra A in Example 2.4 has a d-basis $\{1, a, b, c\}$.

(3) Let A be a unital Archimedean ℓ-algebra over F with a d-basis S and satisfies condition (C). Prove that if H is a convex ℓ-subalgebra of A, then $H \cap S$ is a d-basis for H.

(4) Let V be a vector lattice over F which is a direct sum of maximal convex totally ordered subspaces over F. Prove that V is Archimedean over F if and only if each direct summand is Archimedean over F.

(5) Prove that in Theorem 2.4(2), $a^{n_a-1}a_1 = 0$ or $a^{n_a-1}a_2 = 0$ implies that $a_1 = 0$ or $a_2 = 0$.

(6) Let A be an ℓ-algebra over F and I be a right ℓ-ideal of A. Take $0 < x \in I$. Prove $J = \{a \in A \mid |a| \leq xr \text{ for some } 0 < r \in I\}$ is a right ℓ-ideal of A.

(7) Let G be an ℓ-group and x be a basic element. If $x \leq x_1 + \cdots + x_m$, where $\{x_1, \cdots, x_m\}$ is a disjoint subset of G, then $x \leq x_i$ for some $i = 1, \cdots, m$.

(8) Let R be an ℓ-ring and I be a nilpotent right (left) ℓ-ideal of R. Then $I \subseteq \ell\text{-}N(R)$.

(9) Verify $F^t[G]$ as defined after Definition 2.2 is a vector lattice over F.

(10) Prove that ℓ-field $\mathbb{Q}[\sqrt{2}]$ with the entrywise order may be considered as a twisted group ℓ-algebra over $\mathbb{Q}$.

(11) Prove $F^t[G]$ is an ℓ-domain and its identity element is basic.

(12) Let A be an ℓ-unital ℓ-algebra over F and $1 = a + b$, where $a \wedge b = 0$. Prove that aAa is a convex ℓ-subalgebra of A.

(13) Prove that the relation $\sim$ defined in the proof of Theorem 2.6 is an equivalence relation.

(14) Prove that the operation $(ab)' = a'b'$ defined on G_i in Theorem 2.6 is well-defined, associative, and c_i is the identity element.

(15) Check that t as defined in Theorem 2.6 is a positive twisting function.

(16) Prove that the map φ defined in Theorem 2.6 preserves the addition

on A, that is, for any $a, b \in A$, $\varphi(a + b) = \varphi(a) + \varphi(b)$.

(17) Prove that the map φ defined in Theorem 2.6 preserves the multiplication on A, that is, for any $a, b \in A$, $\varphi(ab) = \varphi(a)\varphi(b)$.

(18) Prove that the φ defined in Theorem 2.6 preserves order, that is, for any $a \in A$, $\varphi(a) \geq 0$ if and only if $a \geq 0$.

(19) Verify that $\{e_{ij}(s) \mid 1 \leq i, j \leq n, s \in H\}$ is a d-basis for the ℓ-algebra $M_n(F^t[H])$.

(20) Prove that the mapping $\phi : B \to M_n(L)$ in Lemma 2.5 is one-to-one and preserves addition and multiplication on B.

(21) Prove that $\varphi|_{E_1} : E_1 \to E_2$ defined in Theorem 2.7 is an ℓ-isomorphism between ℓ-algebras E_1 and E_2.

(22) Verify that $f : K_1^{t_1}[G_1] \to E_1$ is an ℓ-isomorphism between the two ℓ-algebras in Theorem 2.7.

(23) Prove that $K_2 = e_2^{\perp\perp} = \varphi(e_1^{\perp\perp}) = \varphi(K_1)$ in Theorem 2.7.

(24) Suppose that A is a finite-dimensional ℓ-algebra over F. Prove that if the intersection of all the ℓ-prime ℓ-ideals is zero, then there exists a finite number of ℓ-prime ℓ-ideals such that the intersection of them is also zero.

(25) Prove that in Theorem 2.9(3) if $(c_i A)(c_j A) = \{0\}$, then $A_i A_j = \{0\}$.

(26) Let A be an ℓ-algebra over a totally ordered field F and $A_1, \cdots, A_k$ be ℓ-ideals of A. A is the direct sum of $A_1, \cdots, A_k$, denoted by $A = A_1 \oplus \cdots \oplus A_k$, if $A = A_1 + \cdots + A_k$ and $A_i \cap A_j = \{0\}$, $1 \leq i, j \leq n$ and $i \neq j$. Prove Theorem 2.9(4).

(27) Let R be an ℓ-ring and $0 < x \notin \ell\text{-}N(R)$. Prove that if x is basic in R, then $\bar{x} = x + \ell\text{-}N(R)$ is basic in $R/\ell\text{-}N(R)$.

(28) Prove that the H as define in Theorem 2.10 is a convex vector sublattice.

(29) Suppose that R is an ℓ-ring and $a > 0$ is an f-element of R. Prove that if $x_1, \cdots, x_k$ are disjoint elements and

$$ax_1 + \cdots + ax_k = x_1 + \cdots + x_k,$$

then $ax_i = x_i$ for $i = 1, \cdots, k$.

(30) Prove Theorem 2.11.

(31) An ℓ-ring is called a *quasi d-ring* if each nonzero positive element can be written as a sum of disjoint basic elements that are also d-element. Prove that a unital quasi d-ring is ℓ-unital.

(32) Let R be an ℓ-unital ℓ-ring and M, N be ℓ-modules over R. Prove that a module isomorphism from M to N is an ℓ-isomorphism if and only if for any $x, y \in M$, $\varphi(x \vee y) = \varphi(x) \vee \varphi(y)$ or $\varphi(x \wedge y) = \varphi(x) \wedge \varphi(y)$.

(33) Let G be a group and $t : G \times G \rightarrow F \setminus \{0\}$ be a positive twisting function. Prove that for any $g \in G$, $t(g, g^{-1}) = t(g^{-1}, g)$.

(34) Prove that any two statements of $(i), (ii), (iii)$ in Theorem 2.2(2) cannot be true at the same time.

(35) Prove Theorem 2.11.

Chapter 3

Positive derivations on ℓ-rings

In this chapter, we study positive derivations for various ℓ-rings. In section 1 some examples and basic properties are presented. Section 2 is devoted to f-ring and its generalizations. We study positive derivations on matrix ℓ-rings in section 3, and section 4 consists of some results on the kernel of positive derivations of ℓ-rings.

For a ring B, a function $D : B \to B$ is called a *derivation* on B if for any $a, b \in B$

$$D(a + b) = D(a) + D(b) \quad \text{and} \quad D(ab) = aD(b) + D(a)b.$$

If L is an algebra over a field T, then a derivation on L is called a *T-derivation* if T is also a linear transformation, that is, $D(\alpha a) = \alpha D(a)$ for all $\alpha \in T$ and all $a \in L$.

Now let R be a partially ordered ring. A derivation on R is called *positive* if for all $x \in R^+$, $D(x) \geq 0$, and similarly an F-derivation on a partially ordered algebra A over a totally ordered field F is called positive if for all $x \in A^+$, $D(x) \geq 0$.

3.1 Examples and basic properties

The following are some examples of positive derivations. Clearly the map that sends each element to zero is a positive derivation which is called *trivial derivation*. Throughout this chapter F denotes a totally ordered field.

Example 3.1.

(1) Let $R = F[x]$ be the polynomial ℓ-algebra over F with the coordinatewise order. For $f(x) = a_n x^n + \cdots + a_1 x + a_0$ with $a_n \neq 0$, the usual derivative of $f(x)$ is defined as $f'(x) = n a_n x^{n-1} + \cdots + a_1$. Fix

a polynomial $0 < g(x) \in R$. Define $D : R \to R$ by for any $f(x) \in R$, $D(f(x)) = f'(x)g(x)$. It is clear that D is a positive F-derivation on R. On the other hand, if D is a positive F-derivation on R, then it is easily checked that $D(x^n) = nx^{n-1}D(x)$. Hence for any $f(x) \in R$, $D(f(x)) = f'(x)D(x)$ and $D(x)$ is a positive polynomial (Exercise 1).

From Example 1.3(3), $R = F[x]$ can be made into a totally ordered algebra over F in two ways. One total order on R is to define a polynomial positive if the coefficient of the highest power is positive. With respect to this total order, the derivation introduced above is still a positive derivation. Another total order on R is to define a polynomial positive if the coefficient of the lowest power is positive. In this case, for any positive polynomial $f(x)$, $f(x) \leq \alpha 1$, for some $0 < \alpha \in F$, implies that for any positive F-derivation D, $0 \leq D(f(x)) \leq \alpha D(1) = 0$, so $D(f(x)) = 0$. Therefore $D(R) = \{0\}$. Namely the trivial derivation is the only positive F-derivation in this case. In Lemma 3.6, it is shown that for an ℓ-unital ℓ-algebra A over F, each positive derivation on A is an F-derivation. Thus with respect to this total order on R, the trivial derivation is the only positive derivation on R.

(2) For a ring B and $b \in B$, define mapping $D_b : B \to B$ by for any $x \in B$, $D_b(x) = xb - bx$. Then D_b is a derivation on B (Exercise 2) and D_b is called the *inner derivation* determined by b.

Consider 2×2 upper triangular matrix ℓ-algebra $T_2(F)$ over F with the entrywise order. Take $a = (a_{ij}) \in T_2(F)$ with $a_{11} \leq a_{22}$ and $a_{12} = 0$. Then it is easy to see that for each $0 \leq x \in T_2(F)$, $D_a(x) = xa - ax \geq 0$ (Exercise 3). For instance, if $a = e_{22}$, then D_a is a positive derivation.

(3) Let R be an ℓ-ring and $a \in R^+$ with $aR = 0$. Then inner derivation $D_a(x) = xa - ax = xa$ is a positive derivation. Similarly if $Ra = 0$, then $x \to ax$ is a positive derivation on R.

We first show that a positive derivation will map positive nilpotent elements and positive idempotent elements to nilpotent elements.

Lemma 3.1. *Suppose that R is a partially ordered ring and D is a positive derivation on R.*

(1) For $a \in R^+$, if $a^m = 0$ for some positive integer m, then $(D(a))^m = 0$.

(2) For $a \in R^+$, if $a^2 = ka$ for some positive integer k, then $(D(a))^3 = 0$.

Proof. (1) From $a^m = 0$, we have

$$0 = D(0) = D(a^m) = D(a)a^{m-1} + aD(a^{m-1}),$$

and hence $D(a)a^{m-1} = 0$ since $D(a)a^{m-1} \geq 0$ and $aD(a^{m-1}) \geq 0$. It follows that

$$0 = D(0) = D(D(a)a^{m-1}) = D(D(a))a^{m-1} + D(a)D(a^{m-1}),$$

so

$$0 = D(a)D(a^{m-1}) = D(a)(D(a)a^{m-2} + aD(a^{m-2})).$$

Therefore we have $(D(a))^2 a^{m-2} = 0$. Continuing this process we obtain $(D(a))^m = 0$.

(2) From $a^2 = ka$, we have

$$aD(a) + D(a)a - kD(a) = D(a^2) - kD(a) = D(a^2 - ka) = 0.$$

By multiplying the equation on the left by a, we obtain $a^2 D(a) + aD(a)a - kaD(a) = 0$. It follows that $aD(a)a = 0$ since $a^2 = ka$. Then

$$0 = D(aD(a)a) = aD(D(a)a) + D(a)D(a)a$$

implies $(D(a))^2 a = 0$, and hence

$$0 = D((D(a))^2 a) = D(a)^3 + D((D(a))^2)a,$$

so $D(a)^3 = 0$. □

The following result, which will be used later, characterizes minimal ℓ-prime ℓ-ideals for commutative ℓ-rings.

Lemma 3.2. *Let R be a commutative ℓ-ring and P be an ℓ-prime ℓ-ideal of R. Then P is a minimal ℓ-prime ℓ-ideal if and only if for each $0 \leq x \in P$ there exists $0 \leq y \notin P$ such that xy is a nilpotent element.*

Proof. Suppose that $P \neq \{0\}$ is a minimal ℓ-prime ℓ-ideal. Let $0 < x \in P$ and consider the set

$$S = \{x^n a \mid n \geq 1, \ a \in R^+ \setminus P\} \cup (R^+ \setminus P).$$

Then S is an m-system properly containing $R^+ \setminus P$ (Exercise 4). By Theorem 1.30(2), $0 \in S$, and hence $x^n y = 0$ for some $y \in R^+ \setminus P$ and positive integer n. Hence $(xy)^n = 0$. Conversely suppose that P is an ℓ-prime ℓ-ideal which satisfies the given condition and Q is a minimal ℓ-prime ℓ-ideal contained in P. If $Q \neq P$, then there exists $0 \leq x \in P \setminus Q$, and hence xy is nilpotent for some $0 \leq y \notin P$. Then $xy \in Q$ implies that $x \in Q$ or $y \in Q$ by Theorem 1.26, which is a contradiction. Thus $P = Q$ and hence P is a minimal ℓ-prime ℓ-ideal. □

Lemma 3.3. *Let R be an Archimedean ℓ-ring in which the square of each element is positive.*

(1) If $x \in R^+$ and $x^2 = 0$, then $xR = Rx = \{0\}$.

(2) ℓ-$N(R) = \{x \in R \mid |x| \text{ is nilpotent}\}$ and $R^2(\ell\text{-}N(R)) = (\ell\text{-}N(R))R^2 = R(\ell\text{-}N(R))R = \{0\}$.

(3) If R is an f-ring, then $R(\ell\text{-}N(R)) = (\ell\text{-}N(R))R = 0$.

Proof. (1) For any $y \in R^+$ and positive integer n, $(nx - y)^2 \geq 0$ implies

$$n(xy) \leq n(xy) + n(yx) \leq (nx)^2 + y^2 = y^2,$$

and hence $xy = 0$ since R is Archimedean. Similarly $yx = 0$. Thus $xR = Rx = \{0\}$.

(2) We first show that if $x \in R^+$ is nilpotent, then $x^3 = 0$. Suppose that $x^n = 0$ with $n \geq 4$. Then $2n - 4 \geq n$, so $(x^{n-2})^2 = 0$, and hence by (1) $x^{n-1} = xx^{n-2} = 0$. Continuing this process we eventually get $x^3 = 0$.

Now suppose $|x|$ is a nilpotent element and $y \in R^+$. For any positive integer n, $0 \leq (n|x| - y)^2$ implies that $n|x|y \leq n^2|x|^2 + y^2$, and hence

$$nz|x|y \leq n^2 z|x|^2 + zy^2 \text{ and } n|x|yz \leq n^2|x|^2z + y^2z,$$

for any $z \in R^+$. But $(|x|^2)^2 = 0$ implies $z|x|^2 = |x|^2z = 0$ by (1), so that R is Archimedean implies $z|x|y = |x|yz = 0$ for all $y, z \in R^+$. By the definition of ℓ-$N(R)$ and $R^+|x|R^+ = \{0\}$, we have $x \in \ell$-$N(R)$, and hence ℓ-$N(R) = \{x \in R \mid |x| \text{ is nilpotent}\}$. And then $|x|yz = 0$ and $z|x|y = 0$ for any $x \in \ell$-$N(R)$ and $y, z \in R^+$ imply that $(\ell\text{-}N(R))R^2 = \{0\}$ and $R(\ell\text{-}N(R))R = \{0\}$. Similarly, $R^2(\ell\text{-}N(R)) = \{0\}$.

(3) Suppose that R is an f-ring and $0 \leq x \in \ell$-$N(R)$. By (2), $x^3 = 0$. For a positive integer n, consider nx^2 and x. If R is totally ordered, then $nx^2 \leq x$ (Exercise 5). Then $nx^2 \leq x$ is true in an f-ring since it is a subdirect product of totally ordered rings, and hence $x^2 = 0$. Therefore by (1), $xR = Rx = \{0\}$, that is, $R(\ell\text{-}N(R)) = (\ell\text{-}N(R))R = \{0\}$. $\square$

Corollary 3.1. *Let R be an Archimedean ℓ-ring in which the square of each element is positive.*

(1) For a positive derivation D on R, $D(\ell\text{-}N(R)) \subseteq \ell\text{-}N(R)$.

(2) $R/\ell\text{-}N(R)$ is also Archimedean.

Proof. (1) Let $x \in \ell$-$N(R)$. Then $|x|$ is nilpotent by Lemma 3.3, so $D(|x|)$ is also nilpotent by Lemma 3.1. It follows that $D(|x|) \in \ell$-$N(R)$ by Lemma 3.3 again, and since $|D(x)| \leq D(|x|)$ (Exercise 6), $D(x) \in \ell$-$N(R)$. Therefore $D(\ell\text{-}N(R)) \subseteq \ell\text{-}N(R)$.

(2) Take $0 \leq a + \ell\text{-}N(R), 0 \leq b + \ell\text{-}N(R) \in R/\ell\text{-}N(R)$ and suppose that $n(a + \ell\text{-}N(R)) \leq (b + \ell\text{-}N(R))$ for all positive integer n. We may assume $a, b \in R^+$. Then

$$na + \ell\text{-}N(R) = (na + \ell\text{-}N(R)) \wedge (b + \ell\text{-}N(R))$$

implies that $na - na \wedge b = a_n \in \ell\text{-}N(R)$, and hence $na \leq b + a_n$. It follows that $na^3 \leq a^2 b$ since $a^2 a_n \in R^2(\ell\text{-}N(R)) = \{0\}$ by Lemma 3.3. Then R is Archimedean implies that $a^3 = 0$, and hence $a \in \ell\text{-}N(R)$ by Lemma 3.3 again, that is, $a + \ell\text{-}N(R) = 0$. Hence $R/\ell\text{-}N(R)$ is Archimedean. □

3.2 f-ring and its generalizations

Positive derivations on ℓ-rings were first studied for f-rings. We will present the results for f-rings first below.

Lemma 3.4. *Let T be a totally ordered domain and D be a positive derivation on T. Given $a \in T^+$, for any positive integer n,*

$$nD(a^2) \leq a^2 D(a) + D(a)a^2 + D(a).$$

Proof. If $D(a) = 0$, then the inequality is clearly true since $D(a^2) = D(a)a + aD(a) = 0$. Suppose that $D(a) > 0$ and n is a positive integer. If $na \leq a^2$, then

$$nD(a^2) = naD(a) + nD(a)a \leq a^2 D(a) + D(a)a^2.$$

In the following, we consider the case $a^2 < na$. First we show that $2^m a^2 \leq na$ for any positive integer m by mathematical induction. If $aD(a) \leq D(a)a$, then

$$2aD(a) \leq D(a)a + aD(a) = D(a^2) \leq nD(a),$$

and hence $2a^2 D(a) \leq (na)D(a)$. Thus $2a^2 \leq na$ since T is a totally ordered domain. Similarly if $D(a)a \leq aD(a)$, then

$$2D(a)a \leq D(a)a + aD(a) = D(a^2) \leq nD(a)$$

implies $2a^2 \leq na$. Thus $2^m a^2 \leq na$ is true when $m = 1$. Now suppose that $2^k a^2 \leq na$ and we show that $a^{k+1}a^2 \leq na$, $k \geq 1$. If $aD(a) \leq D(a)a$, then

$$2^{k+1} aD(a) \leq 2^k(D(a)a + aD(a)) = 2^k D(a^2) \leq nD(a)$$

implies $2^{k+1}a^2 D(a) \leq (na)D(a)$, so $2^{k+1}a^2 \leq na$. Similarly if $D(a)a \leq aD(a)$, then

$$2^{k+1} D(a)a \leq 2^k(D(a)a + aD(a)) = 2^k D(a^2) \leq nD(a)$$

implies $2^{k+1}D(a)a^2 \leq nD(a)a$, so $2^{k+1}a^2 \leq na$. In any case, $2^{k+1}a^2 \leq na$, and hence by the induction, $2^m a^2 \leq na$ for all positive integer m. Choose m such that $n^2 \leq 2^m$, we get $na^2 \leq a$, and hence $nD(a^2) \leq D(a)$.

Therefore for any case we have proved that $nD(a^2) \leq a^2 D(a) + D(a)a^2 + D(a)$ for $a \in T^+$.   □

We note that Lemma 3.4 is also true for a reduced f-ring since it is a subdirect product of totally ordered domains. We leave the verification of this fact to the reader (Exercise 7).

Theorem 3.1. *Let R be an Archimedean f-ring and D be a positive derivation on R. Then $D(R) \subseteq \ell\text{-}N(R)$ and $D(R^2) = \{0\}$. Thus if $\ell\text{-}N(R) = \{0\}$, then the only positive derivation on R is the trivial derivation.*

Proof. We note that R is commutative by Theorem 1.22. Let P be a minimal ℓ-prime ℓ-ideal of R and $0 \leq x \in P$. From Lemma 3.2, there exists $0 \leq y \notin P$ such that $(xy)^k = 0$ for some positive integer k. Then by Lemma 3.1, $(D(xy))^k = 0$, and hence $D(xy) = xD(y) + D(x)y$ implies that $(D(x)y)^k = 0$. Thus $D(x)y \in P$ and $y \notin P$ imply $D(x) \in P$ by Theorem 1.27. We have proved that $D(P) \subseteq P$ for each minimal ℓ-prime ℓ-ideal P.

Then D induces a positive derivation D_P on $R/P = \{\bar{a} = a + P \mid a \in R\}$ by defining $D_P(\bar{a}) = \overline{D(a)}$ (Exercise 8). Since R/P is a totally ordered domain by Theorem 1.27, using Lemma 3.4, for $a \in R^+$ and any positive integer n,

$$nD_P(\bar{a}^2) \leq \bar{a}^2 D_P(\bar{a}) + D_P(\bar{a})\bar{a}^2 + D_P(\bar{a}),$$

that is, for each minimal ℓ-prime ℓ-ideal P,

$$nD(a^2) + P \leq (a^2 D(a) + D(a)a^2 + D(a)) + P$$

in R/P. Hence

$$nD(a^2) + \ell\text{-}N(R) \leq (a^2 D(a) + D(a)a^2 + D(a)) + \ell\text{-}N(R)$$

in $R/\ell\text{-}N(R)$ since $\ell\text{-}N(R)$ is the intersection of all minimal ℓ-prime ℓ-ideals by Theorem 1.28.

By Corollary 3.1, $R/\ell\text{-}N(R)$ is Archimedean. Then

$$n(D(a^2) + \ell\text{-}N(R)) \leq (a^2 D(a) + D(a)a^2 + D(a)) + \ell\text{-}N(R)$$

implies $D(a^2) + \ell\text{-}N(R) = \{0\}$, that is, $D(a^2) \in \ell\text{-}N(R)$. By Corollary 3.1, $D(D(a^2)) \in \ell\text{-}N(R)$. Hence

$$(D(a))^2 \leq (D(a))^2 + aD(D(a)) = D(aD(a)) \leq D(D(a^2))$$

further implies $(D(a))^2 \in \ell\text{-}N(R)$. Thus $D(a) \in \ell\text{-}N(R)$ for each $a \in R^+$ by Lemma 3.3. It follows that $D(R) \subseteq \ell\text{-}N(R)$.

Finally for all $x, y \in R$, $D(xy) = xD(y) + D(x)y \in R(\ell\text{-}N(R)) + (\ell\text{-}N(R))R = \{0\}$ by Lemma 3.3. Hence $D(R^2) = \{0\}$.   □

Theorem 3.1 was originally proved by P. Colville, G. Davis and K. Keimel [Colville, Davis and Keimel (1977)]. The proof presented here was due to M. Henriksen and F. A. Smith [Henriksen and Smith (1982)] because of elementary and general nature in their proof. We would like to present the proof in [Colville, Davis and Keimel (1977)] based on Theorem 3.2 for Archimedean f-rings whose proof will be omitted, and the reader is referred to [Bigard and Keimel (1969)] for more details. Let R be an f-ring. A *positive orthomorphism* φ of R is an endomorphism of the additive group of R such that for any $x, y \in R$, $x \wedge y = 0 \Rightarrow x \wedge \varphi(y) = 0$. Define

$$\text{Orth}(R) = \{\varphi - \psi \mid \varphi, \psi \text{ are positive orthomorphisms of } R\}.$$

Then $\text{Orth}(R)$ is a partially ordered ring with respect to the positive cone

$$\text{Orth}(R)^+ = \{\varphi \mid \varphi \text{ is a positive orthomorphism of } R\}$$

(Exercise 9).

Theorem 3.2. *Let R be an Archimedean and reduced f-ring. Then $\text{Orth}(R)$ is a unital Archimedean f-ring.*

Another proof of Theorem 3.1 First suppose that $\ell\text{-}N(R) = \{0\}$, that is, R contains no nonzero nilpotent element. Let D be a positive derivation on A and $x \wedge y = 0$ for some $x, y \in R$. Then $xy = 0$ implies that $xD(y) + D(x)y = 0$, and hence $xD(y) = 0 = D(x)y$. Therefore $(x \wedge D(y))^2 = 0$, so $x \wedge D(y) = 0$. Hence D is a positive orthomorphism of R.

For any $a \in R^+$, define $\varphi_a : R \to R$ by $\varphi_a(x) = ax$. Since R is an f-ring, $x \wedge y = 0 \Rightarrow x \wedge \varphi_a(y) = x \wedge ay = 0$. Thus φ_a is a positive orthomorphism. Then D and φ_a commute since $\text{Orth}(R)$ is an Archimedean f-ring implies that it is commutative. For $a, b \in R^+$,

$$D(ab) = D(\varphi_a(b)) = (D\varphi_a)(b) = (\varphi_a D)(b) = \varphi_a(D(b)) = aD(b)$$

and $D(ab) = aD(b) + D(a)b$ imply that $D(a)b = 0$, especially $(D(a))^2 = 0$ when set $b = D(a)$. Thus $D(a) = 0$ for all $a \in R^+$ since R is reduced. Therefore $D(R) = 0$.

For the general case, we consider $\overline{R} = R/\ell\text{-}N(R)$. Then $\overline{R}$ is an Archimedean f-ring with $\ell\text{-}N(\overline{R}) = \{0\}$. By Corollary 3.1, $D(\ell\text{-}N(R)) \subseteq \ell\text{-}N(R)$, so we can define a positive derivation $\overline{D}$ of $\overline{R}$ by $\overline{D}(x + \ell\text{-}N(R)) =$

$D(x) + \ell\text{-}N(R)$. From the above argument, we have $\overline{D} = 0$. Therefore, $D(R) \subseteq \ell\text{-}N(R)$. Since $R(\ell\text{-}N(R)) = (\ell\text{-}N(R))R = \{0\}$ by Lemma 3.3, for any $a, b \in R$, $D(ab) = aD(b) + D(a)b = 0$, so $D(R^2) = \{0\}$. This completes the proof.

A ring B is called *von Neumann regular* if for each $a \in B$ there is an $x \in B$ for which $axa = a$ and B is called *strongly regular* if for each $a \in B$ there is an $x \in B$ for which $a^2x = a$.

Theorem 3.3.

(1) A ring B is strongly regular if and only if B is regular and reduced.

(2) Every regular f-ring is strongly regular.

(3) If D is a positive derivation on a regular f-ring, then $D = 0$.

Proof. (1) Suppose that B is strongly regular. For $a \in B$, if $a^2 = 0$, then there is an $x \in B$ for which $a^2x = a$, so $a = 0$. Thus B is reduced. For $a \in B$, there is an $x \in B$ such that $a^2x = a$. Then $(axa - a)^2 = 0$, so $axa = a$. Conversely if B is regular and reduced, then for each $a \in B$, there is an $x \in B$ for which $axa = a$. Thus $(a^2x - a)^2 = 0$, so $a^2x = a$.

(2) By (1), it is sufficient to show that R is reduced. Let $a \in R$ with $a^2 = 0$. Then there is an $x \in R$ for which $axa = a$, and hence $(ax)^2 = ax$. Since $a \in \ell\text{-}N(R)$ by Theorem 1.27, $ax \in \ell\text{-}N(R)$, so ax is nilpotent. Hence $ax = 0$, so $a = axa = 0$.

(3) We first notice that if L is a totally ordered division ring and D is a positive derivation on L, then for $0 < a \in L$, $a^{-1} > 0$ and $1 = aa^{-1}$ imply that $0 = D(1) = aD(a^{-1}) + D(a)a^{-1}$. Thus $D(a)a^{-1} = 0$ and $D(a) = 0$, so $D = 0$.

Suppose that R is a regular f-ring and D is a positive derivation of R. By (2) R is strongly regular and reduced. Let P be a minimal ℓ-prime ℓ-ideal and $0 \le x \in P$. By Theorem 1.30(3), there is a $0 \le y \notin P$ such that $xy = 0$. Then $xD(y) + D(x)y = 0$, so $D(x)y = 0$. It follows that $D(x) \in P$. Thus $D(P) \subseteq P$ for each minimal ℓ-prime ℓ-ideal P, so D induces a positive derivation D_P on R/P defined by $D_P(x + P) = D(x) + P$. Since R/P is a totally ordered domain by Theorem 1.27(2) and strongly regular, R/P is a totally ordered division ring (Exercise 10). It follows that $D_P = 0$ for each minimal ℓ-prime ℓ-ideal of R, and hence for any $x \in R$, $D(x)$ is contained in each minimal ℓ-prime ℓ-ideal. Therefore $D(x) = 0$ for any $x \in R$ since R is reduced, so $D(R) = 0$. $\qquad\square$

In the following, we consider generalization of Theorem 3.1 to various classes of ℓ-rings. First consider almost f-rings.

Theorem 3.4. *Let R be an Archimedean almost f-ring and D be a positive derivation of R. Then $D(R) \subseteq \ell\text{-}N(R)$ and $D(R^3) = \{0\}$.*

Proof. Since R is an Archimedean almost f-ring, $R/\ell\text{-}N(R)$ is also Archimedean by Corollary 3.1 and $R/\ell\text{-}N(R)$ is an almost f-ring (Exercise 11). From Theorem 1.28, $R/\ell\text{-}N(R)$ is an Archimedean f-ring and reduced, so $\ell\text{-}N(R)$ consists of all the nilpotent elements of R. By Corollary 3.1, $D(\ell\text{-}N(R)) \subseteq \ell\text{-}N(R)$. Thus D induces a positive derivation $\overline{D}$ on $R/\ell\text{-}N(R)$ defined by $\overline{D}(x + \ell\text{-}N(R)) = D(x) + \ell\text{-}N(R)$. It follows from Theorem 3.1 that $\overline{D} = 0$, so $D(R) \subseteq \ell\text{-}N(R)$. Let $x, y, z \in R$, since the square of each element in R is positive, using Lemma 3.3, we have

$$D(xyz) = xD(yz) = x(yD(z) + D(y)z) = xyD(z) + xD(y)z = 0.$$

Hence $D(R^3) = \{0\}$. □

The following example shows that in Theorem 3.4, $D(R^3) = \{0\}$ cannot be replaced by $D(R^2) = \{0\}$ for an Archimedean almost f-ring.

Example 3.2. Let $R = \mathbb{Z} \times \mathbb{Z}$ with the coordinatewise addition and ordering. Define the multiplication by $(a,b)(c,d) = (0, ac)$. Then R is an ℓ-ring (Exercise 12). Since $(a,b) \wedge (c,d) = 0$ implies that $ac = 0$, $(a,b)(c,d) = 0$, and hence R is an almost f-ring. Define $D : R \to R$ by $D(a,b) = (a, 2b)$. Then D is a positive derivation on R (Exercise 12) and $D((1,0)(1,0)) = D(0,1) = (0,2) \neq 0$, so $D(R^2) \neq \{0\}$.

Next we consider positive derivations on Archimedean d-rings.

Lemma 3.5. *Let R be an Archimedean d-ring. Then $\ell\text{-}N(R) = \{a \in R \mid a^3 = 0\}$.*

Proof. We show that for any nilpotent element $a \in R$, $a^3 = 0$. Since $|xy| = |x||y|$ for any $x, y \in R$, we may assume that $a \geq 0$. Suppose that $a^k = 0$ for some positive integer $k \geq 4$, we derive that $a^{k-1} = 0$. Let n be a positive integer and let $(na)^{k-2} \wedge (na)^{k-3} = z_n$. Since $((na)^{k-2} - z_n) \wedge ((na)^{k-3} - z_n) = 0$, $a + a^2$ is a d-element implies that

$$(a + a^2)((na)^{k-2} - z_n) \wedge (a + a^2)((na)^{k-3} - z_n) = 0.$$

Thus

$$a((na)^{k-2} - (na)^{k-2} \wedge (na)^{k-3}) \wedge a^2((na)^{k-3} - (na)^{k-2} \wedge (na)^{k-3}) = 0,$$

that is,

$$(n^{k-2}a^{k-1} - n^{k-2}a^{k-1} \wedge n^{k-3}a^{k-2}) \wedge (n^{k-3}a^{k-1} - n^{k-2}a^k \wedge n^{k-3}a^{k-1}) = 0.$$

Hence $a^k = 0$ implies that

$$(n^{k-2}a^{k-1} - n^{k-2}a^{k-1} \wedge n^{k-3}a^{k-2}) \wedge n^{k-3}a^{k-1} = 0.$$

It follows that

$$(n^{k-2}a^{k-1} - n^{k-2}a^{k-1} \wedge n^{k-3}a^{k-2}) \wedge n^{k-2}a^{k-1} = 0,$$

so $n^{k-2}a^{k-1} - n^{k-2}a^{k-1} \wedge n^{k-3}a^{k-2} = 0$, and $n^{k-2}a^{k-1} \le n^{k-3}a^{k-2}$. Consequently $na^{k-1} \le a^{k-2}$, and hence $a^{k-1} = 0$ since R is Archimedean.

By Theorem 1.25, each element in $\ell\text{-}N(R)$ is nilpotent. By Theorems 1.27 and 1.28, $\ell\text{-}P(R) = \{a \in R \mid a \text{ is a nilpotent}\}$ (Exercise 13). Let $0 \le a, b, c \in \ell\text{-}P(R)$. From the above argument, $abc \le (a+b+c)^3 = 0$ since $a+b+c \in \ell\text{-}P(R)$, that is, $(\ell\text{-}P(R))^3 = 0$. Thus $\ell\text{-}P(R) \subseteq \ell\text{-}N(R)$. It then follows that $\ell\text{-}N(R) = \ell\text{-}P(R) = \{a \in R \mid a^3 = 0\}$. $\square$

Theorem 3.5. *Let R be an Archimedean d-ring. Then $R/\ell\text{-}N(R)$ is a reduced Archimedean f-ring.*

Proof. We only need to show that $R/\ell\text{-}N(R)$ is Archimedean. Suppose that $0 \le a + \ell\text{-}N(R), b + \ell\text{-}N(R) \in R/\ell\text{-}N(R)$ with $n(a + \ell\text{-}N(R)) \le (b + \ell\text{-}N(R))$ for all positive integer n. We may assume that $a, b \in R^+$. Then

$$n(a + \ell\text{-}N(R)) = n(a + \ell\text{-}N(R)) \wedge (b + \ell\text{-}N(R))$$

implies that $na - na \wedge b \in \ell\text{-}N(R)$. By Lemma 3.5, $(na - na \wedge b)^3 = 0$, and by a direct calculation we have

$$0 = n^3 a^3 - n^2(na \wedge b)a^2 - n^2 a(na \wedge b)a + n(na \wedge b)^2 a - n^2 a^2(na \wedge b) +$$
$$n(na \wedge b)a(na \wedge b) + na(na \wedge b)^2 - (na \wedge b)^3.$$

Thus

$$n^3 a^3 \le n^2(na \wedge b)a^2 + n^2 a(na \wedge b)a + n^2 a^2(na \wedge b) + (na \wedge b)^3$$
$$\le n^2 ba^2 + n^2 aba + n^2 a^2 b + b^3$$
$$\le n^2(ba^2 + aba + a^2 b + b^3),$$

and hence $na^3 \le (ba^2 + aba + a^2 b + b^3)$ for all positive integer n. Therefore $a^3 = 0$ since R is Archimedean, so $a \in \ell\text{-}N(R)$ and $a + \ell\text{-}N(R) = 0$. $\square$

Theorem 3.6. *Let R be an Archimedean d-ring and D be a positive derivation on R. Then $D(R) \subseteq \ell\text{-}N(R)$.*

Proof. By Lemma 3.1, for any positive nilpotent element a, $D(a)$ is also nilpotent. Thus $D(\ell\text{-}N(R)) \subseteq \ell\text{-}N(R)$, so D induces a positive derivation $\overline{D}$ on $R/\ell\text{-}N(R)$ defined by $\overline{D}(x+\ell\text{-}N(R)) = D(x)+\ell\text{-}N(R)$ for any $x \in R$. By Theorem 3.5 $R/\ell\text{-}N(R)$ is an Archimedean f-ring and reduced, which implies that $\overline{D} = 0$, so $D(R) \subseteq \ell\text{-}N(R)$. $\qquad\square$

The following is an example of an Archimedean d-ring with a positive derivation D such that $D(R^n) \neq \{0\}$ for any positive integer n.

Example 3.3. Let $A = \mathbb{R}^4$ be the column vector lattice over $\mathbb{R}$ with the coordinatewise addition and ordering, and the multiplication is defined as follows.

$$\begin{pmatrix} \alpha_1 \\ \alpha_2 \\ \alpha_3 \\ \alpha_4 \end{pmatrix} \begin{pmatrix} \beta_1 \\ \beta_2 \\ \beta_3 \\ \beta_4 \end{pmatrix} = \begin{pmatrix} \alpha_1\beta_1 \\ \alpha_1\beta_2 \\ \alpha_3\beta_1 \\ \alpha_3\beta_2 \end{pmatrix}.$$

Then A is an Archimedean d-algebra over $\mathbb{R}$ (Exercise 14) with

$$\ell\text{-}N(A) = \left\{ \begin{pmatrix} 0 \\ \alpha_2 \\ \alpha_3 \\ \alpha_4 \end{pmatrix} \mid \alpha_2, \alpha_3, \alpha_4 \in \mathbb{R} \right\}.$$

Define $D : A \to A$ by

$$D\begin{pmatrix} \alpha_1 \\ \alpha_2 \\ \alpha_3 \\ \alpha_4 \end{pmatrix} = \begin{pmatrix} 0 \\ \alpha_2 \\ 0 \\ 0 \end{pmatrix}.$$

It is straightforward to check that D is a positive derivation on A. For

$$a = \begin{pmatrix} 1 \\ 1 \\ 0 \\ 0 \end{pmatrix}, \quad D(a) = \begin{pmatrix} 0 \\ 1 \\ 0 \\ 0 \end{pmatrix}.$$

Since a is an idempotent element, $D(a^n) = D(a) \neq 0$ for any $n \geq 1$. Thus $D(A^n) \neq \{0\}$ for any positive integer n.

The above example also shows that there are nilpotent elements x in an Archimedean d-ring such that $x^2 \neq 0$. Certainly $x^3 = 0$ by Lemma 3.5.

Now we consider Archimedean ℓ-rings with squares positive. For an ℓ-ring R and $a \in R$,

$$r(a) = \{x \in R \mid |a||x| = 0\} \text{ and } \ell(a) = \{x \in R \mid |x||a| = 0\}$$

are *right* and *left* ℓ-annihilator of a, respectively. Clearly $r(a)$ is a right ℓ-ideal and $\ell(a)$ is a left ℓ-ideal of R.

Theorem 3.7. *Let R be an Archimedean ℓ-ring with squares positive and D be a positive derivation on R. If R contains an idempotent element e with $r(e) \subseteq \ell\text{-}N(R)$ or $\ell(e) \subseteq \ell\text{-}N(R)$, then $D(R) \subseteq \ell\text{-}N(R)$ and $D(R^3) = \{0\}$.*

Proof. First assume that $\ell\text{-}N(R) = \{0\}$. Without loss of generality, we may also assume that $r(e) = \{0\}$. By Lemma 3.3, R is ℓ-reduced. Since $e^2 = e$, $(D(e))^3 = 0$ by Lemma 3.1, so $D(e) = 0$ since R is ℓ-reduced. For $a \in R^+$ and a positive integer n, $0 \leq (ne-a)^2$ implies that $nea \leq n^2e^2 + a^2$, and hence

$$nD(ea) \leq n^2 D(e^2) + D(a^2) = n^2(eD(e) + D(e)e) + D(a^2) = D(a^2).$$

Then R is Archimedean implies $D(ea) = 0$. Thus $eD(a) + D(e)a = 0$, so $eD(a) = 0$. It follows from $r(e) = \{0\}$ that $D(a) = 0$ for each $a \in R^+$. Therefore $D = 0$.

For the general case, by Corollary 3.1, $D(\ell\text{-}N(R)) \subseteq \ell\text{-}N(R)$. Thus D induces a positive derivation $\overline{D}$ on $\overline{R} = R/\ell\text{-}N(R)$ defined by $\overline{D}(x + \ell\text{-}N(R)) = D(x) + \ell\text{-}N(R)$. $\overline{R}$ is also Archimedean by Corollary 3.1. From $r(e) \subseteq \ell\text{-}N(R)$ or $\ell(e) \subseteq \ell\text{-}N(R)$, we have $r(e + \ell\text{-}N(R)) = 0$ or $\ell(e + \ell\text{-}N(R)) = 0$ in $\overline{R}$ (Exercise 15). By the above argument, $\overline{D} = 0$ and hence $D(R) \subseteq \ell\text{-}N(R)$. Using Lemma 3.3 again, we have $D(R^3) = \{0\}$. $\square$

We notice that the ℓ-ring in Example 1.5 satisfies the conditions in Theorem 3.7. Next we show that Theorem 3.3(3) can be generalized to partially ordered rings with squares positive.

Theorem 3.8. *Suppose that R is a partially ordered strongly regular ring in which for each $x \in R$, $x^2 \geq 0$. Then trivial derivation is the only positive derivation on R.*

Proof. Suppose that D is a positive derivation on R. We first show that for each $x \in R^+$, $D(x) = 0$. Since R is strongly regular, there exists a $y \in R$ such that $x^2 y = x$, and hence $xyx = x$ by Theorem 3.3(1). Thus xy is an idempotent and hence $D(xy) = 0$ by Lemma 3.1 and that R is reduced.

Similarly $D(yx) = 0$, so $D(y)x + yD(x) = 0$ and $xD(y)x = -(xy)D(x) \leq 0$ since $xy = (xy)^2 \geq 0$. From $xy^2 \geq 0$, we have

$$D(xy^2) = (xy)D(y) + D(xy)y = (xy)D(y) \geq 0.$$

Multiplying on both sides of the above inequality by x, we obtain $x^2yD(y)x \geq 0$, so $xD(y)x \geq 0$. Hence $xD(y)x = 0$ and $(xy)D(x) = 0$. Then $x = xyx$ implies that $D(x) = (xy)D(x) + D(xy)x = 0$.

Now for any $z \in R$, $z^2 \geq 0$ implies that $D(z^2) = 0$. Suppose $z^2w = z$ for some $w \in R$. Thus, $D(z) = z^2D(w) + D(z^2)w = z^2D(w)$. As in the previous paragraph, wz is idempotent implies that $0 = D(wz) = wD(z) + D(w)z$. Consequently,

$$D(z)z = z^2D(w)z = -z^2wD(z) = -zD(z).$$

Hence

$$0 \leq (zD(z))^2 = z(D(z)z)D(z) = -z^2D(z)^2 \leq 0,$$

so $(zD(z))^2 = 0$ and $zD(z) = 0$ since R is reduced. Since zw is also idempotent, $0 = D(zw) = zD(w) + D(z)w$, so $zD(w) = -D(z)w$. Therefore

$$D(z) = z^2D(w) = z(zD(w)) = -zD(z)w = 0.$$

Therefore $D(z) = 0$ for all $z \in R$. □

For the ℓ-field $F((x))$ of Laurent series with the coordinatewise order in Example 2.1, the usual derivative defined by $D(\sum_{i=n}^{\infty} a_i x^i) = \sum_{i=n}^{\infty} i a_i x^{i-1}$ is a nontrivial positive derivation. Therefore the condition that R is a partially ordered ring with squares positive cannot be omitted in Theorem 3.8.

In the following we show that for a unital finite-dimensional Archimedean ℓ-algebra A over a totally ordered field F with a d-basis, if D is a positive derivation on A, then $D(A) \subseteq \ell\text{-}N(A)$. First we show that for any ℓ-unital ℓ-algebra over F, positive derivation and positive F-derivation coincide.

Lemma 3.6. *Suppose that A is an ℓ-unital ℓ-algebra over F. If D is a positive derivation on A, then D is an F-derivation.*

Proof. Suppose that 1 is the identity element of A. Let $0 < r \in F$. Then $0 < r^{-1} \in F$, and hence $r1 > 0$ and $r^{-1}1 > 0$. Then

$$0 = D(1) = D(r1r^{-1}1) = D(r1)(r^{-1}1) + (r1)D(r^{-1}1)$$

implies $D(r1)r^{-1}1 = 0$ and $D(r1) = 0$. If $r < 0$, then $-r > 0$ and since D is a group-homomorphism, $D(r1) = -D(-r1) = 0$ by above argument. Therefore for any $r \in F$, $D(r1) = 0$, and hence for any $a \in A$, $r \in F$,

$$D(ra) = D((r1)a) = (r1)D(a) + D(r1)a = rD(a),$$

that is, D is an F-derivation. $\square$

Theorem 3.9. *Let A be a unital finite-dimensional Archimedean ℓ-algebra over F with a d-basis and D be a positive derivation on A. Then $D(A) \subseteq \ell\text{-}N(A)$.*

Proof.   From Theorem 2.10, $A = \ell\text{-}N(A) + H$, where H is a convex ℓ-subalgebra of A, and $\ell\text{-}N(A) \cap H = \{0\}$. Suppose that $1 = c_1 + \cdots + c_n$, where $\{c_1, \cdots, c_n\}$ are disjoint basic elements, $n \geq 1$. Then $0 = D(1) = D(c_1) + \cdots + D(c_n)$ implies that $D(c_i) = 0$ for each $i = 1, \cdots, n$. For a basic element $x \in H$, by Lemma 2.6, there exists a basic element y such that $xy = c_i$ and $yx = c_j$ for some i, j. Then $0 = D(c_i) = D(xy) = xD(y) + D(x)y$ implies that $D(x)y = 0$, so $D(x)(yx) = D(x)c_j = 0$. From $yx = c_j$, we have $x = xc_j$. Thus

$$D(x) = D(xc_j) = xD(c_j) + D(x)c_j = 0.$$

Since each strictly positive element in H is a sum of disjoint basic elements in H, we have $D(H) = \{0\}$.

Take a basic element a in $\ell\text{-}N(A)$ and suppose that $D(a) > 0$. Then $D(a) = x_1 + \cdots + x_k$, where $k \geq 1$ and $x_1, \cdots, x_k$ are disjoint basic elements. If some $x_j \notin \ell\text{-}N(A)$, then by Lemma 2.6 again, there exists a basic element z such that $x_j z = c_t$ for some t. Thus $c_t \leq D(a)z \leq D(az)$. Then $az \in \ell\text{-}N(A)$ implies that az is nilpotent, so $D(az)$ is nilpotent by Lemma 3.1. Therefore c_t is nilpotent, which is a contradiction. Then each x_i in $D(a) = x_1 + \cdots + x_k$ belongs to $\ell\text{-}N(A)$, and hence $D(a) \in \ell\text{-}N(A)$. Thus $D(\ell\text{-}N(A)) \subseteq \ell\text{-}N(A)$. Therefore $D(A) \subseteq \ell\text{-}N(A)$. $\square$

3.3   Matrix ℓ-rings

In this section, we consider positive derivations on matrix ℓ-rings and upper triangular matrix ℓ-rings with the entrywise order. For an ℓ-algebra A over a totally ordered field F, an element $u \in A^+$ is called a *strong unit* if for any $x \in A$, there exists $\alpha \in F$ such that $x \leq \alpha u$.

Theorem 3.10. *Let A be a unital ℓ-algebra over a totally ordered field F and D be a positive F-derivation.*

(1) If A contains a strong order u such that $u \leq u^2 \leq \alpha u$ with $1 \leq \alpha < 2$, then D must be the trivial derivation.

(2) If A contains a strong order u such that $u \leq u^2 \leq 2u$, then $(D(x))^2 = 0$ for each $x \in A$.

Proof. (1) We show that $D(x) = 0$ for all $x \in A^+$. Since $u \leq u^2 \leq \alpha u$, we have $0 \leq D(u) \leq uD(u) + D(u)u \leq \alpha D(u)$, so $u^2 D(u)u + uD(u)u^2 \leq \alpha(uD(u)u)$. It follows from $u \leq u^2$ that

$$2uD(u)u \leq u^2 D(u)u + uD(u)u^2 \leq \alpha(uD(u)u),$$

and hence $(\alpha - 2)(uD(u)u) \geq 0$. Thus $uD(u)u \leq 0$ since $\alpha - 2 < 0$. Hence $uD(u)u = 0$.

Suppose that $1 \leq \beta u$ for some $\beta \in F^+$. Then

$$D(u) \leq \beta(uD(u)) \leq \beta^2(uD(u)u) = 0$$

implies that $D(u) = 0$. Hence for each $x \in A^+$, $x \leq \alpha u$ for some $\alpha \in F^+$ implies $D(x) \leq \alpha D(u) = 0$. Thus $D(x) = 0$ for $x \in A^+$ and $D(A) = \{0\}$.

(2) We first show that $(D(u))^2 = 0$. By a similar calculation as in (1), we have

$$2uD(u)u \leq u^2 D(u)u + uD(u)u^2 \leq 2uD(u)u,$$

so $2uD(u)u = u^2 D(u)u + uD(u)u^2$. Thus

$$0 \leq (u^2 - u)D(u)u = uD(u)(u - u^2) \leq 0$$

implies that $(u^2 - u)D(u)u = 0$. Since $(u^2 - u)D(u) \geq 0$ and $\beta u \geq 1$ for some $\beta > 0$ in F, $(u^2 - u)D(u) = 0$. It follows that

$$0 = D((u^2 - u)D(u)) = (u^2 - u)D(D(u)) + D(u^2 - u)D(u),$$

and hence $(u^2-u)D(D(u)) = D(u^2-u)D(u) = 0$. So $(D(u^2)-D(u))D(u) = 0$ implies that

$$(uD(u) + D(u)u)D(u) = u(D(u))^2 + D(u)uD(u) = (D(u))^2.$$

Multiplying the above equation by u from the left, we obtain

$$u^2(D(u))^2 + (uD(u))^2 = u(D(u))^2 \leq u^2(D(u))^2.$$

Consequently, $(uD(u))^2 = 0$ and hence $(D(u))^2 = 0$ since $u \geq \beta 1$. For an arbitrary $x \in A$, $|x| \leq \alpha u$ for some $0 < \alpha \in F$, so

$$|(D(x))^2| \leq |D(x)|^2 \leq (D(|x|))^2 \leq \alpha^2(D(u))^2 = 0$$

implies that $(D(x))^2 = 0$. $\qquad\qquad\square$

Let's consider some applications of Theorem 3.10. For the $n \times n$ matrix algebra $M_n(F)$ over a totally ordered field F, $M_n(F^+)$ is the positive cone of the entrywise order on $M_n(F)$. By Theorem 1.19, for an invertible matrix $f \in M_n(F^+)$, $fM_n(F^+)$ is the positive cone of a lattice order on $M_n(F)$ to make it into an ℓ-algebra over F.

Theorem 3.11. *For any invertible matrix $f \in M_n(F^+)$, the only positive F-derivation on the ℓ-algebra $(M_n(F), fM_n(F^+))$ is the trivial derivation.*

Proof. Suppose that $f = (f_{ij})$. Define $\alpha = \sum_{i=1}^{n} \sum_{j=1}^{n} f_{ij}$, $g = (g_{ij}) \in M_n(F)$ with each $g_{ij} = \alpha^{-1}$, and $u = fg$. For $x \in M_n(F)$, let $0 < \alpha_x \in F$ be greater than each entry in the matrix $f^{-1}x$. Then since $(\alpha\alpha_x)g - f^{-1}x \in M_n(F^+)$,

$$(\alpha\alpha_x)u - x = f((\alpha\alpha_x)g - f^{-1}x) \geq 0$$

with respect to the lattice order $fM_n(F^+)$. Thus $x \leq (\alpha\alpha_x)u$ for each $x \in M_n(F^+)$, so u is a strong unit. As well, a direct calculation shows that $gfg = g$ (Exercise 16) and hence $u^2 = u$. By Theorem 3.10(1), the ℓ-algebra $(M_n(F), fM_n(F^+))$ has no nontrivial positive F-derivation. □

For a totally ordered subfield F of $\mathbb{R}$, each ℓ-algebra $M_n(F)$ over F is ℓ-isomorphic to the ℓ-algebra $(M_n(F), fM_n(F^+))$ for some invertible matrix $f \in M_n(F^+)$ [Steinberg (2010)]. As a direct consequence of this fact and Theorem 3.11, any ℓ-algebra $M_n(F)$ over F has no nontrivial positive F-derivation.

The following example is related to Theorem 3.10(2).

Example 3.4. Consider the following set of upper triangular matrices

$$A = \{ \begin{pmatrix} a & b \\ 0 & a \end{pmatrix} \mid a, b \in \mathbb{R} \}.$$

We leave it to the reader to check that A is an ℓ-algebra over $\mathbb{R}$ with the entrywise order and $u = e_{11} + e_{12} + e_{22}$ is a strong order with $u \leq u^2 \leq 2u$. Clearly D defined by

$$D \begin{pmatrix} a & b \\ 0 & a \end{pmatrix} = \begin{pmatrix} 0 & b \\ 0 & 0 \end{pmatrix}$$

is a positive derivation on A. Since $D \neq 0$, the condition that $u \leq u^2 \leq \alpha u$ for some $1 \leq \alpha < 2$ in Theorem 3.10(1) is not satisfied by A. Clearly $(D(x))^2 = 0$ for any $x \in A$.

For an ℓ-unital ℓ-ring R and the matrix ℓ-ring $M_n(R)$ over R with the entrywise order, we show that positive derivations on R and positive derivations on $M_n(R)$ are in one-to-one correspondence.

For a ring B and the matrix ring $M_n(B)$, if D is a derivation on B, then we may use D to define a derivation D_n on $M_n(B)$ by $D_n(a) = (D(a_{ij}))$, for any $a = (a_{ij}) \in M_n(B)$.

For $a = (a_{ij}), b = (b_{ij}) \in M_n(R)$, clearly $D_n(a + b) = D_n(a) + D_n(b)$. Let $ab = (c_{ij})$, where $c_{ij} = \sum_{1 \le k \le n} a_{ik}b_{kj}$. Then

$$D_n(ab) = (D(c_{ij}))$$
$$= \left(\sum_{1 \le k \le n} (a_{ik}D(b_{kj}) + D(a_{ik})b_{kj}) \right)$$
$$= (a_{ij})(D(b_{kj})) + (D(a_{ik}))(b_{kj})$$
$$= aD_n(b) + D_n(a)b.$$

Thus D_n is indeed a derivation on $M_n(B)$ and D_n is called the *induced derivation* on $M_n(B)$ by D.

Theorem 3.12. *Suppose that R is an ℓ-unital ℓ-ring and $M_n(R)$ is the matrix ℓ-ring over R with the entrywise order.*

(1) *D is a positive derivation on R if and only if D_n is a positive derivation on $M_n(R)$.*

(2) *If D' is a positive derivation on $M_n(R)$, then there exists a positive derivation D on R such that $D' = D_n$.*

Thus positive derivations on $M_n(R)$ and positive derivations on R are in one-to-one correspondence.

Proof. (1) is clear.

(2) Let 1 denote the identity matrix and e_{ij} be the standard matrix units of $M_n(R)$. Since $0 = D'(1) = D'(e_{11}) + \cdots + D'(e_{nn})$ and each $D'(e_{ii}) \ge 0$, we have each $D'(e_{ii}) = 0$. For any $a \in R$,

$$D'(ae_{11}) = D'(ae_{11}e_{11}) = D'(ae_{11})e_{11} + (ae_{11})D'(e_{11}) = D'(ae_{11})e_{11},$$

and similarly $D'(ae_{11}) = e_{11}D'(ae_{11})$. Consequently $D'(ae_{11}) = be_{11}$ for some $b \in R$. Define $D : R \to R$ for any $a \in R$, $D(a) = b$, where $D'(ae_{11}) = be_{11}$. Then D is a positive derivation on R (Exercise 17).

We show that $D_n = D'$. First we notice that for any $1 \le i \le n$, $e_{ii} = e_{i1}e_{1i}$ implies that

$$0 = D'(e_{ii}) = e_{i1}D'(e_{1i}) + D'(e_{i1})e_{1i},$$

so $e_{i1}D'(e_{1i}) = 0$ and $D'(e_{i1})e_{1i} = 0$. Hence $e_{11}D'(e_{1i}) = 0$ and $D'(e_{i1})e_{11} = 0$. Now for $a = (a_{ij}) \in M_n(R)$, suppose that $D(a_{ij}) = b_{ij}$. Then by the definition for D, $D'(a_{ij}e_{11}) = b_{ij}e_{11}$, and $D_n(a) = (D(a_{ij})) = (b_{ij})$. On the other hand,

$$D'(a) = D'\left(\sum_{1 \leq i,j \leq n} a_{ij}e_{ij} \right) = \sum_{1 \leq i,j \leq n} D'(a_{ij}e_{ij})$$

and

$$
\begin{aligned}
D'(a_{ij}e_{ij}) &= D'(e_{i1}(a_{ij}e_{11})e_{1j}) \\
&= D'(e_{i1})((a_{ij}e_{11})e_{1j}) + e_{i1}D'((a_{ij}e_{11})e_{1j}) \\
&= e_{i1}D'((a_{ij}e_{11})e_{1j}) \quad (\text{since } D'(e_{i1})e_{11} = 0) \\
&= e_{i1}D'(a_{ij}e_{11})e_{1j} + e_{i1}(a_{ij}e_{11})D'(e_{1j}) \\
&= e_{i1}D'(a_{ij}e_{11})e_{1j} \\
&= e_{i1}(b_{ij}e_{11})e_{1j} \quad (\text{since } e_{11}D'(e_{1j}) = 0) \\
&= b_{ij}e_{ij}.
\end{aligned}
$$

Therefore $D'(a) = (b_{ij}) = D_n(a)$ for all $a \in M_n(R)$, and hence $D' = D_n$.

Thus the mapping $D \to D_n$ from positive derivations of R to positive derivations of $M_n(R)$ is subjective. It is also injective (Exercise 18). This completes the proof. $\qquad\square$

Next we consider upper triangular matrix ℓ-ring $T_n(R)$ with the entry-wise order over an ℓ-unital commutative ℓ-ring R. In this case each positive derivation on $T_n(R)$ is a sum of an induced positive derivation by a positive derivation on R and a positive inner derivation.

Theorem 3.13. *Let R be an ℓ-unital commutative ℓ-ring and $T_n(R)$ be the upper triangular matrix ℓ-ring with the entrywise order. Suppose that D' is a positive derivation on $T_n(R)$.*

(1) $D' = D_n + D_z$, where D_n is the induced derivation by a positive derivation D on R and D_z is the positive inner derivation determined by $z \in T_n(R)$, where $z \in T_n(R)$ is a diagonal matrix with $z_{11} \leq z_{22} \leq \cdots \leq z_{nn}$.

(2) If $D' = E_n + D_w$, where E_n is the induced derivation by a positive derivation E on R and D_w is the positive inner derivation determined by $w \in T_n(R)$, then $D = E$ and $D_z = D_w$.

Proof. (1) As in Theorem 3.12, $D'(1) = 0$ implies that $D'(e_{ii}) = 0$ for $i = 1, \cdots, n$. For $1 \leq i < j \leq n$,

$$D'(e_{ij}) = D'(e_{ii}e_{ij}) = D'(e_{ii})e_{ij} + e_{ii}D'(e_{ij}) = e_{ii}D'(e_{ij}),$$

and similarly $D'(e_{ij}) = D'(e_{ij})e_{jj}$. Thus for $1 \leq i < j \leq n$, $D'(e_{ij}) = a_{ij}e_{ij}$, where $a_{ij} \in R^+$ since D' is positive.

Suppose that $1 < r < s \leq n$, we claim that $a_{1r} \leq a_{1s}$. In fact, $e_{1s} = e_{1r}e_{rs}$ implies that

$$D'(e_{1s}) = D'(e_{1r}e_{rs}) = e_{1r}D'(e_{rs}) + D'(e_{1r})e_{rs},$$

and hence

$$a_{1s}e_{1s} = e_{1r}(a_{rs}e_{rs}) + (a_{1r}e_{1r})e_{rs}.$$

Hence $a_{1s} = a_{rs} + a_{1r} \geq a_{1r}$.

Define $z = (z_{ij}) \in M_n(R)$ with $z_{ij} = 0$ if $i \neq j$ and $z_{ii} = a_{1i}$ for $i = 1, \cdots, n$. It is straightforward to check that the inner derivation D_z, defined by $D_z(x) = xz - zx$ for any $x \in T_n(R)$, is positive.

Now define $H = D' - D_z$. Then H is also a derivation of $T_n(R)$ (Exercise 19). For $1 < r \leq n$,

$$H(e_{1r}) = D'(e_{1r}) - (e_{1r}z - ze_{1r}) = a_{1r}e_{1r} - a_{1r}e_{1r} + a_{11}e_{1r} = 0,$$

since $a_{11} = D'(e_{11}) = 0$. For $1 < i < j \leq n$,

$$\begin{aligned}
H(e_{ij}) &= e_{ii}H(e_{ij})e_{jj} \\
&= e_{ii}D'(e_{ij})e_{jj} - e_{ii}(e_{ij}z)e_{jj} + e_{ii}(ze_{ij})e_{jj} \\
&= a_{ij}e_{ij} - a_{1j}e_{ij} + a_{1i}e_{ij} \\
&= 0,
\end{aligned}$$

since $a_{1j} = a_{ij} + a_{1i}$, for any $1 < i < j \leq n$. Consequently $H(e_{ij}) = 0$ for $1 \leq i \leq j \leq n$.

Let $r \in R$. For $i = 1, \cdots, n$, $H(re_{ii}) = H(r1)e_{ii} = e_{ii}H(r1)$ implies that $H(r1)$ is a diagonal matrix, and for $1 \leq i \leq j \leq n$, $H(re_{ij}) = H(r1)e_{ij} = e_{ij}H(r1)$ implies that $H(r1)$ is a scalar matrix (Exercise 20). Hence for any $r \in R$, $H(r1) = \bar{r}1$ for some $\bar{r} \in R$. Since $H = D' - D_z$,

$$H(r1) = D'(r1) - ((r1)z - z(r1)) = D'(r1) \geq 0, \quad \text{whenever } r \geq 0.$$

Therefore if $r \in R^+$, then $\bar{r} \in R^+$.

If we define $D : R \to R$ by for any $r \in R$, $D(r) = \bar{r}$, whenever $H(r1) = \bar{r}1$, then D is a positive derivation on R (Exercise 21). And for any $x =$

$(x_{ij}) \in M_n(R)$,

$$(D' - D_z)(x) = H(x)$$

$$= H\left(\sum_{1 \le i \le j \le n} (x_{ij}1)e_{ij}\right)$$

$$= \sum_{1 \le i \le j \le n} H(x_{ij}1)e_{ij} \quad \text{since } H(e_{ij}) = 0$$

$$= \sum_{1 \le i \le j \le n} (\bar{x}_{ij}1)e_{ij}$$

$$= (\bar{x}_{ij})$$

$$= (D(x_{ij}))$$

$$= D_n(x).$$

Therefore we have $D' - D_z = D_n$, that is, $D' = D_n + D_z$.

(2) From $D_n + D_z = E_n + D_w$ and $D_n(e_{ij}) = E_n(e_{ij}) = 0$ for $1 \le i \le j \le n$, we have $D_z(e_{ij}) = D_w(e_{ij})$ for any $1 \le i \le j \le n$, that is,

$$e_{ij}z - ze_{ij} = e_{ij}w - we_{ij} \quad \text{and hence} \quad e_{ij}(z - w) = (z - w)e_{ij}.$$

Thus $z - w$ is a scalar matrix, so $D_z = D_w$ (Exercise 22). Consequently $D_n = E_n$ and $D = E$. $\qquad\square$

Since an Archimedean f-ring is commutative, we have the following consequence of Theorems 3.13 and 3.1.

Corollary 3.2. *Let R be a unital reduced Archimedean f-ring and $T_n(R)$ be the ℓ-ring with the entrywise order. Then each positive derivation on $T_n(R)$ is an inner derivation.*

3.4　Kernel of a positive derivation

Let R be an ℓ-ring and D, E positive derivations on R. The *composition* of D and E is defined as $DE(x) = D(E(x))$ for any $x \in R$. Generally DE is not a derivation (Exercise 23) although DE is still a positive endomorphism of the additive ℓ-group of R. When $D = E$, we use D^2 to denote DD, and $D^n = D^{n-1}D$ for any $n \ge 2$.

Theorem 3.14. *Let R be an ℓ-reduced ℓ-ring and D be a positive derivation.*

(1) If $D^n = 0$ for some positive integer n, then $D = 0$.

(2) For $x \in R^+$, if $D(x^n) = 0$ for some positive integer n, then $D(x) = 0$.

Proof. (1) For $x \in R^+$,

$$D^n(x^n) = 0 \Rightarrow D^{n-1}(x^{n-1}D(x) + D(x^{n-1})x) = 0$$
$$\Rightarrow D^{n-1}(x^{n-1}D(x)) = 0$$
$$\Rightarrow D^{n-2}(D(x^{n-1})D(x) + x^{n-1}D^2(x)) = 0$$
$$\Rightarrow D^{n-2}(D(x^{n-1})D(x)) = 0$$
$$\Rightarrow D^{n-2}(x^{n-2}(D(x))^2) = 0$$

$$\vdots$$

$$\Rightarrow D(x(D(x))^{n-1}) = 0$$
$$\Rightarrow (D(x))^n = 0.$$

Thus $D(x) = 0$ for each $x \in R^+$ since R is ℓ-reduced. Therefore $D = 0$.

(2) The proof of this fact is similar to (1) and Lemma 3.1(1). We leave it as an exercise (Exercise 24). $\square$

For an ℓ-group G, a convex ℓ-subgroup H of G is called a *band* whenever for any subset X of H if X has the least upper bound in G, then the least upper bound of X belongs to H. Clearly G itself and trivial subgroup $\{0\}$ are band. If S is a subset of G, the intersection of all the bands in G containing S is also a band (Exercise 25), which is the smallest band containing S. To construct more bands we prove a general property for ℓ-groups. For a unital f-ring R, $u(R)$ denotes the smallest band containing the units of R.

Theorem 3.15. *Let G be an ℓ-group. For any subset $\{x_i\}$ of G, if $\vee x_i$ exists, then for each $y \in G$, $\vee(y \wedge x_i)$ exists and*

$$y \wedge (\vee x_i) = \vee(y \wedge x_i).$$

Proof. Let $x = \vee x_i$. Then $\vee(x_i - x) = 0$. For any $y \in G$ and any i, $y \wedge x_i \le y \wedge x$, so $y \wedge x$ is an upper bound of $\{y \wedge x_i\}$. Let $z \in G$ with $z \ge y \wedge x_i$, for any i. Since $y \wedge x_i \ge (x_i - x) + y \wedge x$, we have

$$z \ge y \wedge x_i \ge (x_i - x) + y \wedge x,$$

and hence $z - (y \wedge x) \ge x_i - x$ for each i. Therefore $z - (y \wedge x) \ge \vee(x_i - x) = 0$ and $z \ge y \wedge x$. Hence $y \wedge x = \vee(y \wedge x_i)$. $\square$

An immediate corollary of Theorem 3.15 is that $x^{\perp}$ is a band for each x in an ℓ-group.

Lemma 3.7. *Suppose R is a reduced f-ring and D is a positive derivation on R.*

(1) If $a \wedge b = 0$, for $a, b \in R^+$, then $D(a) \wedge D(b) = 0$.
(2) For any $a \in R$, $D(|a|) = |D(a)|$.

Proof. (1) We first notice that in a reduced f-ring, for any $a, b \in R^+$, $a \wedge b = 0$ if and only if $ab = 0$ (Exercise 26). If $a \wedge b = 0$, then $ab = 0$ implies that $aD(b) + D(a)b = 0$, so $aD(b) = 0$. It follows that $D(a)D(b) + aD(D(b)) = 0$, and hence $D(a)D(b) = 0$. Therefore $D(a) \wedge D(b) = 0$.

(2) Let $a \in R$ and $a = a^+ - a^-$. Then $D(a) = D(a^+) - D(a^-)$ and $D(a^+) \wedge D(a^-) = 0$ by (1). Therefore

$$|D(a)| = D(a^+) + D(a^-) = D(a^+ + a^-) = D(|a|).$$

$\square$

For a derivation D on R, the kernel of D is defined as $\mathrm{Ker}D = \{a \in R \mid D(a) = 0\}$.

Theorem 3.16. *Suppose R is a totally ordered domain and D is a positive derivation on R. Then $\mathrm{Ker}D$ is a band and $u(R) \subseteq \mathrm{Ker}D$ if R contains the identity element.*

Proof. Suppose that X is a nonempty subset of $\mathrm{Ker}D$ and $x = \sup X$ in R. For any $a \in X$, $x \geq a$ and D is a positive derivation imply that $D(x) \geq D(a) = 0$. Take an element $0 \neq z \in X$. Then $x - |z| < x$ implies that $x - |z|$ is not an upper bound for X, and hence there exists an element $w \in X$ such that $x - |z| < w$ since R is totally ordered. It follows that $D(x) - D(|z|) \leq D(w)$ and $D(x) \leq 0$. Therefore we must have $D(x) = 0$, that is, $x \in \mathrm{Ker}D$.

Suppose $a \in R$ is a unit. Then $aa^{-1} = 1$ implies that $|a||a^{-1}| = 1$, and $D(|a||a^{-1}|) = D(1) = 0$ implies that $D(|a|)|a^{-1}| + |a|D(|a^{-1}|) = 0$. So $D(|a|)|a^{-1}| = 0$, and hence $D(|a|) = 0$. Therefore $D(a) = 0$, that is, $a \in \mathrm{Ker}D$. Since each unit of R is in $\mathrm{Ker}D$ and $\mathrm{Ker}D$ is a band, we conclude that $u(R) \subseteq \mathrm{Ker}D$. $\square$

An element $b \in R$ is called *almost bounded* if $|b| = \vee(|b| \wedge n1)$, where n runs through all positive integers and 1 is the identity element of R. Let $ab(R)$ denotes the set of almost bounded elements of R.

Theorem 3.17. *Let R be a reduced unital f-ring and D be a positive derivation on R. Then $ab(R) \subseteq \mathrm{Ker}D$.*

Proof. For a reduced unital f-ring, there exist minimal ℓ-prime ℓ-ideals P_i such that $\cap P_i = \{0\}$ by Theorem 1.28. Suppose that $|b| = \vee(|b| \wedge n1)$, where n runs through all positive integers. Consider the following collection of minimal ℓ-prime ℓ-ideals.

$$\mathcal{M} = \{P_j \mid |b| + P_j \leq k1 + P_j \text{ in } R/P_j \text{ for some } k \geq 1\}.$$

We show that $I = \cap P_j = \{0\}$ for $P_j \in \mathcal{M}$. Suppose $I \neq \{0\}$ and take $0 < x \in I$. Then $0 < y = x \wedge 1 \leq 1$ and $y \in I$. For any $P \in \mathcal{M}$,

$$(|b| - y) + P = |b| + P \geq (|b| \wedge n1) + P \text{ in } R/P.$$

For any minimal ℓ-prime ℓ-ideal $Q \notin \mathcal{M}$, $(|b| - y) + Q \geq n1 + Q$ in R/Q for all positive integer n. Otherwise R/Q is a totally ordered domain implies that $(|b| - y) + Q \leq k1 + Q$ for some positive integer k, and hence

$$|b| + Q \leq (y + Q) + (k1 + Q) \leq (k+1)1 + Q,$$

which is a contradiction. Thus $(|b| - y) + J \geq (|b| \wedge n1) + J$ for all positive integer n, and all minimal ℓ-prime ℓ-ideals J. Hence $|b| - y \geq |b| \wedge n1$ in R for all positive integers n, which contradicts with the fact that $|b| = \vee(|b| \wedge n1)$. Therefore we must have $I = \{0\}$.

For each minimal ℓ-prime ℓ-ideal P, if $0 \leq x \in P$, then there exists $0 \leq y \notin P$ such that $xy = 0$, and hence $D(x)y = 0$, so $D(x) \in P$. Thus, as we did before, D induces a positive derivation D_P on R/P by $D_P(a + P) = D(a) + P$ for any $a \in R$. If $P \in \mathcal{M}$, then $|b| + P \leq k1 + P$ in R/P for some positive integer k. It follows that $D_P(|b| + P) = 0$ in R/P, and hence $D(|b|) + P = 0$, that is, $D(|b|) \in P$ for each $P \in \mathcal{M}$. Consequently $D(|b|) = 0$ since $I = \{0\}$, and hence $|b| \in \mathrm{Ker}D$. Therefore $b \in \mathrm{Ker}D$ and $ab(R) \subseteq \mathrm{Ker}D$. □

In a unital f-ring R, an element $a \in R$ is called *bounded* if $|a| \leq n1$ for some positive integer n. It is clear that each bounded element is almost bounded. A unital f-ring is said to have *bounded inversion property* if each element $x \geq 1$ is a unit.

Theorem 3.18. *Let R be a unital f-ring with bounded inversion property. Then the trivial derivation is the only positive derivation on R.*

Proof. Suppose that D is a positive derivation on R. Take $x \in R$ with $x \geq 1$. Then x has the inverse x^{-1}. From $xx^{-1} = 1$, we have $x|x^{-1}| = 1$, so $x^{-1} = |x^{-1}| \geq 0$. Thus $x \geq 1$ implies $1 \geq x^{-1}$, and hence $D(x^{-1}) = 0$. Therefore

$$0 = D(1) = D(xx^{-1}) = xD(x^{-1}) + D(x)x^{-1} = D(x)x^{-1}$$

implies that $D(x)x^{-1} = 0$ and $D(x) = 0$ for any $x \geq 1$. Now for $y \in R^+$, $0 \leq y \leq (1 + y)$, so $0 \leq D(y) \leq D(1 + y) = 0$. Hence $D(y) = 0$ for all $y \in R^+$, so $D = 0$. $\square$

Theorem 3.19. *Let R be an ℓ-ring and D be a positive derivation. Suppose that for $z \in R$, $zD(a) = D(a)z$ for any $a \in R$.*

(1) If R is a domain and $D \neq 0$, then z is contained in the center of R.
(2) If R is a reduced f-ring, then $(az - za) \in KerD$ for every $a \in R$.

Proof. (1) Suppose that z is not in the center of R. We derive a contradiction. For any $u, v \in R$, $[u, v] = uv - vu$ is the commutator of u, v. For all $x, y \in R$, we have $[z, D(xy)] = 0$ by the hypothesis. Since $D(xy) = D(x)y + xD(y)$, we have

$$[z, x]D(y) + D(x)[z, y] = 0.$$

Since z is not in the center, $[z, x_0] \neq 0$ for some $x_0 \in R$, and hence for any $a \in R$, $[z, D(a)] = 0$ implies that $[z, x_0]D(D(a)) = 0$ from the above equation with $x = x_0$, $y = D(a)$. Thus $D(D(a)) = 0$ for all $a \in R$ since R is a domain. Therefore $D^2 = 0$, which is a contradiction by Theorem 3.14. Hence z must be in the center of R.

(2) Since R is reduced, the intersection of minimal ℓ-prime ℓ-ideals is zero. Let P be a minimal ℓ-prime ℓ-ideal. As we did before, D induces a positive derivation D_P on R/P by $D_P(x + P) = D(x) + P$ for any $x \in R$. Since $zD(a) = D(a)z$ for any $a \in R$,

$$(z + P)D_P(a + P) = D_P(a + P)(z + P)$$

for all $a + P$. If $D_P \neq 0$, then R/P is a totally ordered domain implies that $z + P$ is in the center of R/P by (1), so $za - az \in P$ for all $a \in R$. Then $D(za - az) \in P$. If $D_P = 0$, then $D(w) \in P$ for all $w \in R$. Then $D(za - az) \in P$. Therefore for any a, $D(za - az)$ is in every minimal ℓ-prime ℓ-ideal, so we must have $D(za - az) = 0$ for all $a \in R$, and hence $za - az \in KerD$ for all $a \in R$. $\square$

For an ℓ-ring R, let

$$I_0(R) = \{r \in R \mid n|r| \leq x \text{ for some } x \in R^+ \text{ and } n = 1, 2, \cdots \}.$$

Then $I_0(R)$ is an ℓ-ideal of R and R is Archimedean if and only if $I_0(R) = \{0\}$ (Exercise 27).

Theorem 3.20. *Let R be a reduced f-ring and D be a positive derivation on R.*

(1) $D(R^2) \subseteq I_0(R)$.
(2) If R is unital, then $D(R) \subseteq I_0$.
(3) If R is unital with $I_0(R) \subseteq Ker D$, then $D = 0$.

Proof.   (1) For any $x, y \in R$, $|xy| = |x||y| \leq (|x| \vee |y|)^2$ implies that for any positive integer n,

$$n|D(xy)| \leq nD(|xy|)$$
$$\leq nD((|x| \vee |y|)^2)$$
$$\leq (|x| \vee |y|)^2 D(|x| \vee |y|) + D(|x| \vee |y|)(|x| \vee |y|)^2 + D(|x| \vee |y|),$$

by Lemma 3.4 and Exercise 7, and hence $D(xy) \in I_0(R)$. Therefore $D(R^2) \subseteq I_0(R)$.

(2) We first claim that for any $x \in R$, $xD(x) \geq 0$. This fact is clearly true when R is totally ordered, and we leave general case as an exercise (Exercise 28). Then for any positive integer n, $(y - n1)D(y - n1) \geq 0$ implies that $nD(y) \leq yD(y)$ for any $y \in R^+$. Thus $D(y) \in I_0(R)$ for any $y \in R^+$. Therefore $D(R) \subseteq I_0(R)$.

(3) By (2), $D^2(R) = D(D(R)) \subseteq D(I_0(R)) = \{0\}$ implies that $D^2 = 0$. Hence $D = 0$ by Theorem 3.14. $\square$

Exercises

(1) Let $R = F[x]$ be the polynomial ℓ-ring with the coordinatewise order.
    (a) Prove that usual derivative $f'(x)$ is a positive F-derivation on R over the totally ordered field F.
    (b) Prove that if D is a positive F-derivation on R, then for any $f(x) \in R$, $D(f(x)) = f'(x)D(x)$ and $D(x)$ is a positive polynomial in R.
(2) For a ring B and an element $b \in B$, prove the mapping $D_b : B \to B$ defined by $D_b(x) = xb - bx$ is a derivation on B.
(3) Prove that D_a defined in Example 3.1(2) is a positive derivation.
(4) Let R be a commutative ℓ-ring and P be an ℓ-prime ℓ-ideal of R. Define $S = \{x^n a \mid n \geq 0,\ a \in R^+ \setminus P\}$ with $0 < x \in P$. Prove that S is an m-system properly containing $R^+ \setminus P$.
(5) Let R be a totally ordered ring and $x \in R$ with $x^3 = 0$. Then for any positive integer n, $nx^2 \leq x$.
(6) Let R be an ℓ-ring and D be a positive derivation on R. Show $|D(x)| \leq D(|x|)$ for any $x \in R$.
(7) Let R be a reduced f-ring and D be a positive derivation on R. Prove

that for $a \in R^+$ and $n \geq 1$,
$$nD(a^2) \leq a^2 D(a) + D(a)a^2 + D(a).$$

(8) Let R be an ℓ-ring and D be a positive derivation on R. Suppose that I is an ℓ-ideal of R such that $D(I) \subseteq I$. Define $D_I : R/I \to R/I$ by $D_I(\bar{a}) = \overline{D(a)}$, where $\bar{a} = a + I \in R/I$. Prove that D_I is a positive derivation on R/I.

(9) For an ℓ-ring R, prove Orth(R) is a partially ordered ring with the positive cone Orth$(R)^+ = \{\varphi \mid \varphi$ is a positive orthomorphism of $R\}$.

(10) Prove that a strongly regular totally ordered domain is a totally ordered division ring.

(11) Let R be an Archimedean almost f-algebra. Prove that $R/\ell\text{-}N(R)$ is an Archimedean f-algebra and reduced.

(12) Verify that R as defined in Example 3.2 is an ℓ-ring and D is a positive derivation.

(13) Prove that in a d-ring R, $\ell\text{-}P(R) = \{a \in R \mid a$ is nilpotent$\}$.

(14) Verify the ℓ-ring A in Example 3.3 is an Archimedean d-ring.

(15) Let R be an Archimedean ℓ-ring with squares positive and $e \in R^+$ be an idempotent element. Prove that if $r(e) \subseteq \ell\text{-}N(R)$, then $r(e + \ell\text{-}N(R)) = \{0\}$ in $R/\ell\text{-}N(R)$.

(16) Verify $gfg = g$ in Theorem 3.11.

(17) Prove that $D : R \to R$ defined in Theorem 3.12(2) is a positive derivation.

(18) Prove that the mapping $D \to D_n$ from positive derivations of R to positive derivations of $M_n(R)$ in Theorem 3.12 is injective.

(19) For two positive derivations on an ℓ-ring, prove that the sum of them is also a positive derivation.

(20) Prove that for H defined in Theorem 3.13(1), $H(r1)$ is a scalar matrix, for any $r \in R$.

(21) Prove that the function $D : R \to R$ defined in Theorem 3.13(1) by $D(r) = \bar{r}1$, for any $r \in R$, is a positive derivation on R.

(22) Prove that $D_z = D_w$ in Theorem 3.13(2).

(23) Provide an example in which the composition of two positive derivations is not a derivation.

(24) For an ℓ-reduced ℓ-ring R and a positive derivation D on R, prove that if $D(x^n) = 0$ for some $x \in R^+$ and positive integer n, then $D(x) = 0$.

(25) Let G be an ℓ-group. Prove the intersection of a family of bands is also a band.

(26) Let R be a reduced f-ring. Prove that for any $a, b \in R^+$, $a \wedge b = 0$ if and only if $ab = 0$.

(27) Let R be an ℓ-ring. Prove that R is Archimedean if and only if $I_0(R) = \{0\}$.

(28) Let R be a reduced f-ring. Prove that for any $x \in R$, $xD(x) \geq 0$.

(29) Let R be an ℓ-unital ℓ-ring (R may not be commutative) and $T_2(R)$ be the 2×2 upper triangular matrix ℓ-ring with the entrywise order over R. Prove that each positive derivation on $T_2(R)$ is the sum of a derivation induced by a positive derivation on R and an inner derivation on $T_2(R)$.

(30) Let R be an ℓ-unital ℓ-ring (R may not be commutative). Prove that if the trivial derivation is the only positive derivation on R, then each positive derivation on the ℓ-ring $T_n(R)$ with the entrywise order is an inner derivation.

Chapter 4

Some topics on lattice-ordered rings

In this chapter we present some topics of lattice-ordered rings. In section 1, some characterizations of matrix ℓ-rings over ℓ-unital ℓ-rings with the entrywise order are given. In section 2 we study matrix ℓ-rings containing positive cycles. Nonzero f-elements in ℓ-rings could play an important role for the structure of the ℓ-rings. Some topics along this line are presented in section 3. Section 4 is about extending lattice orders on a lattice-ordered Ore domain to its quotient ring. Section 5 contains results on matrix ℓ-algebras over totally ordered integral domains. They generalize results for matrix ℓ-algebras over totally ordered fields. For a unital ℓ-ring in which $1 \not> 0$, 1 still satisfies the definition of f-element given in chapter 1. In section 6, we study d-elements that are not positive. Finally in section 7, we consider lattice-ordered triangular matrices. All lattice orders on 2×2 triangular matrix algebras over totally ordered fields are determined.

4.1 Recognition of matrix ℓ-rings with the entrywise order

In this section, we present some recognition theorems for matrix ℓ-rings with the entrywise order. For an ℓ-unital ℓ-ring R, two right ℓ-ideals I and J are called ℓ-isomorphic if I and J are ℓ-isomorphic right ℓ-modules over R. R is a direct sum of right ℓ-ideals $I_1, \cdots, I_k$, denoted by $R = I_1 \oplus \cdots \oplus I_k$, if $R = I_1 + \cdots + I_k$ and $I_i \cap I_j = \{0\}$ for $1 \leq i, j \leq n$, $i \neq j$.

For a unital ring B, the elements in $\{a_{ij} \mid 1 \leq i, j \leq n\} \subseteq B$ are called matrix units if

$$a_{ij} a_{k\ell} = \delta_{jk} a_{i\ell}, \quad \text{and} \quad a_{11} + \cdots + a_{nn} = 1,$$

where δ_{jk} is called *Kronecker delta* which is defined as

$$\delta_{jk} = \begin{cases} 1, & \text{if } j = k, \\ 0, & \text{if } j \neq k. \end{cases}$$

Lemma 4.1. *Let R be an ℓ-unital ℓ-ring and $\{a_{ij} \mid 1 \leq i, j \leq n\} \subseteq R^+$ be a set of matrix units. Then each a_{ij} is a d-elements of R.*

Proof.   From $a_{11} + \cdots + a_{nn} = 1$ and $0 \leq a_{ii} \leq 1$, $i = 1, \cdots, n$, we have that each a_{ii} is an f-element of R. To see that a_{ij} is a d-element, we just need to show $a_{ij} x \vee 0 = a_{ij}(x \vee 0)$ and $x a_{ij} \vee 0 = (x \vee 0)a_{ij}$, for any $x \in R$. Clearly $a_{ij} x \vee 0 \leq a_{ij}(x \vee 0)$. We show that $a_{ij}(x \vee 0)$ is the sup of $a_{ij}x, 0$. Let $z \geq a_{ij}x, 0$. Then

$$\begin{aligned}
a_{ji}z \geq a_{ji}a_{ij}x, 0 &\Rightarrow a_{ji}z \geq a_{jj}x, 0 \\
&\Rightarrow a_{ji}z \geq a_{jj}x \vee 0 \\
&\Rightarrow a_{ji}z \geq a_{jj}(x \vee 0) \text{ (since } a_{jj} \text{ is an } f\text{-element)} \\
&\Rightarrow a_{ij}a_{ji}z \geq a_{ij}a_{jj}(x \vee 0) \\
&\Rightarrow a_{ii}z \geq a_{ij}(x \vee 0),
\end{aligned}$$

so $z \geq a_{ii}z \geq a_{ij}(x \vee 0)$ since $1 \geq a_{ii}$. Thus $a_{ij}x \vee 0 = a_{ij}(x \vee 0)$. Similarly $x a_{ij} \vee 0 = (x \vee 0)a_{ij}$. Thus each a_{ij} is a d-element of R. $\square$

The following result is fundamental.

Theorem 4.1. *Let R be an ℓ-unital ℓ-ring and $n \geq 2$ be a fixed integer. The following statements are equivalent.*

(1) R is ℓ-isomorphic to a matrix ℓ-ring $M_n(T)$ with the entrywise order, where T is an ℓ-unital ℓ-ring.

(2) R contains a subset of matrix units $\{a_{ij} \mid 1 \leq i, j \leq n\}$ in which each a_{ij} is a d-element of R.

(3) $R_R = I_1 \oplus \ldots \oplus I_n$, where $I_1, \ldots, I_n$ are mutually ℓ-isomorphic right ℓ-ideals of R.

Proof.   (1) $\Rightarrow$ (3) Let $\{e_{ij} \mid 1 \leq i, j \leq n\}$ be the *standard matrix units* in $M_n(T)$, that is, ij^{th} entry in e_{ij} is 1 and other entries in e_{ij} are zero. Define $I_i = e_{ii}M_n(T)$, $i = 1, \ldots, n$. Then $I_1, \ldots, I_n$ are right ℓ-ideals of $M_n(T)$, $M_n(T) = I_1 \oplus \ldots \oplus I_n$ as the direct sum of right ℓ-ideals, and $I_1, \ldots, I_n$ are mutually ℓ-isomorphic right ℓ-modules over $M_n(T)$ (Exercise 1). Since ℓ-rings R and $M_n(T)$ are ℓ-isomorphic, (3) is true.

(3) $\Rightarrow$ (2) Let $1 = a_1 + \ldots + a_n$, where $0 < a_i \in I_i$. Then $a_i < 1$ and $a_i \wedge a_j = 0$ with $i \neq j$ implies that each a_i is an idempotent f-element, and $a_i a_j = 0$ with $i \neq j$. Thus each $a_i R \subseteq I_i$ is a right ℓ-ideal and $R_R = a_1 R \oplus \ldots \oplus a_n R$, so $a_i R = I_i$ for $i = 1, \ldots, n$ (Exercise 2). For any $1 \leq i \leq n$, let $\theta_i : a_1 R \to a_i R$ be an ℓ-isomorphism of the right ℓ-modules over R. Then $0 < b_{i1} = \theta_i(a_1) \in a_i R a_1$. Similarly, $0 < b_{1i} = \theta_i^{-1}(a_i) \in a_1 R a_i$. Hence

$$a_1 = \theta_i^{-1}\theta_i(a_1) = \theta_i^{-1}(b_{i1}) = \theta_i^{-1}(a_i b_{i1}) = \theta_i^{-1}(a_i)b_{i1} = b_{1i}b_{i1},$$

and similarly $b_{i1}b_{1i} = a_i$ for $i = 1, \cdots, n$. Define $a_{ij} = b_{i1}b_{1j}$, $1 \leq i, j \leq n$. Clearly $a_{ij}a_{k\ell} = \delta_{jk}a_{i\ell}$ (Exercise 3), and $a_{ii} = a_i$ implies that $a_{11} + \ldots + a_{nn} = 1$.

Thus $0 \leq a_{ij}, 1 \leq i, j \leq n$, are matrix units, and hence by Lemma 4.1, each a_{ij} is a d-element.

(2) $\Rightarrow$ (1) Define

$$T = \{x \in R \mid a_{ij}x = xa_{ij}, \ 1 \leq i, j \leq n\},$$

which is called the *centralizer* of $\{a_{ij} \mid 1 \leq i, j \leq n\}$. Since each a_{ij} is a d-element, $x \in T$ implies that $a_{ij}|x| = |a_{ij}x| = |xa_{ij}| = |x|a_{ij}$, so $|x| \in T$. Thus T is an ℓ-unital ℓ-subring of R. For an element $x \in R$, define $\alpha_{ij} = \sum_{u=1}^{n} a_{ui}xa_{ju}$ for $i, j = 1, \cdots, n$. For any a_{rs},

$$a_{rs}\alpha_{ij} = a_{rs}a_{si}xa_{js} = a_{ri}xa_{js} \text{ and } \alpha_{ij}a_{rs} = a_{ri}xa_{jr}a_{rs} = a_{ri}xa_{js},$$

so each $\alpha_{ij} \in T$, $1 \leq i, j \leq n$. Also

$$\sum_{i,j=1}^{n} \alpha_{ij}a_{ij} = \sum_{i,j=1}^{n} (\sum_{u=1}^{n} a_{ui}xa_{ju})a_{ij}$$
$$= \sum_{i,j=1}^{n} a_{ii}xa_{jj}$$
$$= (a_{11} + \cdots + a_{nn})x(a_{11} + \cdots + a_{nn})$$
$$= x,$$

that is, $x = \sum_{i,j=1}^{n} \alpha_{ij}a_{ij}$ with $\alpha_{ij} \in T$. Suppose that $x = \sum_{i,j=1}^{n} \beta_{ij}a_{ij}$, where $\beta_{ij} \in T$. Then it is straightforward to check that $\beta_{ij} = \alpha_{ij}$, $i, j = 1, \cdots, n$ (Exercise 4).

Define $\varphi : R \to M_n(T)$ by $\varphi(x) = \sum_{i,j=1}^{n} \alpha_{ij}e_{ij}$, for $x = \sum_{i,j=1}^{n} \alpha_{ij}a_{ij} \in R$. We leave it as an exercise for the reader to verify that φ is one-to-one, onto, and preserves addition (Exercise 5). In the following, we check that φ preserves the multiplication and order. For $x, y \in R$, suppose that $\alpha_{ij} = \sum_{u=1}^{n} a_{ui} x a_{ju}$ and $\alpha'_{ij} = \sum_{u=1}^{n} a_{ui} y a_{ju}$. Then

$$\varphi(x)\varphi(y) = \left(\sum_{i,j=1}^{n} \alpha_{ij} e_{ij} \right) \left(\sum_{i,j=1}^{n} \alpha'_{ij} e_{ij} \right)$$

$$= \sum_{i,j=1}^{n} \left(\sum_{v=1}^{n} \alpha_{iv} \alpha'_{vj} \right) e_{ij},$$

where

$$\sum_{v=1}^{n} \alpha_{iv} \alpha'_{vj} = \sum_{v=1}^{n} (a_{1i} x a_{vv} y a_{j1} + \cdots + a_{ni} x a_{vv} y a_{jn})$$

$$= a_{1i}(xy)a_{j1} + \cdots + a_{ni}(xy)a_{jn}$$

$$= \sum_{u=1}^{n} a_{ui}(xy)a_{ju}.$$

Thus $\varphi(x)\varphi(y) = \varphi(xy)$. For $x = \sum_{i,j=1}^{n} \alpha_{ij} a_{ij} \in R$, where $\alpha_{ij} = \sum_{u=1}^{n} a_{ui} x a_{ju}$, if $\varphi(x) \geq 0$, then each $\alpha_{ij} \geq 0$, so $x \geq 0$. Conversely if $x = \sum \alpha_{ij} a_{ij} \geq 0$, then each $\alpha_{ij} \geq 0$, and hence $\varphi(x) = \sum \alpha_{ij} e_{ij} \geq 0$. Therefore φ is an ℓ-isomorphism between two ℓ-rings. $\qquad \square$

Corollary 4.1. *Let A, B be ℓ-unital ℓ-rings and $f : A \to B$ be an ℓ-homomorphism with $f(1_A) = 1_B$. If A is ℓ-isomorphic to an $n \times n$ $(n \geq 2)$ matrix ℓ-ring with the entrywise order over an ℓ-unital ℓ-ring, then B is also ℓ-isomorphic to an $n \times n$ matrix ℓ-ring with the entrywise order over an ℓ-unital ℓ-ring.*

In particular, if an ℓ-unital ℓ-ring A contains an ℓ-unital ℓ-subring with the same identity which is ℓ-isomorphic to $M_n(S)$ with the entrywise order, where S is an ℓ-unital ℓ-ring, then $A \cong M_n(T)$ with the entrywise order, where T is an ℓ-unital ℓ-ring and $T \supseteq S$.

Proof. Let $\{a_{ij} \mid 1 \leq i, j \leq n\}$ be a set of $n \times n$ matrix units in A. Then $\{f(a_{ij}) \mid 1 \leq i, j \leq n\}$ is a set of $n \times n$ matrix units in B contained in B^+. By Lemma 4.1, each $f(a_{ij})$ is a d-element. Now Theorem 4.1(2) applies. $\square$

We characterize a bit more the centralizer of those matrix units in Theorem 4.1. Suppose that R is an ℓ-unital ℓ-ring and $a \in R^+$ is an f-element and an idempotent. Then aR is an ℓ-subring of R since for any $r \in R$, $|ar| = a|r|$. Define

$$\text{End}_R(aR, aR) = \{\varphi \mid \varphi \text{ is an endomorphism of right } R\text{-module } aR\}.$$

Then $\text{End}_R(aR, aR)$ is a ring with respect to the usual addition and composition of two functions (Exercise 6). For $\theta \in \text{End}_R(aR, aR)$, define $\theta \geq 0$ if $\theta(x) \geq 0$ for each $0 \leq x \in aR$. It is straightforward to check that $\text{End}_R(aR, aR)$ is a partially ordered ring with respect to this order (Exercise 7).

Theorem 4.2. *Let R be an ℓ-unital ℓ-ring and $0 < a \in R$ be an f-element and idempotent.*

(1) $\text{End}_R(aR, aR)$ is an ℓ-ring with respect to the partial order defined above, and $\text{End}_R(aR, aR) \cong aRa$ as ℓ-rings.

(2) Let $0 \leq a_{ij}$, $1 \leq i, j \leq n$, be $n \times n$ matrix units and T be the centralizer of $\{a_{ij} \mid 1 \leq i, j \leq n\}$. Then T and $a_{ii}Ra_{ii}$ are ℓ-isomorphic ℓ-rings.

Proof. (1) We first note that, for $\theta \in \text{End}_R(aR, aR)$, $\theta \geq 0$ if and only if $\theta(a) \geq 0$. In fact, suppose that $\theta(a) \geq 0$. Then for any $0 \leq x \in aR$, $\theta(x) = \theta(ax) = \theta(a)x \geq 0$, so $\theta \geq 0$.

For $\theta \in \text{End}_R(aR, aR)$, define $\theta'(x) = (\theta(a) \vee 0)x$, for $x \in aR$. Clearly $\theta' \in \text{End}_R(aR, aR)$ and $\theta' \geq 0$. Since

$$\begin{aligned}
(\theta' - \theta)(a) &= (\theta(a) \vee 0)a - \theta(a) \\
&= (\theta(a)a \vee 0) - \theta(a) \\
&= (\theta(a^2) \vee 0) - \theta(a) \\
&= (\theta(a) \vee 0) - \theta(a) \\
&\geq 0,
\end{aligned}$$

$\theta' \geq \theta$. Let $\tau \in \text{End}_R(aR, aR)$ with $\tau \geq \theta, 0$. Then $\tau(a) \geq \theta(a), 0$, and hence $\tau(a) \geq \theta(a) \vee 0$. Thus for $0 \leq x \in aR^+$,

$$(\tau - \theta')(x) = (\tau(a) - \theta'(a))x = (\tau(a) - (\theta(a) \vee 0))x \geq 0,$$

so $\tau \geq \theta'$. Therefore $\theta' = \theta \vee 0$ for each $\theta \in \text{End}_R(aR, aR)$. Hence $\text{End}_R(aR, aR)$ is an ℓ-ring.

Now map $\varphi : aRa \to \text{End}_R(aR, aR)$ by $\varphi(x) = \ell_x$, where $\ell_x : aR \to aR$ is the left multiplication by x, that is, for any $z \in R$, $\ell_x(z) = xz$. Then φ is a ring isomorphism. For $x \in aRa$,

$$x \geq 0 \Leftrightarrow \ell_x(a) \geq 0 \Leftrightarrow \ell_x \geq 0,$$

and hence φ is an ℓ-isomorphism between two ℓ-rings.

(2) Define $\varphi : a_{ii}Ra_{ii} \to T$ by for any $x \in a_{ii}Ra_{ii}$, $\varphi(x) = \sum_{u=1}^{n} a_{ui}xa_{iu}$. Then similar to the argument in the proof of Theorem 4.1, φ is an ℓ-isomorphism between two ℓ-rings (Exercise 8). $\square$

Remark 4.1. In the proof of Theorem 4.2(1), it seems we only need to assume that a is a d-element. However the following result shows that each idempotent d-element in an ℓ-unital ℓ-ring must be an f-element.

Lemma 4.2. *Let R be an ℓ-unital ℓ-ring and $a \in R^+$ be a d-element with $a^2 = a$. Then $a \leq 1$ and hence a is an f-element.*

Proof.   Consider $1 \wedge a = b$. We show that $a = b$. Since $a^2 = a$ is a d-element,
$$a = a \wedge a^2 = a(1 \wedge a) = ab = (1 \wedge a)a = ba.$$
From $0 \leq b \leq 1$, b is an f-element, so $1 \wedge a = b$ implies that $b \wedge ba = b^2$, and $b \leq a = ba$ implies that $b = b^2$. Then a direct calculation shows that $(a - b)^2 = -(a - b)$, so $a - b = 0$ and $a = b$.   □

Theorem 4.3. *Let R be an ℓ-unital ℓ-ring and $n \geq 2$ be a fixed integer. Then the following statements are equivalent:*

(1) R is ℓ-isomorphic to a matrix ℓ-ring $M_n(T)$ with the entrywise order, where T is an ℓ-unital ℓ-ring.

(2) There exist positive elements $b, f, g \in R$ such that $f^n = g^n = 0$ and $bg^{n-1} + fbg^{n-2} + f^2bg^{n-3} + \ldots + f^{n-1}b$ is a unit and a d-element.

(3) There exist positive elements $a, f \in R$ such that $f^n = 0$ and $af^{n-1} + faf^{n-2} + f^2af^{n-3} + \ldots + f^{n-1}a = 1$.

(4) There exist positive elements $a, f \in R$ such that $f^n = 0$ and $af^{n-1} + faf^{n-2} + f^2af^{n-3} + \ldots + f^{n-1}a$ is a unit and a d-element.

(5) For any unit and d-element u, there exist positive elements $b, f, g \in R$ such that $f^n = g^n = 0$ and $u = bg^{n-1} + fbg^{n-2} + f^2bg^{n-3} + \ldots + f^{n-1}b$.

Proof.   (1) $\Rightarrow$ (2) Let $f = g = e_{21} + \ldots + e_{n,n-1}$ and $b = e_{1n}$ be in $M_n(T)$. Then a direct calculation shows that $f^n = g^n = 0$ and
$$bg^{n-1} + fbg^{n-2} + f^2bg^{n-3} + \cdots + f^{n-1}b = 1.$$

(2) $\Rightarrow$ (3) Let $u = bg^{n-1} + fbg^{n-2} + f^2bg^{n-3} + \ldots + f^{n-1}b$ and let $v = u^{-1}$. Then
$$fu = fbg^{n-1} + f^2bg^{n-2} + \cdots + f^{n-1}bg = ug$$
implies that $vf = gv$. It follows that $vf^k = g^kv$, $k = 1, \cdots, n - 1$, and hence $u = bg^{n-1} + fbg^{n-2} + f^2bg^{n-3} + \ldots + f^{n-1}b$ implies that
$$\begin{aligned}
1 = uv \\
= bg^{n-1}v + fbg^{n-2}v + f^2bg^{n-3}v + \ldots + f^{n-1}bv \\
= (bv)f^{n-1} + f(bv)f^{n-2} + f^2(bv)f^{n-3} + \ldots + f^{n-1}(bv).
\end{aligned}$$

Since $b \geq 0$, $b = |b| = |(bv)u| = |bv|u$ since u is a d-element, so $bv = |bv| \geq 0$. Let $a = bv$. Then a is positive and $1 = af^{n-1} + faf^{n-2} + f^2af^{n-3} + \ldots + f^{n-1}a$.

(3) $\Rightarrow$ (1) We show that Theorem 4.1(3) is true. For each $t = 1, \ldots, n$, let $g_t = f^{t-1}af^{n-1}$. We first claim that $R = g_1R + g_2R + \ldots + g_nR$ is a direct sum of right ideals of R and $g_1R \cong g_tR$ as right R-modules. Since

$$f^{n-1} = f^{n-1}(af^{n-1} + faf^{n-2} + f^2af^{n-3} + \ldots + f^{n-1}a) = f^{n-1}af^{n-1},$$

af^{n-1} is idempotent. Thus the map $g_1R \to g_tR$ given by left multiplication by f^{t-1} has an inverse map $g_tR \to g_1R$ given by left multiplication by af^{n-t}. Therefore $g_1R \cong g_tR$ as right R-modules.

Suppose that $I = g_1R + \cdots + g_nR$. If $g_1x_1 + \cdots + g_nx_n = 0$ for some $x_i \in R$, then by multiplying the equality from the left by $f^{i-1}af^{n-i}$ in turn, $i = 1, \cdots, n-1$, we have each $g_jx_j = 0$ for $j = 1, \cdots, n$. Thus the sum $g_1R + g_2R + \ldots + g_nR$ is a direct sum. We verify that $I = R$. To this end, we show that $1 \in I$ by showing $f^k \in I$ for each positive integer k. Note $f^{n-1} = g_n \in I$. Suppose that $f^s \in I$ for all positive integers $s > r$. We show $f^r \in I$. In fact, since $f^raf^{n-1} = g_{r+1} \in I$ and $f^{r+1}, \cdots, f^{n-1} \in I$,

$$f^r = f^raf^{n-1} + f^{r+1}af^{n-2} + \cdots + f^{n-1}af^r \in I.$$

Hence $f, \cdots, f^{n-1} \in I$ by the induction, so

$$1 = af^{n-1} + faf^{n-2} + f^2af^{n-3} + \ldots + f^{n-1}a \in I,$$

since $af^{n-1} = g_1 \in I$. Therefore $R = g_1R + \cdots + g_nR$.

We next show that for any $x \in R$, if $x = x_1 + x_2 + \ldots + x_n$, where $x_t \in g_tR$, then $x \geq 0$ if and only if each $x_t \geq 0$, $t = 1, \ldots, n$. It is clear that if each $x_t \geq 0$, then $x \geq 0$. To show if $x \geq 0$, then $x_i \geq 0$, we first consider the identity element 1.

Let $S(a, f)$ be the semigroup generated by a and f with respect to the multiplication of R and let $d_t = f^{t-1}af^{n-t}$, $t = 1, \ldots, n$. We show that for $t = 1, \ldots, n$, $d_t = g_1a_{t1} + g_2a_{t2} + \ldots + g_na_{tn}$, where each of $a_{t1}, a_{t2}, \ldots, a_{tn}$ is a sum of elements from $S(a, f)$. Since $f^{n-1} = f^{n-1}af^{n-1}$, $d_n = f^{n-1}a = g_na$. Suppose that it is true for all d_s with $n \geq s > t \geq 1$. We claim that it is also true for d_t. From $1 = af^{n-1} + faf^{n-2} + f^2af^{n-3} + \ldots + f^{n-1}a$, we have

$$f^{t-1} = f^{t-1}af^{n-1} + f^taf^{n-2} + \ldots + f^{n-1}af^{t-1},$$

so

$$\begin{aligned}
d_t &= f^{t-1}af^{n-t} \\
&= (f^{t-1}af^{n-1} + f^taf^{n-2} + \ldots + f^{n-1}af^{t-1})af^{n-t} \\
&= g_t(af^{n-t}) + d_{t+1}(f^{t-1}af^{n-t}) + \ldots + d_n(f^{t-1}af^{n-t}).
\end{aligned}$$

Thus $d_t = g_1 a_{t1} + g_2 a_{t2} + \ldots + g_n a_{tn}$, where each of $a_{t1}, a_{t2}, \ldots, a_{tn}$ is a sum of elements from $S(a, f)$. Since $1 = d_1 + d_2 + \ldots + d_n$, $1 = g_1 \alpha_1 + g_2 \alpha_2 + \ldots + g_n \alpha_n$, where each of $\alpha_1, \alpha_2, \ldots, \alpha_n$ is a sum of elements in $S(a, f)$, and hence each of $\alpha_1, \alpha_2, \ldots, \alpha_n$ is positive in R. Now for $0 \leq x \in R$,

$$x = g_1 \alpha_1 x + g_2 \alpha_2 x + \ldots + g_n \alpha_n x \text{ with } x_t = g_t \alpha_t x \geq 0, t = 1, \ldots, n.$$

We finally show that each $g_t = f^{t-1} a f^{n-1}$ is a d-element, $t = 1, \ldots, n$. Let $x \in R$, and let $g_t x, 0 \leq z$ for some $z \in R$. Then

$$af^{n-t} g_t x, 0 \leq af^{n-t} z \Rightarrow af^{n-1} x, 0 \leq af^{n-t} z \ (af^{n-1} \text{ is idempotent})$$
$$\Rightarrow af^{n-1}(x \vee 0) \leq af^{n-t} z \ (af^{n-1} \text{ is an } f\text{-element})$$
$$\Rightarrow f^{t-1} af^{n-1}(x \vee 0) \leq f^{t-1} af^{n-t} z$$
$$\Rightarrow g_t(x \vee 0) \leq f^{t-1} af^{n-t} z$$
$$\Rightarrow g_t(x \vee 0) \leq z \ (f^{t-1} af^{n-t} \leq 1).$$

Therefore $(g_t x) \vee 0 = g_t(x \vee 0)$. Similarly we also have $(x g_t \vee 0) = (x \vee 0) g_t$ and we leave the verification as an exercise (Exercise 9). Hence each g_t is a d-element, $t = 1, \ldots, n$.

Let $x \in R$ and $|x| \leq |y|$ for some $y \in g_t R$. Then $y = g_t r$ for some $r \in R$, so $|x| \leq g_t |r|$. Let $|x| = x_1 + \ldots + x_n$ with $x_i \in g_i R$. Then each $x_i \geq 0$ and $0 \leq (-x_1) + \ldots + (g_t|r| - x_t) + \ldots + (-x_n)$ implies that $x_i = 0$ for $i \neq t$. Hence $|x| = x_t \in g_t R$. Since $0 \leq x^+, x^- \leq |x|$, similar argument gives that $x^+, x^- \in g_t R$, and hence $x = x^+ - x^- \in g_t R$. Therefore $g_t R$ is a right ℓ-ideal of $R, t = 1, \ldots, n$. Since $f^{t-1} \geq 0$ and $af^{n-t} \geq 0$, the R-isomorphisms defined before for $g_1 R$ and $g_t R$ are actually now ℓ-isomorphisms over R. Thus Theorem 4.1(3) is true, so (1) is true.

(3) $\Rightarrow$ (4) is clear.

(4) $\Rightarrow$ (5) Given u, let

$$af^{n-1} + faf^{n-2} + f^2 af^{n-3} + \cdots + f^{n-1} a = v.$$

Define $b = av^{-1} u$ and $g = (v^{-1} u)^{-1} f(v^{-1} u)$. Then $g \geq 0$ by Theorem 1.20, $g^n = 0$ and $u = bg^{n-1} + fbg^{n-2} + f^2 bg^{n-3} + \ldots + f^{n-1} b$. Thus (5) is true.

(5) $\Rightarrow$ (2) is clear.

This completes the proof. $\square$

By using three elements with an additional condition, the equation in Theorem 4.3(3) can be shortened.

Lemma 4.3. *Let R be an ℓ-unital ℓ-ring and $n \geq 2$. Then R is ℓ-isomorphic to $M_n(T)$ with the entrywise order, where T is an ℓ-unital ℓ-ring, if and only if R contains positive elements a, b and a d-element f such that $f^n = 0$ and $af^{n-1} + fb = 1$.*

Proof. "⇒" Let $\{e_{ij} \mid 1 \leq i, j \leq n\}$ be the standard $n \times n$ matrix units of R. Take $a = e_{1n}$,

$$b = e_{12} + e_{23} + \cdots + e_{n-1,n} \quad \text{and} \quad f = e_{21} + e_{32} + \cdots + e_{n,n-1}.$$

We leave it as an exercise to verify $f^n = 0$ and $af^{n-1} + fb = 1$ (Exercise 10). Since $(e_{1n} + f)^n = 1$, $e_{1n} + f$ is a d-element by Theorem 1.20, and hence f is a d-element.

"⇐" For $r = 1, \ldots, n$, define $g_r = f^{r-1}af^{n-1}$. Then right ideals $g_1 R, \cdots, g_n R$ are mutually isomorphic, and their sum is a direct sum and equals R. The verification of these facts is similar to the proof of $(3) \Rightarrow (1)$ in Theorem 4.2, so we leave it as an exercise (Exercise 11).

Similar to the proof of Theorem 4.2, we show that Theorem 4.1(3) is true under the given conditions. Let $x = x_1 + \ldots + x_n$, where $x_r \in g_r R$, $r = 1, \ldots, n$. We claim that $x \geq 0$ if and only if each $x_r \geq 0$. As before, we just need to show that 1 is a sum of positive elements. Since $1 = af^{n-1} + fb = g_1 + fb$, we only need to show that f is a sum of positive elements. First, $f^{n-1} = g_n$. Now suppose that for any $n \geq s > r \geq 1$, f^s is a sum of positive elements from $g_1 R, \ldots, g_n R$. Then $f^r = g_{r+1} + f^{r+1}b$ implies that $f^r = y_1 + \ldots + y_n$, where $0 \leq y_r \in g_r R$. Thus it is true that f is a sum of positive elements in $g_1 R, \cdots, g_n R$, and hence 1 is a sum of positive elements of $g_1 R, \ldots, g_n R$.

Since f is a d-element and af^{n-1} is an f-element, each $g_r = f^{r-1}af^{n-1}$ is a d-element, $r = 1, \ldots, n$. Thus each $g_r R$ is a right ℓ-ideal and $R = g_1 R + \cdots + g_n R$ is a direct sum of right ℓ-ideals of R with $g_i R \cong g_j R$, for any i and j. $\qquad\square$

Theorem 4.4. *For an ℓ-unital ℓ-ring R and positive integers m and n, the following conditions are equivalent:*

(1) R is ℓ-isomorphic to a matrix ℓ-ring $M_{m+n}(T)$ with the entrywise order, where T is an ℓ-unital ℓ-ring.

(2) R contains positive elements a, b, and a d-element f such that $f^{m+n} = 0$, and $af^m + f^n b = 1$.

Proof. $(1) \Rightarrow (2)$. Consider the following elements

$$a = e_{1,m+1} + e_{2,m+2} + \ldots + e_{n,m+n}$$
$$b = e_{1,n+1} + e_{2,n+2} + \ldots + e_{m,m+n}$$
$$f = e_{21} + e_{32} + \ldots + e_{m+n,m+n-1}$$

in $M_{m+n}(T)$ with the entrywise order. Then a, b, f are all positive, $f^{m+n} = 0$, and $af^m + f^n b = 1$ (Exercise 12). Let $e = f + e_{1,m+n}$. Then that $e \geq 0$

and $e^{m+n} = 1$, where 1 is the identity matrix, implies that e is a d-element by Theorem 1.20, so f is also a d-element since $0 \le f \le e$.

(2) $\Rightarrow$ (1). Suppose that there exist positive elements a, b, f such that $f^{m+n} = 0$, $1 = af^m + f^n b$, and f is a d-element. Then

$$
\begin{aligned}
1 &= af^{m-1}(1 - f^n b)f + f^n b + f^{n-1}bf - (1 - af^m)f^{n-1}bf \\
&= (af^{m-1}a)f^{m+1} + f^{n-1}(fb + bf - (fb)f^{n-1}(bf)) \\
&= a'f^{m+1} + f^{n-1}b',
\end{aligned}
$$

where $a' = af^{m-1}a \ge 0$ and $b' = fb + bf - (fb)f^{n-1}(bf)$. Now we use the condition that f is a d-element. Since

$$
1 = |1| = |a'f^{m+1} + f^{n-1}b'| \le a'f^{m+1} + |f^{n-1}b'| = a'f^{m+1} + f^{n-1}|b'|,
$$

we have $1 = x + y$, where $0 \le x \le a'f^{m+1}$ and $0 \le y \le f^{n-1}|b'|$. Then $xf^{n-1} = 0$ and $f^{m+1}y = 0$, and hence $1 = a'f^{m+1} + f^{n-1}b'$ implies that $x = xa'f^{m+1}$ and $y = f^{n-1}b'y$. Let $a_1 = xa'$ and $b_1 = |b'y|$. Then $y = |y| = f^{n-1}|b'y| = f^{n-1}b_1$, so $1 = x + y = a_1f^{m+1} + f^{n-1}b_1$ with $a_1 \ge 0$, $b_1 \ge 0$. Continuing the above procedure, we have

$$
\begin{aligned}
1 &= a_1 f^m (1 - f^{n-1}b_1)f + f^{n-1}b_1 + f^{n-2}b_1 f - (1 - a_1 f^{m+1})f^{n-2}b_1 f \\
&= (a_1 f^m a_1)f^{m+2} + f^{n-2}(fb_1 + b_1 f - (fb_1)f^{n-2}(b_1 f)) \\
&= a''f^{m+2} + f^{n-2}b'',
\end{aligned}
$$

where $a'' = a_1 f^m a_1 \ge 0$ and $b'' = fb_1 + b_1 f - (fb_1)f^{n-2}(b_1 f)$. Similar to the argument before, $1 = a_2 f^{m+2} + f^{n-2}b_2$ with $a_1 \ge 0$ and $b_2 \ge 0$. Repeating this process, we will eventually arrive at $1 = a_{n-1}f^{m+n-1} + fb_{n-1}$ with $a_{n-1} \ge 0$ and $b_{n-1} \ge 0$. Now by Lemma 4.3, R is ℓ-isomorphic to the ℓ-ring $M_{m+n}(T)$ with the entrywise order, where T is an ℓ-unital ℓ-ring. $\square$

Another characterization of matrix ℓ-rings with entrywise order is given below.

Theorem 4.5. *For an ℓ-unital ℓ-ring R and $n \ge 2$, the following are equivalent:*

(1) $R \cong M_n(T)$ with the entrywise order, where T is an ℓ-unital ℓ-ring.
(2) There exist positive elements $x, y \in R$ such that $x^{n-1} \ne 0$, $x^n = y^2 = 0$, $x + y$ has the positive inverse and $\ell(x^{n-1}) \cap Ry = \{0\}$, where $\ell(x^{n-1}) = \{a \in R \mid |a|x^{n-1} = 0\}$.

Proof. (1) $\Rightarrow$ (2) Let $x = e_{12} + \ldots + e_{n-1,n}$ and $y = e_{n1}$. Then $x^{n-1} \ne 0$, $x^n = y^2 = 0$. Since $(x + y)^n = 1$, $(x + y)^{-1} = (x + y)^{n-1} > 0$. Since $x^{n-1} = e_{1n}$, it is clear that $\ell(x^{n-1}) \cap Ry = \{0\}$.

(2) $\Rightarrow$ (1) Let r be the inverse of $x + y$. We note that since $x + y$ and r are both positive, they are d-elements by Theorem 1.20, and hence x, y are also d-elements. Since x is a d-element, $\ell(x^{n-1}) = \{a \in R \mid ax^{n-1} = 0\}$.

Define $a_{ij} = r^{n-i}(ry)x^{n-j}$, $i, j = 1, \cdots, n$. We show that $\{a_{ij} \mid 1 \leq i, j \leq n\}$ is a set of $n \times n$ matrix units of R by two steps.

(i) $yr^k y = 0$ and $yr^k x^j = 0$ for $2 \leq k \leq n, k \leq j \leq n$.

First we show that it is true for $k = 2$. From that r is an inverse of $x+y$, we have $1 = rx + ry$, so $y = yrx + yry$ and $(1 - yr)y = yrx \in Rx \cap Ry$. Since $x^n = 0$, $Rx \subseteq \ell(x^{n-1})$. Thus $Rx \cap Ry \subseteq \ell(x^{n-1}) \cap Ry = \{0\}$ implies that $yrx = 0$. By $1 = rx + ry$ and $x^n = 0$, we have $x^{n-1} = ryx^{n-1}$, and hence

$$yr^2 yx^{n-1} = (yr)(ryx^{n-1}) = yrx^{n-1} = 0,$$

since $n \geq 2$. Therefore $yr^2 y \in \ell(x^{n-1}) \cap Ry = \{0\}$ implies that $yr^2 y = 0$. Then $yr = yr^2 x + yr^2 y = yr^2 x$ implies that $yr^2 x^2 = yrx = 0$, and hence for any $j \geq 2$, $yr^2 x^j = 0$. Hence (i) is true when $k = 2$.

Now suppose that $yr^i y = 0, yr^i x^j = 0$ for $2 \leq i < k, i \leq j$. We prove that $yr^k y = 0, yr^k x^j = 0$ for $k \leq j$. From $x^{n-1} = rx^n + ryx^{n-1} = ryx^{n-1}$ and the inductive assumption, we have

$$yr^k yx^{n-1} = yr^{k-1}(ryx^{n-1}) = yr^{k-1}x^{n-1} = 0,$$

so $yr^k y \in \ell(x^{n-1}) \cap Ry = \{0\}$ implies that $yr^k y = 0$. Finally from $yr^{k-1} = yr^k x + yr^k y = yr^k x$ and $yr^{k-1}x^j = 0$, we have $yr^k x^{j+1} = 0$. Thus $yr^k x^j = yr^k x^{j+1} + yr^k yx^j = 0$ for any $k \leq j$. Therefore (i) is true.

(ii) $ryx^i r^j ry = \delta_{ij} ry$ for all $0 \leq i, j \leq n - 1$, where δ_{ij} is the Kroneker delta.

From $yry = y$ and $yr^k y = 0$ for $k \geq 2$ by (i), the equation is true when $i = 0$. Now

$$1 = xr + yr \Rightarrow x^{i-1}r^{i-1} = x^i r^i + x^{i-1}yr^i, \ 1 \leq i \leq n - 1$$
$$\Rightarrow x^{i-1}r^i y = x^i r^{i+1}y + x^{i-1}yr^{i+1}y = x^i r^{i+1}y,$$

since $yr^{i+1}y = 0$ by (i). Thus $x^{i-1}r^{i-1}(ry) = x^i r^i(ry)$, for any $i = 1, \cdots, n - 1$. Then

$$x^{n-1}r^{n-1}(ry) = x^{n-2}r^{n-2}(ry) = \cdots = xr(ry) = ry \ (\star),$$

and hence $(ry)x^i r^i(ry) = (ry)^2 = ry$ for $1 \leq i \leq n - 1$. So the equation (ii) is true when $1 \leq i = j \leq n - 1$.

For $i > j$,

$$x^i r^j ry = x^{i-j}x^j r^j(ry) = x^{i-j}(ry) \text{ by } (\star),$$

so $x^{i-j}ry = 0$ since $xry = (1-xr)x \in Rx \cap Ry = \{0\}$. Hence the equation (ii) is true for this case.

For $1 \le i < j$ and $0 < t \le i$, $1 = xr + yr$ implies

$$x^{i-t}r^{i-t}r^{j-i}ry = x^{i-t+1}r^{i-t+1}r^{j-i}ry + x^{i-t}yr^{j-t+1}y$$
$$= x^{i-t+1}r^{i-t+1}r^{j-i}ry$$

since $2 \le j - t + 1 \le n$ implies $yr^{j-t+1}y = 0$ by (i). Let $t = 1, \cdots, i$ in the above equation, we have

$$x^i r^i r^{j-i}ry = x^{i-1}r^{i-1}r^{j-i}ry = \cdots = r^{j-i}ry,$$

and hence $ryx^i r^j ry = ryr^{j-i}ry = 0$ since $2 \le j - i + 1 \le n$. Thus the equation (ii) is true also for this case.

By (ii), $a_{ij}a_{rs} = r^{n-i}(ry)x^{n-j}r^{n-r}(ry)x^{n-s} = \delta_{jr}a_{is}$. Since $1 = rx + ry$, $r^{n-i}x^{n-i} = r^{n-i+1}x^{n-i+1} + r^{n-i}(ry)x^{n-i}$ implies

$$\sum_{i=1}^{n} a_{ii} = \sum_{i=1}^{n-1} a_{ii} + ry$$
$$= \sum_{i=1}^{n-1} r^{n-i}(ry)x^{n-i} + ry$$
$$= \sum_{i=1}^{n-1} (r^{n-i}x^{n-i} - r^{n-i+1}x^{n-i+1}) + ry$$
$$= -r^n x^n + rx + ry$$
$$= 1.$$

Hence $a_{ij}, 1 \le i, j \le n$, are $n \times n$ matrix units.

Finally since x, y and r are all d-elements, each $a_{ij} = r^{n-i}(ry)x^{n-j}$ is a d-element, so Theorem 4.1 applies.

We also note that since $Ry = R(ry)$ and $a_{nn} = ry$, Ry is a left ℓ-ideal of R. $\qquad\square$

The following result is an immediate consequence of Theorem 4.5.

Corollary 4.2. *Let R be an ℓ-unital ℓ-ring. If there exist positive elements $x, y \in R$ such that $x^2 = y^2 = 0$ and $x + y$ has the positive inverse, then $R \cong M_2(T)$ with the entrywise order, where T is an ℓ-unital ℓ-ring.*

Proof. We just need to show that $\ell(x) \cap Ry = \{0\}$. If $a \in \ell(x) \cap Ry$, then $a = by$ for some $b \in R$, so $|a| = |b|y$ since y is a d-element. It follows from $y^2 = 0$ that $|a|(x + y) = 0$. Thus $|a| = 0$, so $\ell(x) \cap Ry = \{0\}$. $\qquad\square$

Let F be a totally ordered field and $M_n(F)$ ($n \geq 2$) be an $n \times n$ matrix ℓ-ring. Using previous results, various conditions may be obtained such that $M_n(F)$ is ℓ-isomorphic to $M_n(F)$ with the entrywise order. We state one such result below.

Theorem 4.6. *Let F be a totally ordered field and let $M_n(F)$ ($n \geq 2$) be an ℓ-algebra. If there are positive elements a, f such that*

$$f^n = 0 \text{ and } 1 = af^{n-1} + faf^{n-2} + \cdots + f^{n-1}a,$$

then $M_n(F)$ is ℓ-isomorphic to the ℓ-algebra $M_n(F)$ with the entrywise order.

Proof.   By Theorem 4.3(3), $M_n(F)$ contains a set of $n \times n$ matrix units $\{a_{ij} \mid 1 \leq i, j \leq n\}$ and each a_{ij} is a d-element. Let S be the centralizer of a_{ij}. Then $F1 \subseteq S$. Since a_{ij}, $1 \leq i, j \leq n$, are linearly independent, they form a basis for $M_n(F)$ as a vector space over F, so each standard matrix unit e_{rs} is a linear combination of a_{ij} over F. Thus each matrix in S is in the centralizer of e_{rs}, $1 \leq r, s \leq n$, and hence each matrix in S must be scalar matrix. Therefore $S = F1$ and $M_n(F)$ is ℓ-isomorphic to $M_n(F)$ with the entrywise order by the proof of Theorem 4.3. $\square$

4.2   Positive cycles

In this section, we consider the structure of ℓ-unital ℓ-rings with positive elements of finite order. For a unital ring R, an element e is said to have *finite order* if $e^m = 1$ for some positive integer m. For an element e with finite order, the *order* of e is the smallest positive integer n such that $e^n = 1$.

Lemma 4.4. *Let R be a unital ℓ-ring with a positive element e of order $n \geq 2$ and M be a maximal convex totally ordered subgroup of its additive ℓ-group. Then $e^i M e^j$ is a maximal convex totally ordered subgroup for $1 \leq i, j \leq n$.*

Proof.   Clearly $e^i M e^j$ is a totally ordered subgroup. Suppose that $0 < b \leq e^i a e^j$ for some $0 \leq a \in M$. Then $0 \leq e^{n-i} b e^{n-j} \leq a$ implies that $e^{n-i} b e^{n-j} \in M$, so $b \in e^i M e^j$. Therefore $e^i M e^j$ is convex. Assume that $e^i M e^j \subseteq N$ for some convex totally ordered subgroup N. Then from $M \subseteq e^{n-i} N e^{n-j}$ and $e^{n-i} N e^{n-j}$ is convex totally ordered, we have $M = e^{n-i} N e^{n-j}$, and hence $e^i M e^j = N$. $\square$

Theorem 4.7. *Let R be a unital ℓ-ring with a positive element e of order $n \geq 2$. Suppose that R satisfies the following conditions.*

(1) R contains a basic element $a \leq 1$ such that $a \wedge (1 - a) = 0$.
(2) $1 \in \sum_{i,j=1}^{n} e^i M e^j$, where $M = a^{\perp\perp}$.

Let $k \geq 2$ be the smallest positive integer with $e^k a = a e^k$. Then $B = \{e^i a e^{n-j} \mid 1 \leq i, j \leq k\}$ is a disjoint set of basic d-elements and also a set of matrix units. Therefore R is ℓ-isomorphic to the matrix ℓ-ring $M_k(T)$ with the entrywise order, where T is the centralizer of B in R.

Proof.   We first note that $e^n = 1$ implies that e is a d-element by Theorem 1.20 and $a \wedge (1 - a) = 0$ implies that $a(1 - a) = 0$, since $a, 1 - a$ are f-element, so $a = a^2$.

Since a is basic, M is a maximal convex totally ordered subgroup by Corollary 1.1. We claim that the sum

$$(eMe + \cdots + eMe^n) + \cdots + (e^k Me + \cdots + e^k Me^n)$$

is a direct sum. Since each $e^i M e^j$ is a maximal convex totally ordered subgroup by Lemma 4.4, by Theorem 1.16, any two terms in the sum are either disjoint or equal. Consider the following array and we claim that any two different terms cannot be equal.

$$
\begin{array}{cccc}
eMe & eMe^2 & \cdots & eMe^n \\
e^2 Me & e^2 Me^2 & \cdots & e^2 Me^n \\
\vdots & \vdots & \cdots & \vdots \\
e^k Me & e^k Me^2 & \cdots & e^k Me^n
\end{array}
$$

Suppose that for some positive integer m with $1 \leq m < n$, $e^m M = M$. Then for any $0 \leq x \in M$, $e^m x$ and x are comparable. If $x < e^m x$, then $x < e^m x < e^{2m} x < \cdots < e^{nm} x = x$, which is a contradiction. Similarly $e^m x \not\leq x$. Thus we must have $e^m x = x$, and hence for each $z \in M$, $e^m z = z$. From $1 \in \sum_{i,j=1}^{n} e^i M e^j$, $1 = \sum_{i,j} e^i x_{ij} e^j$, where $x_{ij} \in M$, and hence

$$e^m = e^m \sum_{i,j} e^i x_{ij} e^j = \sum_{i,j} e^i (e^m x_{ij}) e^j = \sum_{i,j} e^i x_{ij} e^j = 1,$$

which is a contradiction. Hence n is the smallest positive integer such that $e^n M = M$, and similarly n is the smallest positive integer such that $Me^n = M$. Therefore any two terms in the same row or column of the above array are different. Suppose that for $1 \leq s < k, 1 \leq t < n$, $e^s Me^t = M$, then $e^s M = Me^{n-t}$. Similar to the above proof, $e^s a = a e^{n-t}$. We show that $Me^s = Me^{n-t}$. If $Me^s \neq Me^{n-t}$, then $Me^s \cap Me^{n-t} = \{0\}$, and hence

$ae^s \wedge ae^{n-t} = 0$. Thus a is an f-element implies that $ae^s a \wedge ae^{n-t} = 0$, so since e is a d-element, we have

$$(a^2 \wedge a)e^{n-t} = a^2 e^{n-t} \wedge ae^{n-t} = ae^s a \wedge ae^{n-t} = 0.$$

It follows that $a^2 \wedge a = 0$, and hence $a = 0$ since $a^2 = a$, which is a contradiction. Thus $Me^s = Me^{n-t}$, then $Me^{s+t} = M$ implies $n \mid (s+t)$, and hence $s+t = n$ since $s+t < 2n$. Hence $e^s M = Me^s$, which contradicts the fact that $1 \le s < k$ and k is the smallest positive integer satisfying $e^k M = Me^k$. This proves that $e^s Me^t \ne M$ for $1 \le s < k, 1 \le t < n$. Therefore the sum

$$\sum_{1 \le i \le k, 1 \le j \le n} e^i Me^j$$

is a direct sum (Exercise 13).

We next show that $1 = a + eae^{n-1} + \ldots + e^{k-1}ae^{n-k+1}$. Since 1 is a sum of disjoint basic elements and $1 = a + (1-a)$ implies that $1 = e^i e^{n-i} = e^i ae^{n-i} + e^i(1-a)e^{n-i}$, $1 \le i < k$, we have $a, eae^{n-1}, \ldots, e^{k-1}ae^{n-k+1}$ are all in the sum for 1 (Exercise 14). By condition (2), each basic element in the sum for 1 is equal to $c = e^s xe^t \le 1$ for some $0 < x \in M$ and $1 \le s \le k, 1 \le t \le n$. Then $e^s xe^t e^s xe^t = e^s xe^t$ implies that $xe^{s+t}x = x$. Suppose $xe^v x = x$ with $0 < v < n$. Since $M \cap Me^v = \{0\}$, $x \wedge xe^v = 0$. If $x \le a$, then x is an f-element, so $x \wedge xe^v x = 0$ implies that $x = 0$, which is impossible. Hence $a < x$. Since $x = e^{n-s} ce^{n-t}$ is a d-element, $x \wedge xe^v = 0$ implies that $x^2 \wedge xe^v x = 0$, so $x^2 \wedge x = 0$. Then $a = a^2 \wedge a \le x^2 \wedge x = 0$, which is a contradiction. Therefore there is no positive integer $v < n$ such that $xe^v x = x$. It follows from $xe^{s+t}x = x$ that we must have $s + t = n$, and hence $c = e^s xe^{n-s}$, $1 \le s \le k$ and x is idempotent. From

$$1 = c + (1-c) = e^s xe^{n-s} + (1 - e^s xe^{n-s})$$

and

$$c \wedge (1-c) = e^s xe^{n-s} \wedge (1 - e^s xe^{n-s}) = 0,$$

we have $1 = x + (1-x)$ and $x \wedge (1-x) = 0$. Since we also have $1 = a + (1-a)$ with $a \wedge (1-a) = 0$, we must have $x = a$ since $a, x \in M$. Therefore $c = e^s ae^{n-s}$, $1 \le s \le k$, and hence $1 = a + eae^{n-1} + \ldots + e^{k-1}ae^{n-k+1}$. Then $\{a, eae^{n-1}, \ldots, e^{k-1}ae^{n-k+1}\}$ is a disjoint set implies that $e^i ae^{n-i} e^j ae^{n-j} = 0$ for $i \ne j$.

For $1 \le i \le k, 1 \le j \le k$, define $c_{ij} = e^i ae^{n-j}$. It is clear that each c_{ij} is d-element since a is an f-element and e is a d-element, and $\{c_{ij} \mid 1 \le i, j \le k\}$ is a set of $k \times k$ matrix units, that is, $c_{ij}c_{rs} = \delta_{jr}c_{is}$, where δ_{jr} is the Kronecker delta, and $c_{11} + \ldots + c_{kk} = 1$ (Exercise 15). Let $T = \{x \in R \mid xc_{ij} = c_{ij}x, 1 \leq i, j \leq k\}$ be the centralizer of $\{c_{ij} \mid 1 \leq i, j \leq k\}$ in R. By Theorem 4.1, R is ℓ-isomorphic to the matrix ℓ-ring $M_k(T)$ with the entrywise order. This completes the proof. $\qquad\square$

For a ring R and $x \in R$, we define $i(x) = \{a \in R \mid ax = xa = a\}$. Clearly $i(x)$ is a subring of R and if R is an algebra over F then $i(x)$ is a subalgebra of R over F.

Theorem 4.8. *Let A be a unital finite-dimensional Archimedean ℓ-algebra over a totally ordered field F. Suppose that A contains a positive element e with order $n \geq 2$ and $dim_F i(e) = 1$. Then A is ℓ-isomorphic to $M_k(F[G])$ as the ℓ-algebra over F with the entrywise order, where $k \mid n$, G is a finite cyclic group of order n/k, and $F[G]$ is the group ℓ-algebra of G over F with the coordinatewise order.*

Proof. We first show conditions in Theorem 4.7 are satisfied and then we determine the ℓ-unital ℓ-ring T in Theorem 4.7. Since A is finite-dimensional and Archimedean over F, by Corollary 1.3, A is a finite direct sum of maximal convex totally ordered subspaces over F. Then 1 is a sum of disjoint basic elements, and hence there exists a basic element a such that $a \leq 1$ and $a \wedge (1 - a) = 0$, that is, condition (1) in Theorem 4.7 is satisfied.

Let $M = a^{\perp\perp}$ and $x = \sum_{i,j=1}^{n} e^i a e^j$. Then $ex = xe = x$ implies that $x \in i(e)$, so $x = \alpha(1 + e + \cdots + e^{n-1})$ for some $0 < \alpha \in F$ since $dim_F i(e) = 1$ and $1 + e + \cdots + e^{n-1} \in i(e)$. It follows that $1 \leq \alpha^{-1}x \in \sum_{i,j=1}^{n} e^i M e^j$, and hence $1 \in \sum_{i,j=1}^{n} e^i M e^j$, that is, condition (2) in Theorem 4.7 is also satisfied. We note that above arguments have actually proved $A = \sum_{i,j=1}^{n} e^i M e^j$. Otherwise there is a maximal convex totally ordered subspace N that is not contained in $H = \sum_{i,j=1}^{n} e^i M e^j$, then $H \cap J = \{0\}$, where $J = \sum_{i,j=1}^{n} e^i N e^j$ (Exercise 16). On the other hand, by a similar argument we have $1 \in J$, which is a contradiction. Therefore $A = \sum_{i,j=1}^{n} e^i M e^j$.

Suppose that k is the smallest positive integer with $e^k a = a e^k$ and $k \geq 2$. By Theorem 4.7, $\{c_{ij} = e^i a e^{n-j} \mid 1 \leq i, j \leq k\}$ is a disjoint set of basic d-elements and a set of $k \times k$ matrix units, and $A \cong M_k(T)$ with the entrywise order, where

$$T = \{x \in A \mid xc_{ij} = c_{ij}x, 1 \leq i, j \leq k\}$$

is an ℓ-unital ℓ-ring. Also from the proof of Theorem 4.7, $A = \sum_{1 \leq i \leq k, 1 \leq j \leq n} e^i M e^j$ is a direct sum as a vector lattice.

We prove that $\dim_F M = 1$. Let $u, v \in M$ be linearly independent over F. Define $x = \sum_{1 \leq i \leq k, 1 \leq j \leq n} e^i u e^j$ and $y = \sum_{1 \leq i \leq k, 1 \leq j \leq n}^n e^i v e^j$. If $\alpha x + \beta y = 0$ for some $\alpha, \beta \in F$, then

$$\sum_{1 \leq i \leq k, 1 \leq j \leq n} e^i (\alpha u + \beta v) e^j = 0,$$

and hence $\alpha u + \beta v = 0$, so $\alpha = \beta = 0$. It follows that x and y are linearly independent. On the other hand, $x, y \in i(e)$ implies that they are linearly dependent since $\dim_F i(e) = 1$. This contradiction shows that $\dim_F M = 1$, and hence $M = Fa$.

Since $e^k a = ae^k$, $e^k, e^{2k}, \cdots, e^{\ell k} \in T$, where $n = \ell k$. We prove that

$$T = \{\alpha_0 + \alpha_1 e^k + \cdots + \alpha_{\ell-1} e^{(\ell-1)k} \mid \alpha_i \in F, \ 0 \leq i \leq \ell - 1\}.$$

Suppose that $x \in T$. Since A is a direct sum of $e^i (Fa) e^j$, $1 \leq i \leq k, 1 \leq j \leq n$, a direct calculation shows that $x = \sum_{1 \leq i, j \leq k} v_{ij} c_{ij}$, where $v_{ij} \in F + Fe^k + \cdots + Fe^{(\ell-1)k}$, then $\{c_{ij}\}$ is a set of $k \times k$ matrix units implies that $v_{ij} = 0$ if $i \neq j$ and $v_{11} = \cdots = v_{kk}$. Hence

$$x = v_{11}(c_{11} + \cdots + c_{kk}) = v_{11} 1 = v_{11}$$

(Exercise 17), so $T = F + Fe^k + \cdots + Fe^{(\ell-1)k}$. For an element $x = \alpha_0 + \alpha_1 e^k + \cdots + \alpha_{\ell-1} e^{(\ell-1)k}$ in T, it is clear that $x \geq 0$ if and only if each $\alpha_i \geq 0$, and hence T is a group ℓ-algebra of a finite cyclic group of order $\ell = n/k$ over F with the coordinatewise order.

If $k = 1$, that is, $ea = ae$, then A is ℓ-isomorphic to the group ℓ-algebra $F[G]$ of a cyclic group of order n over F. The verification of this fact is left to the reader (Exercise 78). This completes the proof of Theorem 4.8. $\quad\square$

An *n-cycle* $(i_1 i_2 \cdots i_n)$ on the set $\{1, \cdots, n\}$ is a permutation which sends $i_1 \to i_2, \cdots, i_{n-1} \to i_n$, and $i_n \to i_1$. The permutation matrix $e_{i_1 i_2} + \cdots + e_{i_{n-1} i_n} + e_{i_n i_1}$, where $(i_1 i_2 \cdots i_n)$ is an n-cycle, is called an *n-cycle* in matrix ring $M_n(R)$ over a unital ring R.

Lemma 4.5. *Let T be a unital ring and e be an n-cycle in $M_n(T)$. For $x \in M_n(T)$, if $ex = xe$, then $x = \alpha_0 1 + \alpha_1 e + \cdots + \alpha_{n-1} e^{n-1}$ for some $\alpha_i \in T$, where 1 is the identity matrix.*

Proof. First we assume that $e = e_{12} + e_{23} + \cdots + e_{n1}$. Let $x = (x_{ij})$. For $1 \leq k \leq n$, a direct calculation shows that

$$e^k = e_{1,k+1} + e_{2,k+2} + \cdots + e_{n-k,n} + \cdots + e_{n,k}$$

and

$$x_{1,k} = x_{2,k+1} = \cdots = x_{n-k,n-1} = x_{n-k+1,n} = \cdots = x_{n,k-1}$$

(Exercise 18), and hence $x = x_{11}1 + x_{12}e + x_{13}e^2 + \cdots + x_{1n}e^{n-1}$.

Now suppose that $e' = e_{i_1 i_2} + e_{i_2 i_3} + \cdots + e_{i_n i_1}$. Define $d = e_{1 i_1} + e_{2 i_2} + \cdots + e_{n i_n}$. Then $d^{-1} = e_{i_1 1} + e_{i_2 2} + \cdots + e_{i_n n}$ and $de'd^{-1} = e$. If $xe' = e'x$, then $(dxd^{-1})(de'd^{-1}) = (de'd^{-1})(dxd^{-1})$, and hence, by previous argument, there exist $\alpha_0, \cdots, \alpha_{n-1} \in T$ such that

$$
\begin{aligned}
dxd^{-1} &= \alpha_0 1 + \alpha_1 e + \cdots + \alpha_{n-1}e^{n-1} \\
&= \alpha_0 1 + \alpha_1(de'd^{-1}) + \cdots + \alpha_{n-1}(de'd^{-1})^{n-1} \\
&= d(\alpha_0 1 + \alpha_1 e' + \cdots + \alpha_{n-1}(e')^{n-1})d^{-1},
\end{aligned}
$$

since each entry in d is either 0 or 1 implies $\alpha d = d\alpha$ for $\alpha \in T$. Therefore $x = \alpha_0 + \alpha_1 e' + \cdots + \alpha_{n-1}(e')^{n-1}$. $\qquad\square$

Theorem 4.9. *Let T be a unital totally ordered ring, and $R = M_n(T)$ ($n \geq 2$) be an ℓ-ring and f-bimodule over T with respect to left and right scalar multiplication. Assume that R is a direct sum of convex totally ordered subgroups and contains a positive n-cycle. Then R is ℓ-isomorphic to the ℓ-ring $M_n(T)$ with the entrywise order.*

Proof. Let e be a positive n-cycle. Since each entry in e is either 1 or 0, for any $\alpha \in T$, $\alpha e = e\alpha$. From $1 = e^n > 0$, 1 is a sum of disjoint basic elements, so there is a basic element a such that $a \leq 1$ and $a \wedge (1-a) = 0$. Hence the condition (1) in Theorem 4.7 is satisfied.

Let $M = a^{\perp\perp}$ and $H = \sum_{i,j=1}^{n} e^i M e^j$. We claim that $R = H$. If $R \neq H$, then there exists a maximal convex totally ordered subgroup N that is not in the sum of H, and hence $H \cap J = \{0\}$, where $J = \sum_{i,j=1}^{n} e^i N e^j$. Take $0 < x \in N$ and consider $z = \sum_{i,j=1}^{n} e^i x e^j$. Then $ez = ze = z$, so $z = \alpha(1 + e + \ldots + e^{n-1})$ for some $0 < \alpha \in T$ by Lemma 4.5. On the other hand, if $w = \sum_{i,j=1}^{n} e^i a e^j$, then $ew = we = w$, so $w = \beta(1 + e + \ldots + e^{n-1})$ for some $0 < \beta \in T$. Thus $\beta 1 \in H$ and $\alpha 1 \in J$, which implies that $0 < \min\{\beta, \alpha\}1 \in H \cap J$, which is a contradiction. Therefore we must have $R = H = \sum_{i,j=1}^{n} e^i M e^j$, so $1 \in \sum_{i,j=1}^{n} e^i M e^j$, and condition (2) in Theorem 4.7 is satisfied.

Suppose that $e^i a = ae^i$ for some $1 \leq i \leq n-1$, then $R = \sum_{i,j=1}^{n} e^i M e^j$ implies that e^i is in the center of R, which is a contradiction (Exercise 19). Hence n is the smallest positive integer with $e^n a = ae^n$, then by Theorem 4.7, R is ℓ-isomorphic to the ℓ-ring $M_n(S)$ with the entrywise order, where S is the centralizer of $\{c_{ij} = e^i a e^{n-j} \mid 1 \leq i, j \leq n\}$ in R. We show that S consists of all scalar matrices over T.

Let $0 < \alpha \in T$. To show that $\alpha 1 \in S$, it is sufficient to show that $(\alpha 1)a = a(\alpha 1)$ since $\alpha 1$ commutes with e. From $1 = a + eae^{n-1} + \ldots + e^{n-1}ae$

and $\alpha 1 = 1\alpha$, we have

$$\alpha a + \alpha e a e^{n-1} + \ldots + \alpha e^{n-1} a e = a\alpha + eae^{n-1}\alpha + \ldots + e^{n-1}ae\alpha.$$

Since $M_n(T)$ is an f-bimodule over T, $a \wedge e^i a e^{n-i} = 0$ for any $i = 1, \ldots, n-1$ implies that $\alpha a \wedge e^i a e^{n-i} \alpha = 0$, so

$$
\begin{aligned}
\alpha a &= \alpha a \wedge (\alpha a + \alpha e a e^{n-1} + \ldots + \alpha e^{n-1} a e) \\
&= \alpha a \wedge (a\alpha + eae^{n-1}\alpha + \ldots + e^{n-1}ae\alpha) \\
&\leq (\alpha a \wedge a\alpha) + (\alpha a \wedge eae^{n-1}\alpha) + \ldots + (\alpha a \wedge e^{n-1}ae\alpha) \\
&= \alpha a \wedge a\alpha \\
&\leq a\alpha.
\end{aligned}
$$

Similarly, $a\alpha \leq \alpha a$, so $\alpha a = a\alpha$. Thus $T1 \subseteq S$.

Now let $0 < x \in S$. Then x commutes with e since

$$e = e1 = ea + e^2 a e^{n-1} + \ldots + ae = c_{1n} + c_{21} + c_{32} + \ldots + c_{n,n-1},$$

and hence $x = \alpha_0 1 + \alpha_1 e + \ldots + \alpha_{n-1} e^{n-1}$ for some $\alpha_i \in T$, $i = 0, \ldots, n-1$ by Lemma 4.5. Since $a = c_{nn}$, $xa = ax$, and hence

$$\alpha_0 a + \alpha_1 e a + \ldots + \alpha_{n-1} e^{n-1} a = \alpha_0 a + \alpha_1 ae + \ldots + \alpha_{n-1} ae^{n-1}$$

implies that $\alpha_i e^i a = 0$ for any $i = 1, \cdots, n-1$. Hence $\alpha_i a = 0$ for $i = 1, \cdots, n$ since $e^n = 1$. We claim that each $\alpha_i = 0$, $i = 1, \cdots, n$. Suppose that $\alpha_k \neq 0$ for some k. We may assume that $\alpha_k > 0$. Since $a \wedge (1 - a) = 0$ and $M_n(T)$ is an f-bimodule over T, we have $a \wedge \alpha_k(1 - a) = a \wedge \alpha_k 1 = 0$, and hence $a \wedge 1 = 0$, which is a contradiction. Thus $\alpha_i = 0$ for $i = 1, \cdots, n$.

Therefore $x = \alpha_0 1 \in T1$. This proves that $S = T1$, that is, S consists of all the scalar matrices over T. Since for any $\alpha \in T$, $\alpha \geq 0$ in T if and only if $\alpha 1 \geq 0$ in S, S and T are ℓ-isomorphic ℓ-rings. Therefore $M_n(T)$ is ℓ-isomorphic to the ℓ-ring $M_n(T)$ with the entrywise order. This completes the proof. $\qquad\square$

A unital domain R is called a *left (right) Ore domain* if R can be embedded in a division ring Q such that

$$Q = \{a^{-1}x \mid a, x \in R, a \neq 0\} \ (Q = \{xa^{-1} \mid a, x \in R, a \neq 0\}).$$

The Q is called the *classical left (right) quotient ring* of R. Theorem 4.9 is true when R is a unital ℓ-simple totally ordered left (right) Ore domain. We will first prove the following result.

Lemma 4.6. *Suppose that R is a unital totally ordered ring.*

(1) R contains a unique maximal (left, right) ℓ-ideal.

(2) If R is a domain, then the unique maximal left (right) ℓ-ideal of R is a maximal ℓ-ideal.

(3) If R is ℓ-simple, then R is a domain and R and {0} are the only left (right) ℓ-ideals of R.

Proof.   (1) Since R is unital, by Zorn's Lemma R contains a maximal ℓ-ideal. Suppose that M, N are maximal ideals and $M \neq N$. Then $R = M + N$ implies that $1 = x + y$ for some $x \in M$ and $y \in N$. It follows from $1 > 0$ that $1 = |1| = |x + y| \leq |x| + |y|$, and hence $1 = a + b$ for some $0 \leq a \leq |x|$ and $0 \leq b \leq |y|$. Hence we have $a \in M$ and $b \in N$. Now that R is totally ordered implies that $a \leq b$ or $b \leq a$, so $1 \leq 2b$ or $2a$ implies that $1 \in N$ or $1 \in M$, which is a contradiction. Similarly there is a unique maximal left ℓ-ideal and a unique maximal right ℓ-ideal.

(2) Let I be the unique maximal left ℓ-ideal. Consider the ℓ-ideal $\langle I \rangle$ generated by I. Then

$$\langle I \rangle = \{x \in R \mid |x| \leq ar, a \in I^+, r \in R^+\}.$$

If $\langle I \rangle = R$, then $1 \leq ar$ for some $a \in I^+$ and $r \in R^+$, and hence $ra \leq (ra)^2$. Then that R is a totally ordered domain implies $1 \leq ra \in I$, so $1 \in I$, which is a contradiction. Therefore $\langle I \rangle \neq R$, so $\langle I \rangle$ is contained in a maximal left ℓ-ideal by a standard argument using Zorn's Lemma, and hence $\langle I \rangle \subseteq I$. Thus $I = \langle I \rangle$ is an ℓ-ideal.

(3) If R is ℓ-simple, then ℓ-$N(R) = \{0\}$, and R is reduced by Theorem 1.28. It follows that R is a domain since R is totally ordered. Let $I \neq \{0\}$ be a left ℓ-ideal of R. Then $\langle I \rangle = R$. By a similar argument in (2), we must have $1 \in I$, so $I = R$. $\square$

Corollary 4.3. *Let R be a unital ℓ-simple totally ordered left (right) Ore domain and $M_n(R)$ ($n \geq 2$) be an ℓ-ring and f-bimodule over R with respect to left and right scalar multiplication. If ℓ-ring $M_n(R)$ contains a positive n-cycle, then it is ℓ-isomorphic to the ℓ-ring $M_n(R)$ with the entrywise order.*

Proof.   Let Q be the classical left quotient ring of R. Then $M_n(R) \subseteq M_n(Q)$. Consider $M_n(R)$ ($M_n(Q)$) as a left module over R (Q) by left scalar multiplication. Since $M_n(Q)$ is an n^2-dimensional vector space over the division ring Q and matrices in $M_n(R)$ that are linearly independent over R are also linearly independent over Q (Exercise 20), $M_n(R)$ has at most n^2 linearly independent matrices over R. Suppose that $\{f_i \mid i \in I\} \subseteq M_n(R)$

is a disjoint set. Then it is linearly independent over R by a similar proof of Theorem 1.13 since R is a domain and $M_n(R)$ is a left f-module over R (Exercise 21). Therefore R does not contain any infinite set of disjoint elements, so condition (C) in Theorem 1.15 is satisfied.

We next show that $M_n(R)$ contains no maximal convex totally ordered subgroup that is bounded above. Suppose that M is a maximal convex totally ordered subgroup of $M_n(R)$ and $0 < a \in M_n(R)$ such that $x \leq a$ for all $x \in M$. For $0 < y \in M$, $y^{\perp\perp} = M$. Since $M_n(R)$ is a left f-module over R, for any $0 < \alpha \in R$, $\alpha y \in y^{\perp\perp}$ implies that $\alpha y \leq a$ for all $0 \leq \alpha \in R$. Thus for all $0 < \alpha \in R$,

$$\alpha \sum_{i,j=1}^{n} e^i y e^j = \alpha\beta(1 + e + \ldots + e^{n-1}) \leq \sum_{i,j=1}^{n} e^i a e^j = \gamma(1 + e + \ldots + e^{n-1}),$$

for some $0 < \beta, \gamma \in R$. Therefore $\alpha\beta \leq \gamma$ for all $\alpha \in R$. Let I be the left ℓ-ideal generated by β in R. By Lemma 4.6, $I = R$, so $\gamma \leq \delta\beta$ for some $\delta \in R^+$, which contradicts with $\alpha\beta \leq \gamma$ for all $\alpha \in R^+$. Hence $M_n(R)$ has no maximal convex totally ordered subgroup bounded above. By Theorem 1.17, $M_n(R)$ is a direct sum of convex totally ordered subgroups, so Theorem 4.9 applies. This completes the proof. $\square$

Let's consider two important cases that Corollary 4.3 applies. For a totally ordered subring R of $\mathbb{R}$, each $0 < \alpha \in R$ is less than $k1$ for some positive integer k. Thus an ℓ-algebra $M_n(R)$ over R is an f-bimodule over R, so if $M_n(R)$ contains a positive n-cycle, then it is ℓ-isomorphic to the ℓ-algebra $M_n(R)$ with the entrywise order. If R is a totally ordered division ring, then each ℓ-module over R is an f-module, so if $M_n(R)$ is an ℓ-ring and ℓ-bimodule over R and contains a positive n-cycle, then $M_n(R)$ is ℓ-isomorphic to the ℓ-ring $M_n(R)$ with the entrywise order.

4.3 Nonzero f-elements in ℓ-rings

For an ℓ-ring R with nonzero f-elements, for instance, an ℓ-unital ℓ-ring, properties of R are affected by $f(R)$. In this section we present some results in this direction. Recall that $f(R) = \{a \in R \mid |a|$ is an f-element of $R\}$. For an ℓ-ring R, let $U_f = U_f(R)$ be the upper bound of $f(R)$, that is, $U_f = \{x \in R \mid |x| \geq a$, for each $a \in f(R)\}$.

Lemma 4.7. *Let R be an ℓ-ring and $f(R) \neq 0$ be totally ordered. Then $R = U_f \cup (f(R) \oplus f(R)^{\perp})$, where the direct sum is regarded as the direct sum of convex ℓ-subgroups, and $U_f \cap (f(R) \oplus f(R)^{\perp}) = \emptyset$.*

Proof. Suppose that $a \in R$ and $a \notin U_f$. Then there exists an f-element $b > 0$ such that $|a| \not\geq b$, and hence $|a| \wedge b < b$. Consider $a_1 = |a| - |a| \wedge b$ and $b_1 = b - |a| \wedge b$. We have $a_1 \wedge b_1 = 0$, so $(a_1 \wedge e) \wedge b_1 = 0$, where $0 < e \in f(R)$. Since $0 < b_1, a_1 \wedge e \in f(R)$ and $f(R)$ is totally ordered, we must have $a_1 \wedge e = 0$, that is, $a_1 \in f(R)^{\perp}$. Thus $|a| = (|a| \wedge b) + a_1 \in f(R) + f(R)^{\perp}$. It follows from $0 \leq a^+, a^- \leq |a|$ that $a^+, a^- \in f(R) + f(R)^{\perp}$, and hence $a = a^+ - a^- \in f(R) + f(R)^{\perp}$. It is clear that $f(R) \cap f(R)^{\perp} = \{0\}$.

Suppose that $x \in U_f \cap (f(R) \oplus f(R)^{\perp})$. Then $x = x_1 + y_1, x_1 \in f(R), y_1 \in f(R)^{\perp}$. Thus $2|x_1| \leq |x| \leq |x_1| + |y_1|$, and hence $x_1 = 0$. Therefore $x \in f(R)^{\perp}$ which implies $f(R) = \{0\}$, which is a contradiction. Therefore we have $U_f \cap (f(R) \oplus f(R)^{\perp}) = \emptyset$. $\qquad\qquad\square$

We provide an application of the decomposition in Lemma 4.7. For an ℓ-ring R, an element e is called f-*superunit* if e is an f-element and for any $x \in R^+$, $ex \geq x$ and $xe \geq x$.

For a ring B and an element $a \in B$, if $ab = ba = nb$ for some integer n and all $b \in B$, then a is called an n-*fier* and n is said to have an n-fier a in B. Define $K = \{n \in \mathbb{Z} \mid n \text{ has an } n\text{-fier in } B\}$. Then K is an ideal of $\mathbb{Z}$ (Exercise 22). The ideal K is called the *modal ideal* of B and its nonnegative generator is called the *mode* of B.

Lemma 4.8.

(1) If R is an f-ring with mode $k > 0$, then R has a unique k-fier $x \geq 0$.
(2) Let R be an ℓ-ring with an f-superunit and $f(R)$ is totally ordered. Then mode of R and the mode of $f(R)$ are the same, and if k is the mode of R, then the k-fier of R is equal to the k-fier of $f(R)$.

Proof. (1) If x, y both are k-fier, then $kx = xy = ky$ implies that $x = y$. Thus there is only one k-fier x. From $xb = bx = kb$ for any $b \in R$ and R is an f-ring, we have $|x|a = a|x| = ka$ for each $a \in R^+$, and hence $|x|b = b|x| = kb$ for each $b \in R$. Therefore the uniqueness of x implies that $x = |x| \geq 0$.

(2) Let e be an f-superunit of R, n be the mode of R, and m be the mode of $f(R)$. If x is an m-fier of $f(R)$, then $xa = ax = ma$ for each $a \in f(R)$, especially $xe = ex = me$. Thus for each $b \in R$, $(bx)e = (mb)e$, so $bx = mb$ since e is an f-superunit. Similarly $xb = mb$. Therefore $n \mid m$. Now let y be an n-fier in R. We show that $y \in f(R)$. By Lemma 4.7, $R = U_f \cup (f(R) \oplus f(R)^{\perp})$. If $y \in U_f$, then we have $ne = |ey| = e|y| \geq |y| \geq (n+1)e$, which is a contradiction. Thus $y = z + w$ with $z \in f(R)$ and

$w \in f(R)^{\perp}$ implies that $ne = ey = ez + ew$, and hence $ew = 0$. Therefore $w = 0$ and $y \in f(R)$. Hence $m \mid n$, so $n = m$.

Suppose that k is the mode of R, by the above argument, a k-fier of R is also a k-fier of $f(R)$, so by (1) R has a unique k-fier $x \geq 0$. $\quad\square$

Let $\overline{R} = \{(n, r) \mid n \in \mathbb{Z}, r \in R\}$. Then $\overline{R}$ becomes a ring having identity element $(1, 0)$ with respect to the coordinatewise addition and the multiplication

$$(n, r)(m, s) = (nm, ns + mr + rs).$$

It is well known that $a \to (0, a)$ is a one-to-one ring homomorphism from R to $\overline{R}$. So R may be considered as a subring of $\overline{R}$.

Suppose that $k > 0$ is the mode of R and x is the unique k-fier of R. Let $I(k, x) = \{n(k, -x) \mid n \in \mathbb{Z}\}$. Then $I(k, x)$ is an ideal of $\overline{R}$ (Exercise 24). Define $R_1 = \overline{R}/I(k, x)$ as the quotient ring with identity $\overline{(1, 0)}$ and $a \to \overline{(0, a)}$ from R to R_1 which is a one-to-one ring homomorphism from R to R_1. Hence R can be considered as a subring of R_1.

We need the following result in the proof of Theorem 4.10.

Lemma 4.9. *Let R be an ℓ-ring with $a \in R$ and $0 \leq e \in f(R)$. Then $|ae + ea| = |a|e + e|a|$.*

Proof. Since $a^+ \wedge a^- = 0$,

$$ea^+ \wedge ea^- = ea^+ \wedge a^- e = a^+ e \wedge a^- e = a^+ e \wedge ea^- = 0,$$

so $(a^+ e + ea^+) \wedge (a^- e + ea^-) = 0$. Thus

$$
\begin{aligned}
|ae + ea| &= |a^+ e - a^- e + ea^+ - ea^-| \\
&= |(a^+ e + ea^+) - (a^- e + ea^-)| \\
&= (a^+ e + ea^+) + (a^- e + ea^-) \\
&= |a|e + e|a|.
\end{aligned}
$$
$\quad\square$

Theorem 4.10. *Let R be an ℓ-ring with an f-superunit and $f(R)$ be totally ordered. Suppose that R_1 is defined as above. Then R can be embedded in an ℓ-unital ℓ-rings R_1 such that $f(R) \subseteq f(R_1)$ and $f(R_1)$ is totally ordered. Moreover if R is ℓ-simple or R is squares positive, so is R_1.*

Proof. Before we proceed with the proof, we comment that the following proof works for $\overline{R}$ and we leave the verification of it to the reader.

By Lemma 4.7, $R = U_f \cup (f(R) \oplus f(R)^{\perp})$. Let e be an f-superunit and $A = \{ne + ae \mid n \in \mathbb{Z}, a \in R\}$ be the subring generated by e and Re. We first show that A is an ℓ-subring of R. Let $ne + ae \in A$, where $n \in \mathbb{Z}$ and $a \in R$. We consider two cases.

(1) $a \in f(R) \oplus f(R)^{\perp}$. Suppose that $a = b + c$ with $b \in f(R), c \in f(R)^{\perp}$. Then

$$(ne + ae)^{+} = (ne + be + ce)^{+} = (ne + be)^{+} + (ce)^{+} = (ne + be)^{+} + c^{+}e$$

and $(ne + be)^{+} = ne + be$ or 0 since $f(R)$ is totally ordered. Therefore $(ne + ae)^{+} = ne + be + c^{+}e$ or $c^{+}e$, so $(ne + ae)^{+} \in A$.

(2) $a \in U_f$. Since $a = a^{+} - a^{-}$ and $a^{+} \wedge a^{-} = 0$, we must have one of $a^{+}, a^{-} \in U_f$, but not both of them (Exercise 23). Suppose that $a^{+} \in U_f$. Then $a^{+} \wedge a^{-} = 0$ implies that $a^{-} \in f(R)^{\perp}$. Since e is an f-superunit, $ne + a^{+}e \geq ne + a^{+} \geq 0$, and hence

$$0 \leq (ne + a^{+}e) \wedge a^{-}e \leq (|n|e + a^{+}e) \wedge a^{-}e \leq (|n|e \wedge a^{-}e) + (a^{+}e \wedge a^{-}e) = 0,$$

since $a^{-}e \in f(R)^{\perp}$ and $|n|e \in f(R)$. Hence $(ne + a^{+}e) \wedge a^{-}e = 0$ and

$$(ne + ae)^{+} = (ne + a^{+}e - a^{-}e)^{+} = ne + a^{+}e \in A.$$

If $a^{-} \in U_f$, then $a^{+} \in f(R)^{\perp}$. Since $-ne + a^{-}e \geq -ne + a^{-} \geq 0$,

$$0 \leq (-ne + a^{-}e) \wedge a^{+}e \leq (|n|e + a^{-}e) \wedge a^{+}e \leq (|n|e \wedge a^{+}e) + (a^{-}e \wedge a^{+}e) = 0,$$

so $(-ne + a^{-}e) \wedge a^{+}e = 0$ and

$$(ne + ae)^{+} = (ne + a^{+}e - a^{-}e)^{+} = (a^{+}e - (-ne + a^{-}e))^{+} = a^{+}e \in A.$$

Thus in any case, $(ne + ae)^{+} \in A$, and hence A is an ℓ-subring of R.

Define $\varphi : R_1 \to A$ by $\varphi(\overline{(n, a)}) = ne + ae$. It is left to the reader to check that φ is a well-defined isomorphism between two additive groups of R_1 and A (Exercise 25). Now we define an element $\overline{(n, a)} \geq 0$ if $\varphi(\overline{(n, a)}) = ne + ae \geq 0$ in A. Since A is an ℓ-ring, R_1 becomes an ℓ-group with respect to its addition, and φ becomes an ℓ-isomorphism of two additive ℓ-groups.

We show that the product of two positive elements of R_1 is also positive. Suppose that $\overline{(n, a)}, \overline{(m, b)} \geq 0$. Then $ne + ae, me + be \geq 0$ in A, and hence e is an f-superunit implies that $ne + ea, me + eb \geq 0$. Thus

$$(ne + ea)(me + be) = e(nme + mae + nbe + abe) \geq 0,$$

implies that

$$\varphi(\overline{(n, a)(m, b)}) = \varphi(\overline{(nm, ma + nb + ab)}) = nme + (ma + nb + ab)e \geq 0.$$

Hence the product of $\overline{(n, a)}$ and $\overline{(m, b)}$ is positive in R_1. Therefore R_1 is an ℓ-unital ℓ-ring.

We may directly check that an element $\overline{(n,a)} \in R_1$ is in $f(R_1)$ if and only if $ne + ae \in f(R)$. Suppose that $\overline{(n,a)} \in R_1$ is an f-element, and $x, y \in R$ with $x \wedge y = 0$. Then $xe \wedge ye = 0$ implies that $\overline{(0,x)} \wedge \overline{(0,y)} = 0 \in R_1$. Then

$$\overline{(0,x)}\,\overline{(n,a)} \wedge \overline{(0,y)} = 0 \Rightarrow \overline{(0, nx + xa)} \wedge \overline{(0,y)} = 0$$
$$\Rightarrow (nx + xa)e \wedge ye = 0$$
$$\Rightarrow x(ne + ae) \wedge y = 0 \ (e \text{ is an } f\text{-superunit}).$$

Now we show that $(ne + ae)x \wedge y = 0$. From $x \wedge y = 0$ and e is an f-element, we have $ex \wedge y = 0$, so $\overline{(0,ex)} \wedge \overline{(0,y)} = 0$ implies

$$\overline{(n,a)}\,\overline{(0,ex)} \wedge \overline{(0,y)} = 0 \Rightarrow \overline{(0, nex + aex)} \wedge \overline{(0,y)} = 0$$
$$\Rightarrow (nex + aex)e \wedge ye = 0$$
$$\Rightarrow (nex + aex) \wedge y = 0 \ (e \text{ is an } f\text{-superunit}).$$
$$\Rightarrow (ne + ae)x \wedge y = 0.$$

Thus $ne + ae \in f(R)$. Similarly to show that if $ne + ae \in f(R)$, then $\overline{(n,a)} \in f(R_1)$. We leave the verification of it to the reader (Exercise 26).

Now $a \to \overline{(0,a)}$ is a one-to-one ring homomorphism with $a \geq 0$ in R if and only if $\overline{(0,a)} \geq 0$ in R_1. Thus we may consider R as an ℓ-subring of R_1 and write $R_1 = \mathbb{Z} + R$. Then $f(R) \subseteq f(R_1)$ and $f(R_1) = \{n + a \mid a \in f(R)\}$ is totally ordered.

Suppose that R is ℓ-simple. For an ℓ-ideal I of R_1. That $I \cap R$ is an ℓ-ideal of R implies that $I \cap R = R$ or $I \cap R = \{0\}$. If $I \cap R = R$, then $e \in I$ and $1 \leq e$ implies that $1 \in I$, and hence $I = R_1$. If $I \cap R = \{0\}$, then $IR = \{0\}$, and hence $I = \{0\}$ since $e \in R$ is an f-superunit of R. Therefore R_1 is ℓ-simple.

Finally we suppose that R has squares positive. For $n + a \in R_1 = \mathbb{Z} + R$, first assume that $a \in U_f$. For any $0 \leq m \in \mathbb{Z}$, $0 \leq (me \pm a)^2$ and Lemma 4.9 yield

$$me|a| + m|a|e = |mea + mae| \leq m^2 e^2 + a^2 \leq m|a|e + a^2,$$

since $me \leq |a|$. Hence $m|a| \leq me|a| \leq a^2$. Therefore, for any $n \in \mathbb{Z}$, $(n + a)^2 = n^2 + 2na + a^2 \geq 0$ in R_1. Now suppose that $a = x + y$ with $x \in f(R)$ and $y \in f(R)^{\perp}$. Then

$$(n + a)^2 = n^2 + 2na + a^2 = n^2 + 2nx + 2ny + x^2 + xy + yx + y^2.$$

Since $n + x \in f(R_1)$, $n^2 + 2nx + x^2 = (n + x)^2 \geq 0$. For any $0 \leq d \in f(R)$,

$$|y|d \leq |y|d + d|y| = |yd + dy| \leq y^2 + d^2$$

implies that

$$|y|d = |y|d \wedge (y^2 + d^2)$$
$$\leq (|y|d \wedge y^2) + (|y|d \wedge d^2)$$
$$= (|y|d \wedge y^2) + (|y| \wedge d)d$$
$$= |y|d \wedge y^2 \quad (|y| \wedge d = 0)$$
$$\leq y^2.$$

Similarly $d|y| \leq y^2$. Hence we have

$$-6ny \leq (6ny)e \leq y^2, -3xy \leq y^2, \text{ and } -3yx \leq y^2,$$

so $-(6ny + 3xy + 3yx) \leq 3y^2$. Therefore $-(2ny + xy + yx) \leq y^2$. It then follows that

$$(n + a)^2 = (n^2 + 2nx + x^2) + (2ny + xy + yx + y^2) \geq 0.$$

This completes the proof that R_1 is squares positive. $\qquad\square$

Lemma 4.10.

(1) For an ℓ-ring R with ℓ-$N(R) = \{0\}$, if $a \in f(R)$ and $a^2 = 0$, then $a = 0$.

(2) Let R be an ℓ-unital ℓ-reduced ℓ-ring. For an ℓ-prime ℓ-ideal P, $f(R/P)$ is a totally ordered domain.

(3) For a totally ordered ring, any two $(right, left)$ ℓ-ideals are comparable. For an f-ring and an ℓ-prime ℓ-ideal P, if I, J are ℓ-ideals containing P, then $I \subseteq J$ or $J \subseteq I$.

(4) For an ℓ-ring R with an f-superunit, if $f(R)$ is totally ordered, then R has a unique maximal $(right, left)$ ℓ-ideal.

Proof. (1) Since $f(R)$ is an f-ring, we have $|a|^2 = |a^2| = 0$, so we may assume that $a \geq 0$. For $x \in R^+$, $(ax - xa)^+ \wedge (ax - xa)^- = 0$ implies that

$$axa = axa \wedge axa = (ax - xa)^+ a \wedge a(ax - xa)^- = 0$$

since a is an f-element of R. Thus $aR^+a = \{0\}$, and hence $a = 0$ since ℓ-$N(R) = \{0\}$.

(2) Let $\overline{R} = R/P$, and for each $x \in R$ write $\overline{x} = x + P \in R/P$. Suppose that $\overline{a} \wedge \overline{b} = 0$ for some $\overline{a}, \overline{b} \in f(\overline{R})$. Then $a \wedge b = c \in P$, and hence $(a - c) \wedge (b - c) = 0$. It follows that

$$((a - c) \wedge 1) \wedge ((b - c) \wedge 1) = 0.$$

However $(a-c)\wedge 1, (b-c)\wedge 1 \in f(R)$ implies that $((b-c)\wedge 1)((a-c)\wedge 1) = 0$, so

$$[((a - c) \wedge 1)x((b - c) \wedge 1)]^2 = 0 \text{ for each } x \in R^+.$$

Then that R is ℓ-reduced implies $((a - c) \wedge 1)x((b - c) \wedge 1) = 0$ for each $x \in R^+$. Therefore in $\overline{R}$, we have $(\overline{a} \wedge \overline{1})\overline{x}(\overline{b} \wedge \overline{1}) = 0$ for each $\overline{x} \in (\overline{R})^+$, and hence $\overline{a} \wedge \overline{1} = 0$ or $\overline{b} \wedge \overline{1} = 0$ since $\overline{R}$ is ℓ-prime. Consequently $\overline{a} = 0$ or $\overline{b} = 0$ since $\overline{a}, \overline{b} \in f(\overline{R})$. Thus $f(\overline{R})$ is totally ordered (Exercise 1.12). By (1) $f(\overline{R})$ contains no nonzero nilpotent element, then $f(\overline{R})$ is totally ordered implies that it is a domain.

(3) Suppose that R is totally ordered and I, J are ℓ-ideals. If $I \nsubseteq J$, then there exists $0 < a \in I \setminus J$, so for any $0 \le b \in J$, $b \le a$. Thus $b \in I$ for each $b \in J^+$, and hence $J \subseteq I$.

For an f-ring R, by Theorem 1.27, R/P is totally ordered, and hence $I/P \subseteq J/P$ or $J/P \subseteq I/P$. Thus $I \subseteq J$ or $J \subseteq I$.

(4) Let e be an f-superunit of R. By Zorn's Lemma, R has a maximal ℓ-ideal. Suppose that M, N are maximal ℓ-ideals. If $M + N = R = U_f \cup (f(R) \oplus f(R)^\perp)$, then $e = x + y$ for some $0 \le x \in M$ and $0 \le y \in N$, and hence $x, y \in f(R) \oplus f(R)^\perp$ since $M \cap U_f = N \cap U_f = \emptyset$. Suppose that $x = a + b, y = c + d$ with $a, c \in f(R)$, $b, d \in f(R)^\perp$. Then $e = a + c$ and $b + d = 0$. Since $M \cap f(R)$ and $N \cap f(R)$ are ℓ-deals of $f(R)$, by (3) they are comparable. Thus $e \in M \cap f(R)$ or $N \cap f(R)$, and hence $M = R$ or $N = R$, which is a contradiction. Consequently $M + N \ne R$, and hence $M = M + N = N$.

Similar argument shows that there exists a unique maximal left ℓ-ideal and a unique maximal right ℓ-ideal. $\qquad\square$

For a general ℓ-ring, Lemma 4.10(3) is not true. For instance, let $R = \mathbb{R}[x, y]$ be the polynomial ℓ-ring in two variables over $\mathbb{R}$ with the coordinatewise order. Then R is a domain and xR, yR are ℓ-ideals that are not comparable.

For an ℓ-ring R, $f(R)$ is called *dense* if for any nonzero ℓ-ideal I, $I \cap f(R) \ne \{0\}$.

Theorem 4.11. *Let R be an ℓ-reduced Archimedean ℓ-ring such that $f(R)$ is dense. Then R is ℓ-isomorphic to a subdirect product of ℓ-simple ℓ-rings with f-superunits.*

Proof. By Theorem 1.30, R is a subdirect product of ℓ-domains, and hence R contains ℓ-ideals I_α such that $\cap I_\alpha = \{0\}$ and each $R_\alpha = R/I_\alpha$ is an ℓ-domain. Take $0 < a \in f(R)$. Since R is Archimedean, there exists positive integer n such that $na^2 \not\leq a$, and hence either $na^2 \geq a$ or na^2 and a are not comparable. Now $0 < a \in f(R)$ implies that $a_\alpha = a + I_\alpha \in f(R_\alpha)$ for each α (Exercise 27). Since $f(R_\alpha)$ is totally ordered, there is at least one α such that $na_\alpha^2 > a_\alpha$. Then that R_α is an ℓ-domain implies that na_α is an f-superunit of R_α. Define $\Gamma = \{\alpha \mid R_\alpha \text{ has an } f\text{-superunit}\}$. The above argument shows $\Gamma \neq \emptyset$. Let $I = \cap I_\alpha$, $\alpha \in \Gamma$. We show that $I = \{0\}$ by showing $I \cap f(R) = \{0\}$. Let $0 \leq e \in I \cap f(R)$ and $b = (ke^2 - e)^+$, where k is a positive integer. For each $\alpha \in \Gamma$, $e \in I$ implies that $e_\alpha = 0$, so $b_\alpha = 0$. If $b_\beta \neq 0$ for some $\beta \notin \Gamma$, then $b_\beta = (ke_\beta^2 - e_\beta)^+ > 0$ and $ke_\beta^2 - e_\beta \in f(R_\beta)$, which is totally ordered, implies that $ke_\beta^2 - e_\beta > 0$. It follows that ke_β is an f-superunit of R_β, which contradicts with $\beta \notin \Gamma$. Therefore $b_\alpha = 0$ for all α, and hence $b = (ke^2 - e)^+ = 0$. Consequently $ke^2 \leq e$ for all positive integer k and $e^2 = 0$ since R is Archimedean, so $e = 0$ and $I \cap f(R) = \{0\}$. Hence $I = \{0\}$.

By Lemma 4.10(4), for each $\alpha \in \Gamma$, R_α contains a unique maximal ℓ-ideal denoted by M_α/I_α, where M_α is a maximal ℓ-ideal of R. Now R/M_α is an ℓ-simple ℓ-ring with f-superunits. Suppose that $M = \cap M_\alpha, \alpha \in \Gamma$. We claim that $M = \{0\}$. Let $0 \leq a \in M \cap f(R)$. Suppose that e_α is an f-superunit of R_α for each $\alpha \in \Gamma$. We have $ne_\alpha a_\alpha \leq e_\alpha$ for any positive integer n since $ne_\alpha a_\alpha \in M_\alpha$ and $ne_\alpha a_\alpha$, e_α are comparable. Hence $ne_\alpha a_\alpha^2 \leq e_\alpha a_\alpha$, and hence $na_\alpha^2 \leq a_\alpha$ for each $\alpha \in \Gamma$. Therefore $na^2 \leq a$ for all positive integer n. Then R is Archimedean implies that $a = 0$, and hence $M \cap f(R) = \{0\}$. Thus $M = \{0\}$ and R is a subdirect product of R/M_α, $\alpha \in \Gamma$. $\qquad\square$

The condition that $f(R)$ is dense in Theorem 4.11 cannot be omitted. For instance, in the polynomial ℓ-ring $R = \mathbb{R}[x]$ with the coordinatewise order, $f(R)$ is not dense and xR is the unique maximal ℓ-ideal of R. The following is a direct consequence of Theorem 4.11.

Corollary 4.4. *Let R be an ℓ-reduced Archimedean ℓ-ring such that $f(R)$ is dense. If R satisfies descending chain condition on ℓ-ideals, then R is ℓ-isomorphic to a finite direct sum of ℓ-simple ℓ-rings with f-superunits.*

Proof. By Theorem 4.11, there are maximal ℓ-ideals M_α such that $\cap_\alpha M_\alpha = \{0\}$. Since R satisfies descending chain condition on ℓ-ideals, similar to Exercise 2.24, there are finitely many maximal ℓ-ideals $M_1, \cdots, M_k$ such that $M_1 \cap \cdots \cap M_k = \{0\}$. By the same argument used in the proof of Theorem 2.8, R is ℓ-isomorphic to $R/M_1 \oplus \cdots \oplus R/M_k$. $\qquad\square$

For an ℓ-reduced f-ring, Corollary 4.4 is true without assuming it is Archimedean, the reader is referred to [Birkhoff and Pierce (1956)] for more details.

Let R be an ℓ-ring. An ℓ-ideal I of R is called an ℓ-*annihilator* ℓ-*ideal* if $I = \ell(X)$ and $I = r(Y)$ for some $X, Y \subseteq R$, where

$$\ell(X) = \{r \in R \mid |r||x| = 0, \forall x \in X\},$$

and

$$r(Y) = \{r \in R \mid |y||r| = 0, \forall y \in Y\}.$$

Lemma 4.11. *Suppose that R is an ℓ-ring with ℓ-$N(R) = \{0\}$. If R satisfies ascending chain condition on ℓ-annihilator ℓ-ideals, then there are a finite number of ℓ-prime ℓ-ideals with zero intersection.*

Proof. We first note that for an ℓ-ideal I, $\ell(I)$ and $r(I)$ are ℓ-ideals, and since ℓ-$N(R) = \{0\}$, $\ell(I) = r(I)$ (Exercise 28). We show that each ℓ-annihilator ℓ-ideal contains the product of a finite number of ℓ-prime ℓ-ideals. Suppose not, then, by ascending chain condition on ℓ-annihilator ℓ-ideals, there exists a maximal ℓ-annihilator ℓ-ideal I such that I does not contain any product of a finite number of ℓ-prime ℓ-ideals. In particular, I is not ℓ-prime, and hence there are ℓ-ideals J and K such that $JK \subseteq I$ with $J \nsubseteq I$ and $K \nsubseteq I$. Let $A = I + J$ and $B = I + K$ and let $B' = r(\ell(I)A)$ and $A' = \ell(B'r(I))$. Then A' and B' are ℓ-annihilator ℓ-ideal (Exercise 29) properly containing I, and hence A' and B' contain the product of ℓ-prime ℓ-ideals. Since $A'B'r(I) = \{0\}$, $A'B' \subseteq \ell(r(I)) = I$, so I contains the product of ℓ-prime ℓ-ideals, which is a contradiction. Then each ℓ-annihilator ℓ-ideal contains a product of a finite number of ℓ-prime ℓ-ideals. Since $\ell(R) = \{0\}$ is an ℓ-annihilator ℓ-ideal, there exist ℓ-prime ℓ-ideals $P_1 \cdots P_k = \{0\}$. Then $P_1 \cap \cdots \cap P_k = \{0\}$ since ℓ-$N(R) = \{0\}$. $\square$

Corollary 4.5. *Let R be an ℓ-unital ℓ-reduced Archimedean ℓ-ring such that $f(R)$ is dense. If R satisfies ascending chain condition on ℓ-annihilator ℓ-ideals, then R is ℓ-isomorphic to a finite direct sum of ℓ-unital ℓ-simple ℓ-rings.*

Proof. By Theorem 4.11, there are maximal ℓ-ideals M_α such that $\cap_\alpha M_\alpha = \{0\}$. By Lemma 4.11, there are ℓ-prime ℓ-ideals $P_1, \cdots, P_k$ such that $P_1 \cdots P_k = \{0\}$. Each P_i is contained in a maximal ℓ-ideal M_i, $i = 1, \cdots, k$. Let M be a maximal ℓ-ideal of R. Then $P_j \subseteq M$ for some j. By Lemma 4.10(2) $f(R/P_j)$ is totally ordered, and hence by Lemma

4.10(4), R/P_j contains a unique maximal ℓ-ideal, so $M/P_j = M_j/P_j$. Thus we have $M = M_j$, that is, R contains only a finite number of maximal ℓ-ideals. Thus there exist maximal ℓ-ideals $M_1, \cdots, M_n$ such that $M_1 \cap \cdots \cap M_n = \{0\}$, so R is ℓ-isomorphic to a direct sum of ℓ-unital ℓ-simple ℓ-rings $R/M_1, \cdots, R/M_n$. □

Similarly to Corollary 4.4, for an f-ring, Archimedean condition is not necessary in Corollary 4.5 [Anderson (1962)].

An open question on ℓ-rings with squares positive posted by J. Diem asks whether or not an ℓ-prime ℓ-ring with squares positive is an ℓ-domain. The question seems simple, however it is still unsolved. In the following we present some conditions that ensure the assertion true. It is easy to verify that if $R \neq \{0\}$ is an ℓ-ring, then R is an ℓ-domain if and only if R is ℓ-prime and ℓ-reduced (Exercise 30).

Theorem 4.12. *Suppose that R is an ℓ-prime ℓ-ring with squares positive.*

(1) If R is Archimedean, then R is an ℓ-domain.
(2) If disjoint elements of R commute, then R is an ℓ-domain.
(3) If $f(R) \neq \{0\}$, then R is a domain.
(4) If R contains a nonzero idempotent element that is in the center of R, then R is an ℓ-domain.

Proof.   (1) Suppose that $x \in R^+$ with $x^2 = 0$. By Lemma 3.3, $xR = \{0\}$, so $x = 0$ by Lemma 1.26(2). Therefore R is an ℓ-domain.

(2) Suppose that $a \in R^+$ with $a^2 = 0$. We show that for any $z \in R^+$, $aza = 0$, so $aR^+a = \{0\}$. Then R is ℓ-prime implies that $a = 0$.

If $(az - za)^+ = 0$, then $az \leq za$, and hence $aza \leq za^2 = 0$ implies that $aza = 0$. If $(az - za)^- = 0$, then $az \geq za$ implies that $a^2z \geq aza$, so $aza = 0$. In the following we assume that $(az-za)^+ \neq 0$ and $(az-za)^- \neq 0$. Since $(az - za)^+ \wedge (az - za)^- = 0$,

$$(az - za)^+(az - za)^- = (az - za)^-(az - za)^+ \leq (za)(az) = 0,$$

so $(az - za)^+(az - za)^- = 0$. Thus for any $y \in R^+$,

$$[(az - za)^- y(az - za)^+]^2 = 0.$$

Then R has squares positive and $a^2 = 0$ imply that $a(az-za)^- y(az-za)^+ = 0$. Since R is ℓ-prime and $(az - za)^+ \neq 0$, $a(az - za)^- = 0$. We also have $a(az - za)^+ \leq a^2z = 0$. Therefore

$$aza = -a(az - za) = a(az - za)^- - a(az - za)^+ = 0.$$

Therefore, $aza = 0$ for any $z \in R^+$.

(3) By Lemma 4.10(1), for any $0 < z \in f(R)$, $z^2 \neq 0$. Let $0 < e \in f(R)$ and $x \in R^+$ with $x^2 = 0$. We show that $x = 0$. Suppose that $x \neq 0$. We derive a contradiction. From $(e - x)^2 \geq 0$, we have $ex + xe \leq e^2$. It follows that $xex \leq xe^2 \leq e^3$, and hence $xex \in f(R)$. Thus $xex = 0$ by Lemma 4.10 since $(xex)^2 = 0$, so $(xe)^2 = 0$ and $xe \in f(R)$ further imply that $xe = 0$. Similarly $ex = 0$.

For any $y, z \in R^+$, $(eyx)^2 = (xze)^2 = 0$ and R has squares positive, and hence $xze^2yx = 0$. Fix z first, since R is ℓ-prime and $x \neq 0$, $xze^2 = 0$. Now since $z \in R^+$ is arbitrary and $x \neq 0$, we must have $e^2 = 0$, which is a contradiction. Therefore for any $x \in R^+$, $x^2 = 0$ implies $x = 0$, and hence R is ℓ-reduced. Thus R is an ℓ-domain.

Take $w \in R$ with $w^2 = 0$ and $0 < e \in f(R)$. Then $(e \pm w)^2 \geq 0$ implies that $|we + ew| \leq e^2$. By Lemma 4.9,

$$|w|e \leq (|w|e + e|w|) = |we + ew| \leq e^2,$$

and hence $|w|e = |w|e \wedge e^2 = (|w| \wedge e)e$, so $(|w| - |w| \wedge e)e = 0$. It follows from that R is an ℓ-domain that $|w| = |w| \wedge e \leq e$. Hence $w \in f(R)$. Consequently $|w|^2 = |w^2| = 0$ since $f(R)$ is an f-ring. Therefore $|w| = 0$ and $w = 0$, that is, R is reduced. Finally suppose that $a, b \in R$ with $ab = 0$. Then $a^2b^2 = 0$ implies that $a^2 = 0$ or $b^2 = 0$, and hence $a = 0$ or $b = 0$. Therefore R is a domain.

(4) Let $e = e^2 \neq 0$ be in the center of R. We first show that Re is an ℓ-subring of R. Suppose $x \in R^+$ and $xe = 0$. Since $xRe = (xe)R = \{0\}$ and R is ℓ-prime, $x = 0$. For $a, b \in R$,

$$(ae \vee be)e \geq (ae \vee be) \text{ and } [(ae \vee be)e - (ae \vee be)]e = 0.$$

By the above argument, $(ae \vee be)e - (ae \vee be) = 0$, and hence $ae \vee be \in Re$. Similarly $ae \wedge be \in Re$. Thus Re is an ℓ-ring with squares positive. We leave it to the reader to check that Re is also an ℓ-prime ℓ-ring (Exercise 31). Since e is the identity element in Re, by (3) Re is a domain. Let $x \in R^+$ with $x^2 = 0$. Then $(xe)^2 = 0$ implies that $xe = 0$, and hence $x = 0$ by previous argument. Therefore R is ℓ-reduced, so R is an ℓ-domain. $\square$

The ℓ-ring in Example 1.5 is a commutative ℓ-ring with squares positive. It contains a nonzero idempotent and contains no nonzero f-element.

Consider polynomial $p(x) = x^2$. An ℓ-ring R with squares positive is an ℓ-ring that satisfies $p(a) \geq 0$ for all $a \in R$. We may just call that R is an ℓ-ring with *polynomial constraint* $p(x) \geq 0$ or $p(x)^- = 0$. We may also use polynomials with two variables. For instance, an ℓ-ring with squares positive is an ℓ-ring with the polynomial constraint

$$f(x,y) = -(xy + yx) + x^2 + y^2 \geq 0.$$

That means for any $a, b \in R$, $f(a,b) \geq 0$.

ℓ-rings and ℓ-algebras with polynomial constraints were first systematically studied by S. Steinberg. Because of introductory nature of the book, we are not going to present general topic on ℓ-rings with polynomial constraints, and the reader is refereed to [Steinberg (2010)] for more detail. The interested reader may begin by reading [Steinberg (1983)] first.

In the following, we present a few examples to show some ideas of generalizing results on ℓ-rings with squares positive to ℓ-rings with more general polynomial constraints.

The key ingredient in the proof of Theorem 4.12(3) is stated in the following result.

Lemma 4.12. *Let R be an ℓ-prime ℓ-ring with $f(R) \neq \{0\}$. If there exists $0 < e \in f(R)$ such that for any $a \in R^+$ with $a^2 = 0$, $ae, ea \in f(R)$, then R is an ℓ-domain.*

Proof. We just need to show that R is ℓ-reduced. Suppose that $a \in R^+$ with $a^2 = 0$. Then $(a \wedge ae)^2 = 0$ and $a \wedge ae \in f(R)$ imply that $a \wedge ae = 0$ by Lemma 4.10(1). Since $e \in f(R)$, $ae \wedge ae = 0$, so $ae = 0$. Similarly $ea = 0$. Take $x \in R^+$, then $(exa)^2 = (exa)(exa) = 0$, and hence by previous argument, we have $e^2xa = 0$. Therefore $e^2R^+a = \{0\}$. Since R is ℓ-prime and $e^2 \neq 0$ by Theorem 4.10(1), we must have $a = 0$. Hence R is ℓ-reduced, so R is an ℓ-domain. $\square$

Let's look at some examples.

Example 4.1.

(1) Let R be an ℓ-prime ℓ-ring with $f(R) \neq \{0\}$. If R satisfies the polynomial constraint

$$f(x,y) = -(xy + yx) + (x^{2n} + y^{2n}) \geq 0, \ n \geq 1,$$

then R is an ℓ-domain.

In fact, take $0 < e \in f(R)$, and let $a \in R^+$ and $a^2 = 0$. Then $f(a,e) \geq 0$ implies that $ae + ea \leq e^{2n}$. Thus $ae, ea \in f(R)$, so R is an ℓ-domain by Lemma 4.12. There are ℓ-rings that satisfy the above polynomial constraint (Exercise 81).

(2) Let R be an ℓ-prime ℓ-ring with $0 < e \in f(R)$ in the center of R. If R satisfies $x^{2n} \geq 0$ for some fixed positive integer n, then R is an ℓ-domain.

Let $a \in R^+$, $a^2 = 0$. Since $ea = ae$, we have

$$(e - a)^{2n} = e^{2n} - (2n)e^{2n-1}a \geq 0,$$

so $(2n)e^{2n-1}a \leq e^{2n} \in f(R)$. Thus $(2n)e^{2n-1}a \in f(R)$, so by Lemma 4.12, R is an ℓ-domain.

(3) Let R be an ℓ-prime ℓ-ring with $f(R) \neq \{0\}$. If R satisfies $x^3 \geq 0$ or $x^3 \leq 0$, that is, for any $a \in R$, $a^3 \geq 0$ or $a^3 \leq 0$, then R is an ℓ-domain. Take $0 < e \in f(R)$. Let $a \in R^+$ with $a^2 = 0$. First we claim that $(e - a)^3 \leq 0$ is not possible. If $(e - a)^3 \leq 0$, then

$$(e - a)^3 = e^3 - ae^2 - eae - e^2a + aea \leq 0$$

implies that $e^3 + aea \leq ae^2 + eae + e^2a$. Since $(a \wedge e)^2 = 0$ and $a \wedge e \in f(R)$, $a \wedge e = 0$ by Lemma 4.10, so

$$ae^2 \wedge e^3 = 0, \ eae \wedge e^3 = 0, \ e^2a \wedge e^3 = 0,$$

we must have $e^3 = 0$, which is a contradiction.

Thus we must have $(e - a)^3 \geq 0$. Then

$$(e - a)^3 = e^3 - ae^2 - eae - e^2a + aea \geq 0$$

implies that $ae^2 + eae + e^2a \leq e^3 + aea$. Similarly

$$ae^2 \wedge e^3 = 0, \ eae \wedge e^3 = 0, \ ea \wedge e^3 = 0,$$

so $ae^2 + eae + e^2a \leq e^3 + aea$ implies $ae^2 + eae + e^2a \leq aea$. Thus $eaea \leq aea^2 = 0$ and $aeae \leq a^2ea = 0$. Hence $ae^2 + eae + e^2a \leq e^3 + aea$ implies that $ae^3, e^3a \leq e^4$. Therefore e^3a and ae^3 are f-elements, so R must be an ℓ-domain by Lemma 4.12.

Let R be a unital ℓ-ring with squares positive. An important property of R is that the inverse a^{-1} of a positive invertible element a is also positive since $a^{-1} = a(a^{-1})^2$. Thus each positive invertible element of R is a d-element by Theorem 1.20(2). As a direct consequence of this fact, in a lattice-ordered division ring R with squares positive, each positive element is a d-element, that is, R is a d-ring, and hence R is a totally ordered division ring.

For reader's convenience, we present a direct proof for the fact that a lattice-ordered division ring with squares positive must be totally ordered. Let $x \in R$ and $a = x^+ + 1$, $b = x^- + 1$. Since $a^{-1} > 0$ and $b^{-1} > 0$,

$$0 \leq a^{-1}(ax^+b \wedge ax^-b)b^{-1} \leq a^{-1}ax^+bb^{-1} \wedge a^{-1}ax^-bb^{-1} = x^+ \wedge x^- = 0,$$

so $a^{-1}(ax^+b \wedge ax^-b)b^{-1} = 0$ and $ax^+b \wedge ax^-b = 0$. It follows that

$$x^+x^- = x^+x^- \wedge x^+x^- \leq ax^+b \wedge ax^-b = 0,$$

and hence $x^+x^- = 0$. Therefore $x^+ = 0$ or $x^- = 0$, that is, R is totally ordered.

In 1956, G. Birkhoff and R. Pierce proved that an ℓ-field with squares positive must be totally ordered [Birkhoff and Pierce (1956)]. Their elementary proof didn't use the commutative condition for multiplication. Therefore, as pointed out by S. Steinberg in 1970, G. Birkhoff and R. Pierce have proved that a lattice-ordered division ring with squares positive is totally ordered [Steinberg (1970)], although they didn't precisely state the result. S. Steinberg also generalized this result to ℓ-rings satisfying minimal condition on right (left) ideals as stated in Theorem 4.13. The reader is referred to [Steinberg (1970)] for more details.

Theorem 4.13. *Let R be an ℓ-ring with squares positive and an identity element. If R has the minimal condition on right ideals, then R is an f-ring.*

As a direct consequence of Theorem 4.13, for an ℓ-prime ℓ-ring R with squares positive and an identity element, if R has the minimal condition on right ideals, then R is totally ordered since R is a domain by Theorem 4.12.

The following result gives the conditions for ℓ-rings with zero ℓ-radical to become an f-ring.

Theorem 4.14. *Let R be a nonzero ℓ-ring with ℓ-$N(R) = \{0\}$. Then R is an f-ring if and only if $f(R) \neq \{0\}$ and $f(R)^{\perp} = \{0\}$.*

Proof. We just need to show if $f(R) \neq \{0\}$ and $f(R)^{\perp} = \{0\}$, then R is an f-ring. We first show that R is ℓ-reduced. Suppose that $x^2 = 0$ for some $x \in R^+$. Take $0 < e \in f(R)$. Then $(x \wedge e)^2 = 0$ implies $x \wedge e = 0$ by Lemma 4.10. Thus $x \in f(R)^{\perp}$, so $x = 0$.

Next we claim that R is an almost f-ring. Suppose first that $a \wedge b = 0$ with $b \in f(R)$. Since b is an f-element, $ab \wedge b = 0$. Thus for any $0 < e \in f(R)$, $(ab \wedge e) \wedge b = 0$ implies that $b(ab \wedge e) = 0$, and hence

$$0 \leq (ab \wedge e)^2 \leq ab(ab \wedge e) = 0.$$

Thus $(ab \wedge e)^2 = 0$. It follows from Lemma 4.10 that $ab \wedge e = 0$, so $ab \in f(R)^{\perp}$. Therefore $ab = 0$.

Now consider $x \wedge y = 0$ for some $x, y \in R$. Take $0 < e \in f(R)$. Let $x_1 = x - x \wedge e$ and $e_1 = e - x \wedge e$. Then $y \wedge (x \wedge e) = 0$ and previous argument implies that $y(x \wedge e) = 0$, so $ye = ye_1$. Similarly we have $ye(x \wedge e) = 0$, $x_1 e_1 = 0$ and $e_1 x_1 = 0$ (Exercise 32). Thus

$$yex = yex_1 = ye_1 x_1 = 0,$$

and hence $(xye)^2 = 0$ implies that $xye = 0$ since R is ℓ-reduced. It follows that $(xy \wedge e)^2 = 0$, so $xy \wedge e = 0$, that is, $xy \in f(R)^{\perp}$. Therefore $xy = 0$ and R is an almost f-ring. By Theorem 1.28, R is an f-ring because of $\ell\text{-}N(R) = \{0\}$. $\qquad\square$

For an ℓ-ring R. An element $a \in R^+$ is called a *weak unit* if for any $b \in R^+$, $a \wedge b = 0 \Rightarrow b = 0$. If a nonzero ℓ-ring R with $\ell\text{-}N(R) = \{0\}$ contains a weak unit which is an f-element, then R is an f-ring by Theorem 4.14. However, if $\ell\text{-}N(R) \neq \{0\}$, this is not true as shown in Exercise 1.47.

Corollary 4.6. *Suppose that R is an Archimedean ℓ-ring. If R contains a weak unit $e \in f(R)$ with $\ell(e) = \{0\}$ or $r(e) = \{0\}$, then R is an f-ring.*

Proof. Let $x \in R^+$ with $x^2 = 0$. Then $(x \wedge e), e \in f(R)$ implies that $(e - n(x \wedge e))^2 \geq 0$ for any positive integer n. Since $(x \wedge e)^2 = 0$,

$$n(x \wedge e)e \leq (n(x \wedge e))^2 + e^2 = e^2,$$

and hence $(x \wedge e)e = 0$ since R is Archimedean. It follows from $\ell(e) = \{0\}$ that $x \wedge e = 0$, so $x = 0$ since e is a weak unit. Therefore R is ℓ-reduced and by Theorem 4.14, R is an f-ring. The proof is similar if $r(e) = \{0\}$. $\square$

We study the relation between weak units and equation $x^+ x^- = 0$ in ℓ-rings.

Lemma 4.13. *Let R be an ℓ-ring.*

(1) Suppose there exists a weak unit $e \in f(R)$ with $\ell(e) = \{0\}$ or $r(e) = \{0\}$. If $a \in R^+$ and $(a \wedge e)^2 = 0$, then $a \leq e$.

(2) Suppose there exists an element $e \in f(R)$ with $\ell(e) = \{0\}$ (or $r(e) = \{0\}$), and for any $a \in R^+$, $(a \wedge e)^2 = 0$ implies $a \in f(R)$. Then for any $x \in R, y \in f(R)$, $x \wedge y = 0$ implies $xy = 0$ (or $yx = 0$).

Proof. (1) Since $a \wedge e \leq a \wedge 2e \leq 2(a \wedge e)$, we have $(a \wedge 2e)^2 = 0$, and since $a \wedge 2e, e \in f(R)$ that is an f-ring, we have

$$(a \wedge 2e)e \leq (a \wedge 2e)^2 + e^2 = e^2.$$

Thus $((a \wedge 2e) \vee e - e)e = 0$. It follows from $\ell(e) = \{0\}$ that $(a \wedge 2e) \vee e = e$, and hence

$$((a - e) \wedge e) \vee 0 = 0 \text{ and } ((a - e) \vee 0) \wedge e = 0.$$

Then e is a weak unit implies that $(a - e) \vee 0 = 0$. Therefore $a \le e$.

(2) From $x \wedge y = 0$, we have $(x \wedge e) \wedge y = 0$, so $(x \wedge e)y = y(x \wedge e) = 0$ since they both belong to $f(R)$. Let $x_1 = x - (x \wedge e)$ and $e_1 = e - (x \wedge e)$. Then

$$\begin{aligned} x_1 \wedge e_1 = 0 &\Rightarrow e_1 x_1 \wedge e_1 = 0 \\ &\Rightarrow (e_1 x_1 \wedge e) \wedge e_1 = 0 \\ &\Rightarrow (e_1 x_1 \wedge e)e_1 = 0 \\ &\Rightarrow (e_1 x_1 \wedge e)^2 = 0. \end{aligned}$$

By the assumption, $e_1 x_1 \in f(R)$, and hence $e_1 x_1 \wedge y = 0$ implies that $y e_1 x_1 = 0$ and $y e x_1 = 0$ (Exercise 33). Consequently $x_1 y e \in f(R)$ since $(x_1 y e)^2 = 0$. From that $x_1 \wedge e_1 = 0$ and $e, y \in f(R)$, we have $x_1 y e \wedge e_1 = 0$, so $x_1 y e e_1 = 0$ and $x y e^2 = 0$. Therefore $xy = 0$ since $\ell(e) = \{0\}$. Similarly if $r(e) = \{0\}$, then $yx = 0$. □

Theorem 4.15. *Suppose that R is an ℓ-ring and $e > 0$ is an f-element with $\ell(e) = \{0\}$ or $r(e) = \{0\}$. Then the following statements are equivalent.*

(1) For any $x \in R$, $x^+ ex^- = 0$.
(2) e is a weak unit.
(3) For any $a \in R^+$, if $(a \wedge e)^2 = 0$, then $a \in f(R)$.

Proof. (1) $\Rightarrow$ (2) Assume that $\ell(e) = \{0\}$. Suppose that $a \wedge e = 0$ for some $a \in R$. Let $x = a - e$. Then $x^+ = a$ and $x^- = e$. Thus we have $ae^2 = 0$, so $a = 0$ since $\ell(e) = \{0\}$. Therefore e is a weak unit. A similar argument works for $r(e) = \{0\}$.

(2) $\Rightarrow$ (3) By Lemma 4.13(1).

(3) $\Rightarrow$ (1) Suppose $x \wedge y = 0$ for some $x, y \in R$ and $r(e) = \{0\}$. Let $x_1 = x - (x \wedge e)$ and $e_1 = e - (x \wedge e)$. Then $x_1 \wedge e_1 = 0$. By Lemma 4.13(2) $e_1 x_1 = 0$, and hence $(x_1 e_1)^2 = 0$ and $x_1 e_1 \in f(R)$ by the assumption. Since $x_1 \wedge y = 0$, $x_1 e_1 \wedge y = 0$ and $x_1 e_1 y = 0$ by Lemma 4.13(2) again. Thus $x e y = x_1 e y = x_1 e_1 y = 0$. We leave it to the reader to verify that it is also true when $\ell(e) = \{0\}$. □

Corollary 4.7. *Let R be an ℓ-ring and $0 < e \in f(R)$ with $\ell(e) = r(e) = \{0\}$. The following statements are equivalent.*

(1) $x^+x^- = 0$ *for all* $x \in R$.

(2) e *is a weak unit.*

(3) *For any* $a \in R^+$, *if* $(a \wedge e)^2 = 0$, *then* $a \in f(R)$.

Proof. (1) $\Rightarrow$ (2) If $x \wedge e = 0$, then $xe = 0$ and $x = 0$ by $\ell(e) = \{0\}$. Hence e is a weak unit.

(2) $\Rightarrow$ (3) By Theorem 4.15.

(3) $\Rightarrow$ (1) Suppose that $x \wedge y = 0$. Then $xey = 0$ by Theorem 4.15, so $(eyx)^2 = 0$ implies that $eyx \in f(R)$ by the assumption. Since $f(R)$ has squares positive, $e^2(eyx) \le e^4 + (eyx)^2 = e^4$, and hence $e^3yx = e^3yx \wedge e^4 = e^3(yx \wedge e)$. It follows from $r(e) = \{0\}$ that $yx = yx \wedge e \le e$. Similarly $yex = 0$ implies that $xy \le e$. Suppose

$$x_1 = x - x \wedge e, \ e_1 = e - x \wedge e, \ y_1 = y - y \wedge e, \ e_2 = e - y \wedge e.$$

Then $x_1 \wedge e_1 = y_1 \wedge e_2 = 0$. By Lemma 4.13, $x_1e_1 = e_1x_1 = y_1e_2 = e_2y_1 = 0$. Since $(x \wedge e)y = 0$, $e_1xy = e_1x_1y = 0$ and $exy = (x \wedge e)xy \le x^2y$. From $xy \le e$ and $xey = 0$, we have $x^2y^2 = 0$, and hence $exy^2 \le x^2y^2 = 0$ and $xy^2 = 0$ by $r(e) = 0$. Since $x \wedge (y \wedge e) = 0$ implies that $x(y \wedge e) = 0$, we have

$$xy_1 = xy, \ xye_2 = xy_1e_2 = 0, \ \text{and} \ xye = xy(y \wedge e) \le xy^2 = 0.$$

Hence $xye = 0$ and $xy = 0$ by $\ell(e) = 0$. This completes the proof. $\square$

4.4 Quotient rings of lattice-ordered Ore domains

Let R be a lattice-ordered integral domain and Q be its quotient field. It is still an open question whether or not the lattice order on R can be extended to Q. As an example, consider polynomial ℓ-ring $\mathbb{R}[x]$ with the coordinate-wise order. We still don't know if this lattice order can be extended to the field of rational functions over $\mathbb{R}$. In this section, we provide some conditions to extend lattice orders on R to Q, and more generally we work on lattice-ordered Ore domains.

An arbitrary domain R is called a *left (right) Ore domain* if for given nonzero elements $x, a \in R$, $Rx \cap Ra \ne \{0\}$ ($xR \cap aR \ne \{0\}$). A *classical left (right) quotient ring* of a domain R is a ring Q which contains R as a subring such that every nonzero element of R is invertible in Q and

$$Q = \{a^{-1}x \mid x, a \in R, a \ne 0\} \ (Q = \{xa^{-1} \mid x, a \in R, a \ne 0\}).$$

Theorem 4.16. *For a domain R, R has a classical left (right) quotient ring if and only if R is a left (right) Ore domain.*

Proof. "$\Rightarrow$" Let $x, a \in R$ and $a \neq 0$, $x \neq 0$. Then $xa^{-1} \in Q$ implies that there exist $y, b \in R$ with $b \neq 0$ such that $b^{-1}y = xa^{-1}$. Then $y \neq 0$. It follows that $bx = ya \neq 0$, that is, $Rx \cap Ra \neq \{0\}$, so R is a left Ore domain.

"$\Leftarrow$" Suppose $S = \{a \in R \mid a \neq 0\}$ and consider $R \times S$. Define the relation on the set $R \times S$ by

$$(r, s) \sim (r', s') \quad \text{if } s_1 r = s_2 r', s_1 s = s_2 s' \text{ for some } s_1, s_2 \in S.$$

We show that $\sim$ is an equivalence relation on $R \times S$. Clearly $\sim$ is reflexive and symmetric. Suppose $(r, s) \sim (r', s')$ and $(r', s') \sim (r'', s'')$. Then there exist $s_1, s_2, s_3, s_4 \in S$ such that $s_1 r = s_2 r'$, $s_1 s = s_2 s'$ and $s_3 r' = s_4 r''$, $s_3 s' = s_4 s''$. Since R is a left Ore domain, there exist $z_1, z_2 \in S$ such that $z_1 s_2 = z_2 s_3$, and hence

$$z_1 s_1 r = z_1 s_2 r' = z_2 s_3 r' = z_2 s_4 r'' \quad \text{and} \quad z_1 s_1 s = z_1 s_2 s' = z_2 s_3 s' = z_2 s_4 s'',$$

and $z_1 s_1 \neq 0$, $z_2 s_4 \neq 0$. Hence $(r, s) \sim (r'', s'')$. Therefore $\sim$ is an equivalence relation on $R \times S$. Let r/s be the equivalence class of (r, s) and $Q = \{r/s \mid r \in R, s \in S\}$.

Define the addition and multiplication in Q as follows.

$$r/s + r'/s' = (s_1 r + s_2 r')/s_2 s', \quad \text{where } s_1 s = s_2 s' \text{ and } s_1, s_2 \in S.$$

$$(r/s)(r'/s') = (r_1 r')/(s_1 s), \quad \text{where } s_1 r = r_1 s' \text{ and } r_1 \in R, s_1 \in S.$$

We first notice that the definition of the addition is independent of choice of $s_1, s_2 \in S$. Suppose we also have $t_1 s = t_2 s'$ for some $t_1, t_2 \in S$. By left Ore condition and $s_1 s, t_1 s \in S$, we have $w, w' \in S$ such that $w(s_1 s) = w'(t_1 s)$, and hence $w(s_2 s') = w'(t_2 s')$. Thus we have $ws_1 = w't_1$ and $ws_2 = w't_2$, so $w(s_1 r + s_2 r') = w'(t_1 r + t_2 r')$. It follows that

$$(s_1 r + s_2 r', s_2 s') \sim (t_1 r + t_2 r', t_2 s'),$$

that is,

$$(s_1 r + s_2 r')/s_2 s' = (t_1 r + t_2 r')/t_2 s'.$$

Similarly the definition of the multiplication is independent of the choice of $r_1 \in R$ and $s_1 \in S$. In fact, if $t_1 r = u_1 s'$ for some $u_1 \in R$, $t_1 \in S$, then, by left Ore condition, $zs_1 = z't_1$ for some $z, z' \in S$, and hence $z'u_1 s' = z't_1 r = zs_1 r = zr_1 s'$. It follows that $z'u_1 = zr_1$ and $(r_1 r', s_1 s) \sim (u_1 r', t_1 s)$. Therefore $(r_1 r')/(s_1 s) = (u_1 r')/(t_1 s)$.

To see that the addition is well defined, suppose that $r/s = a/b$ and $r'/s' = c/d$. Then there exist $t_1, t_2, t_3, t_4 \in S$ such that $t_1 r = t_2 a, t_1 s = t_2 b$, and $t_3 r' = t_4 c, t_3 s' = t_4 d$. By left Ore condition, there exist $z, z' \in S$ such that $z(t_1 s) = z'(t_3 s')$, and hence $z t_2 b = z' t_4 d$. Hence we have

$$r/s + r'/s' = (zt_1 r + z't_3 r')/z't_3 s' = (zt_2 a + z't_4 c)/z't_4 d = a/b + c/d.$$

We leave it to the reader to verify that the multiplication is also well defined and Q becomes a ring with respect to the operations (Exercise 35). Clearly $0/s$ is the zero element in Q for any $s \in S$, and s/s is the identity element in Q for any $s \in S$. For any $0 \neq r/s \in Q$, $r \neq 0$, so s/r is the inverse of r/s, and hence Q is a division ring.

Define $\varphi : R \to Q$ for any $r \in R$, $\varphi(r) = (sr)/s$ for any $s \in S$. Clearly φ is a homomorphism between two rings (Exercise 36). Suppose that $\varphi(r) = 0/s$. Then $(sr)/s = 0/s$, so $r = 0$, namely, φ is one-to-one. Hence we may identify R with $\varphi(R)$ and consider R as a subring of Q. For any $r/s \in Q$, $r/s = (s/ss)(sr/s) = \varphi(s)^{-1}\varphi(r) = s^{-1}r$, that is, Q is the classical left quotient ring of R. $\qquad\square$

Let R be a left Ore domain and an ℓ-ring. We say that its classical left quotient ring Q is an ℓ-ring extension of R if Q can be made into an ℓ-ring such that R is an ℓ-subring of Q.

Theorem 4.17. *Let R be a left Ore domain and an ℓ-ring with $f(R) \neq \{0\}$. If for each nonzero element a of R, $Ra \cap f(R) \neq \{0\}$, then its classical left quotient ring Q can be made into an ℓ-ring extension of R, and Q is certainly a lattice-ordered division ring. Moreover, if R is Archimedean, then Q is also Archimedean.*

Proof. Since R is a domain, $f(R) = \{a \in R \mid |a| \text{ is an } f\text{-element}\}$ is totally ordered. We also notice that for any $x, y \in R$, $x \neq 0$, there exist z, w with $0 < w \in f(R)$ such that $zx = wy$. In fact, by left Ore condition, there exist z_1, z_2, $z_2 \neq 0$, such that $z_1 x = z_2 y$. Then $z_2 \neq 0$ implies that $Rz_2 \cap f(R) \neq \{0\}$, and hence there exists z_3 such that $z_3 z_2 = w > 0$ and $w \in f(R)$. Let $z = z_3 z_1$. Then we have $zx = wy$.

Suppose that $Q = \{a^{-1}b \mid a, b \in R, a \neq 0\}$ is the classical left quotient ring of R. For $a \neq 0$, $Ra \cap f(R) \neq \{0\}$ implies that there exists a_1 such that $a_1 a = c_1 > 0$ and $c_1 \in f(R)$. Then $a^{-1}b$ can be written as $a^{-1}b = c_1^{-1}(a_1 b)$. Thus each element q of Q can be expressed as $q = c^{-1}b$ with $0 < c \in f(R)$, $b \in R$. Then we define $q \geq 0$ in Q if $b \geq 0$ in R, that is, define the positive cone of Q as follows:

$$P = \{q \in Q \mid q = c^{-1}b, \text{ where } 0 < c \in f(R), b \in R^+\}.$$

We first show that this definition is independent of the representations of elements in Q. Suppose that $c^{-1}b = c_1^{-1}b_1$ with $b \in R^+$, $0 < c, c_1 \in f(R)$. We will derive that $b_1 \in R^+$. By the fact proved in the previous paragraph, $wc_1 = zc$ for some $0 < w \in f(R)$ and $0 \neq z \in R$, and hence $wc_1 = |wc_1| = |zc| = |z|c$ since c is an f-element. It follows that $wb_1 = |z|cc_1^{-1}b_1 = |z|b \in R^+$, so $w(b_1 \wedge 0) = wb_1 \wedge 0 = 0$ since w is an f-element. Thus $b_1 \wedge 0 = 0$, that is, $b_1 \geq 0$ in R.

It is routine to check that $P + P \subseteq P$, $PP \subseteq P$, and $P \cap -P = \{0\}$. We leave the verification of these facts as an exercise (Exercise 37). For $r \in R^+$, $r = c^{-1}(cr)$ for $0 < c \in f(R)$, so $r \in P$. Hence $R^+ \subseteq P$. Now let $q = s^{-1}r \in P \cap R$ with $0 < s \in f(R)$. Then $r \in R^+$, and hence $sq \in R^+$. Since s is an f-element, we must have $q \in R^+$. Therefore $R^+ = P \cap R$.

We show that P is a lattice order on Q. Let $q = c^{-1}b \in Q$ with $0 < c \in f(R)$. We claim that $q^* = c^{-1}b^+ = c^{-1}(b \vee 0)$ is the least upper bound of q and 0 in Q. First we show that q^* is well-defined.

Suppose we also have $q = d^{-1}e$ with $0 < d \in f(R)$ and $e \in R$. Then $wc = zd$ for some $0 < w \in f(R)$ and $z \neq 0$, and hence $z > 0$ since $zd = |zd| = |z|d$ implies that $z = |z|$. In Q, let $z^{-1} = d(wc)^{-1} = x^{-1}y$ for some $0 < x \in f(R)$ and $y \in R$. Then similarly we have $y > 0$ in R since $xd = y(wc)$ and $0 < xd, 0 < wc$ are in $f(R)$, and hence $z^{-1} > 0$ in Q. Suppose that $f = e \vee 0$ in R. We show that f is also the least upper bound of e and 0 in Q. Let $q \in Q$ with $q \geq e, 0$. Suppose that $q = r^{-1}s$ with $0 < r \in f(R)$ and $s \in R^+$. We have

$$s = rq \geq re, 0 \Rightarrow s \geq re \vee 0 = r(e \vee 0) = rf,$$

since $r \in f(R)$, and hence $q = r^{-1}s \geq f$ in Q since $r^{-1} \geq 0$ in Q. Therefore $f = e \vee 0$ in Q. We claim that $zf = ze \vee 0$ in R. Clearly $zf \geq ze, 0$. Let $u \in R$ and $u \geq ze, 0$. We have $z^{-1}u \geq e, 0$ since $z^{-1} > 0$ in Q, so $z^{-1}u \geq f$ since $f = e \vee 0$ in Q. It follows that $u \geq zf$, and hence $zf = ze \vee 0$. From $q = c^{-1}b = d^{-1}e$, we have $q = (wc)^{-1}(wb) = (zd)^{-1}(ze)$, so $wb = ze$ implies that $w(b \vee 0) = (wb) \vee 0 = (ze) \vee 0 = z(e \vee 0)$. Hence

$$c^{-1}(b \vee 0) = (wc)^{-1}(w(b \vee 0)) = (zd)^{-1}(z(e \vee 0)) = d^{-1}(e \vee 0),$$

so q^* is well defined.

Now let $p = g^{-1}t \in Q$, where $0 < g \in f(R), t \in R$, with $p \geq q, 0$ in Q. Then $t \in R^+$, and $p - q = g^{-1}t - c^{-1}b \in P$. Thus there exist $z \in R$, $0 < w \in f(R)$ such that $zg = wc = e$, so $|z|g = |zg| = wc = e$ since c, g, w are all f-elements. Then $e \in f(R)$ and

$$p - q = g^{-1}t - c^{-1}b = e^{-1}(|z|t - wb) \in P$$

implies $|z|t - wb \in R^+$. Since $|z|t \geq wb$ and $|z|t \geq 0$ in R, $|z|t \geq wb \vee 0 = w(b \vee 0) = wb^+$ since again w is an f-element of R. Therefore $p - q^* = g^{-1}t - c^{-1}b^+ = e^{-1}(|z|t - wb^+) \in P$, that is, $p \geq q^*$ in Q. Hence in Q, $q^* = q \vee 0$, and hence P defines a lattice order on Q and (Q, P) becomes an ℓ-ring. For an element $x \in R$, $x = a^{-1}(ax)$, where $0 < a \in f(R)$. Then $x^* = a^{-1}(ax)^+ = a^{-1}(ax^+) = x^+$. Therefore Q is an ℓ-ring extension of R.

It is clear that if R is Archimedean, then Q is also Archimedean. We omit the proof and leave the verification to the reader. $\square$

Let B be a unital ring and M be a left B-module. Then $_BM$ is said to be *finite-dimensional* over B provided M does not contain the direct sum of an infinite number of nonzero B-submodules of M. Certainly for a vector space over a division ring, this definition coincides with the usual meaning of finite-dimensional vector space.

Theorem 4.18. *Let R be an ℓ-unital ℓ-ring and a domain. If $_{f(R)}R$ is finite-dimensional, then R is a left Ore domain and its classical left quotient ring can be made into an ℓ-ring extension of R.*

Proof. Since $_{f(R)}R$ is finite-dimensional, $_RR$ is also finite-dimensional, that is, R does not contain the direct sum of an infinite number of nonzero left ideals of R. We show that R is a left Ore domain by verifying $Ra \cap Rb \neq \{0\}$ for any $a, b \in R \setminus \{0\}$.

Suppose that $\mathcal{M}$ is the family of nonzero left ideals I which contains two nonzero left ideals J and H such that $J \cap H = \{0\}$. We claim that not every nonzero left ideal belongs to $\mathcal{M}$. Suppose not. R is a direct sum of two nonzero left ideals I_1, I_1', then $I_1' \in \mathcal{M}$ implies that I_1' is a direct sum of two nonzero left ideals I_2, I_2', so R is a direct sum of I_1, I_2, I_2'. Continuing this process, we get a family of nonzero left ideals $\{I_k\}_{k=1}^{\infty}$ such that R is a direct sum of them, which is a contradiction. Thus there exists at least a nonzero left ideal $J \notin \mathcal{M}$. Take $0 \neq z \in J$. Then Raz, Rbz are nonzero left ideals contained in J, and hence $Raz \cap Rbz \neq \{0\}$ since $J \notin \mathcal{M}$. Hence there exist $c, d \in R$ such that $caz = dbz \neq 0$, so $ca = db \neq 0$. Therefore $Ra \cap Rb \neq \{0\}$.

Let $0 \neq a \in R$. Since $\sum_{i=1}^{\infty} f(R)a^i$ is not a direct sum over $f(R)$, there exist positive integers $i_1, \cdots, i_n$ and $f_{i_1}, \cdots, f_{i_n} \in f(R)$ such that
$$0 \neq f_{i_1}a^{i_1} + \cdots + f_{i_n}a^{i_n} \in f(R).$$
Thus $Ra \cap f(R) \neq \{0\}$. Now Theorem 4.17 applies. $\square$

The following result gives a sufficient condition such that the quotient field Q of a lattice-ordered integral domain R can be made into an ℓ-ring extension of R. For an ℓ-unital lattice-ordered integral domain R, R is called *algebraic* over $f(R)$ if for any $a \in R$, there exists a nonzero polynomial $g(x) \in f(R)[x]$ such that $g(a) = 0$.

Theorem 4.19. *Let R be an ℓ-unital lattice-ordered integral domain. If R is algebraic over $f(R)$, then the quotient field of R can be made into an ℓ-ring extension of R.*

Proof. For $0 \neq a \in R$, there exists a nonzero polynomial $g(x) \in f(R)[x]$ such that $g(a) = 0$. Suppose that $g(x) = \alpha_n x^n + \cdots + \alpha_1 x + \alpha_0$ with $\alpha_i \in f(R)$. Then $g(a) = \alpha_n a^n + \cdots + \alpha_1 a + \alpha_0 = 0$. We may assume $\alpha_0 \neq 0$. Then $\alpha_n a^n + \cdots + \alpha_1 a = -\alpha_0 \in Ra \cap f(R)$ implies that $Ra \cap f(R) \neq \{0\}$. By Theorem 4.17, the quotient field of R can be made into an ℓ-ring extension of R. □

Lattice-ordered division rings were first constructed around 1989 by J. Dauns [Dauns (1989)] and R. Redfield [Redfield (1989)] independently. Theorem 4.17 provides us a method to construct lattice-ordered division rings. Let's consider an example. For general construction of Ore domains, the reader is referred to [Lam (1999)].

Example 4.2. Let F be a totally ordered field, and let σ be an order-preserving injective ring endomorphism of F, that is, σ is an injective endomorphism of F with $\sigma(F^+) \subseteq F^+$. Certainly σ is not the identity mapping. Let $R = F[x; \sigma]$ be the *skew polynomial ring* over F in one variable x. The elements of R are left polynomials of the form $\sum_{i=0}^n a_i x^i$, where $a_i \in F$, with the usual addition, and the multiplication defined by

$$(\sum a_i x^i)(\sum b_j x^j) = \sum a_i \sigma^i(b_j) x^{i+j}.$$

Then R is a noncommutative domain.

We claim that R is a left Ore domain. We begin by noting that Euclidean Division Algorithm is valid in R for one-sided division, that is, if $f(x), g(x)$ are left polynomials in R with $f(x) \neq 0$, then there are unique $q(x)$ and $r(x)$ in R such that

$$g(x) = q(x)f(x) + r(x), \quad \text{with } r(x) = 0 \text{ or } \deg r(x) < \deg f(x).$$

The verification of this fact is left as an exercise (Exercise 38). Then any left ideal of R is a principle left ideal generated by any polynomial in it with the least degree. Now take $f, g \in R \setminus \{0\}$. If $Rf \cap Rg = \{0\}$, we get a contradiction. Since $Rf + Rg$ is a left ideal, $Rf + Rg = Rh$ for some $h \in R$, and hence $h = tf + sg$ and $g = rh$ for some $r, s, t \in R$. It follows that

$$g = rh = rtf + rsg \Rightarrow (rt)f = (1 - rs)g \in Rf \cap Rg,$$

so $rtf = 0$. Thus $tf = 0$ and $h = sg \in Rg$, and hence $Rf \subseteq Rh \subseteq Rg$. Therefore $Rf = \{0\}$, which is contradiction. Hence for any $f, g \in R \setminus \{0\}$, $Rf \cap Rg \neq \{0\}$, that is, R is a left Ore domain.

Consider the subring $F[x^2; \sigma] = \{\sum_{i=0}^{n} a_i x^{2i} \mid a_i \in F\}$ of R. Totally order $F[x^2; \sigma]$ by saying a left polynomial positive if the coefficient of the lowest term is positive in F. Then $F[x^2; \sigma]$ is a totally ordered domain. Since each element in R can be uniquely expressed as $f + gx$, where f and $g \in F[x^2; \sigma]$, we may order R by $f + gx \geq 0$ if $f \geq 0$ and $g \geq 0$ in $F[x^2; \sigma]$. Then it is easily checked that R becomes an ℓ-ring with $f(R) = F[x^2; \sigma]$ (Exercise 39). Given an element $\sum_{i=0}^{n} a_i x^i \in R$, we denote $\sum_{i=0}^{n} \sigma(a_i) x^i$ by $\sigma(\sum_{i=0}^{n} a_i x^i)$. Let $0 \neq a \in R$. We show that $Ra \cap F[x^2; \sigma] \neq \{0\}$. Suppose $a = f + gx$, where f and $g \in F[x^2; \sigma]$. If $g = 0$, then $a = f \in F[x^2; \sigma]$. If $f = 0$, then $xa = x(gx) = \sigma(g)x^2 \in F[x^2; \sigma]$. Now suppose that $f \neq 0$ and $g \neq 0$. Then $\sigma(f) \neq 0$. Since $F[x^2; \sigma]$ is also a left Ore domain, there exist $h \neq 0$ and $k \neq 0$ in $F[x^2; \sigma]$ such that $h\sigma(f) = kg$. Let $b = -k + hx \in F[x; \sigma]$. Then

$$
\begin{aligned}
0 \neq ba \\
&= (-k + hx)(f + gx) \\
&= -kf + hxf - kgx + hxgx \\
&= (-kf + h\sigma(g)x^2) + (h\sigma(f) - kg)x \\
&= (-kf + h\sigma(g)x^2) \in F[x^2; \sigma].
\end{aligned}
$$

Thus, $Ra \cap f(R) \neq \{0\}$, for any $0 \neq a \in R$, so by Theorem 4.17, the classical left quotient ring of R can be made into an ℓ-ring extension of R.

For the polynomial ℓ-ring $R = \mathbb{R}[x]$ with the entrywise order, $f(R) = \mathbb{R}$ and R is not algebraic over $\mathbb{R}$, so Theorem 4.19 cannot apply to this situation. In the following we present some thoughts that may be useful in further study of this problem. We notice that ℓ-ring $R = \mathbb{R}[x]$ with the entrywise order is an Archimedean ℓ-ring in which x is a d-element and satisfies condition (C) in Theorem 1.15.

Theorem 4.20. *The entrywise order on $R = \mathbb{R}[x]$ cannot be extended to an Archimedean lattice order on its quotient field Q such that x is a d-element of Q and Q satisfies condition (C) in Theorem 1.15.*

Proof. Suppose that the lattice order on R can be extended to a lattice order on Q that satisfies all three conditions. We derive a contradiction. We first show that $S = \{x^i \mid i \in \mathbb{Z}\}$ is a basis. Since x is a d-element in

Q, $x^{-1} > 0$ by Theorem 1.20(2), and hence x^{-1} is also a d-element. It follows that for any $i, j \in \mathbb{Z}$, $i < j$, $x^i \wedge x^j = x^i(1 \wedge x^{j-i}) = 0$. Hence S is disjoint. Let $0 \le a, b \le x^i$ for some $i \in \mathbb{Z}$. By multiplying x^{-i} to each side, we get $0 \le ax^{-i}, bx^{-i} \le 1$, so $ax^{-i}, bx^{-i} \in f(Q)$ which is totally ordered. Therefore ax^{-i}, bx^{-i} are comparable and hence a, b are comparable. This proves that x^i is a basic element for any $i \in \mathbb{Z}$. Now we show that $S^{\perp} = \{0\}$. Suppose that $0 < z \in S^{\perp}$. Since $z \in Q$, there exists $0 \ne w \in R$ such that $z|w| > 0$. Let $|w| = \alpha_n x^n + \cdots + \alpha_1 x + \alpha_0 \in R$ with $\alpha_n \ne 0$, and $z|w| = \beta_m x^m + \cdots + \beta_1 x + \beta_0 \in R$ with $\beta_m \ne 0$. Then $\alpha_i \ge 0$, $i = 1, \cdots, n$, and $\beta_j \ge 0$, $j = 1, \cdots, m$. It follows from $z \in S^{\perp}$ that $z \wedge x^{m-k} = 0$ for $k = 0, \cdots, n$, and then that each element in S is a d-element implies $zx^k \wedge x^m = 0$, $k = 0, \cdots, n$. Thus $z(\alpha_k x^k) \wedge x^m = 0$ since R is an f-module over $\mathbb{R}$. From Theorem 1.5(7),

$$
\begin{aligned}
0 &\le z|w| \wedge x^m \\
&= [z(\alpha_n x^n) + \cdots + z(\alpha_1 x) + z(\alpha_0 1)] \wedge x^m \\
&\le z(\alpha_n x^n) \wedge x^m + \cdots + z(\alpha_1 x) \wedge x^m + z(\alpha_0 1) \wedge x^m \\
&= 0.
\end{aligned}
$$

Hence $z|w| \ge \beta_m x^m$ and $z|w| \wedge x^m = 0$ imply that $\beta_m x^m = 0$, which is a contradiction. Therefore $S^{\perp} = \{0\}$, and hence S is a basis, actually S is a d-basis defined in chapter 2.

We prove that $f(Q) = \mathbb{R}$. Certainly $\mathbb{R} \subseteq f(Q)$. Suppose that $0 < q \in f(Q)$. We show that $q \in \mathbb{R}$. Let $q = f(x)/g(x)$ with $g(x) \ne 0$. Then $f(x) = qg(x)$ and $|f(x)| = q|g(x)|$ since q is an f-element. Suppose that

$$
|f(x)| = \alpha_n x^{k_n} + \cdots + \alpha_1 x^{k_1}, k_n > \cdots > k_1 \ge 0 \text{ and } \alpha_i > 0, i = 1, \cdots, n,
$$

and

$$
|g(x)| = \beta_m x^{j_m} + \cdots + \beta_1 x^{j_1}, j_m > \cdots > j_1 \ge 0 \text{ and } \beta_i > 0, i = 1, \cdots, m.
$$

If some x^{k_i} is not in the sum for $|g(x)|$, then $x^{k_i} \wedge |g(x)| = 0$ implies that $x^{k_i} \wedge q|g(x)| = 0$, so $x^{k_i} \wedge |f(x)|$ and $x^{k_i} = 0$, which is a contradiction. On the other hand, if some x^{j_t} is not in the sum for $|f(x)|$, then $x^{j_t} \wedge |f(x)| = 0$ implies that $x^{j_t} \wedge q|g(x)| = 0$, and hence $x^{j_t} \wedge q\beta_t x^{j_t} = 0$. Hence $q\beta_t x^{j_t} = 0$, which is a contradiction. Therefore we have $k_n = j_m, \cdots, k_1 = j_1$, so $|g(x)| = \beta_m x^{k_n} + \cdots + \beta_1 x^{k_1}$ and we have

$$
|f(x)| = \alpha_n x^{k_n} + \cdots + \alpha_1 x^{k_1} = q|g(x)| = q\beta_m x^{k_n} + \cdots + q\beta_1 x^{k_1}.
$$

By Exercise 2.7, we must have $\alpha_n x^{k_n} = q\beta_m x^{k_n}$ and it follows that $\alpha_n = q\beta_m$, and hence $q = (\beta_m^{-1}\alpha_n) \in \mathbb{R}$. Therefore $f(Q) = \mathbb{R}$.

Since Q is Archimedean, for $0 < q \in (x^i)^{\perp\perp}$ there exists positive integer n such that $0 < q \le nx^i$, and hence $0 < qx^{-i} \le n$. It follows that $qx^{-i} \in f(Q) = \mathbb{R}$, so $q \in \mathbb{R}x^i$. Hence $(x^i)^{\perp\perp} = \mathbb{R}x^i$ for each $i \in \mathbb{Z}$.

If Q satisfies condition (C), then Q is a direct sum of $(x^i)^{\perp\perp} = \mathbb{R}x^i$ by Theorem 1.17, $i \in \mathbb{Z}$. Then

$$\frac{1}{1+x} = \alpha_{k_1}x^{k_1} + \alpha_{k_2}x^{k_2} + \cdots + \alpha_{k_n}x^{k_n}, \; k_1 < k_2 < \ldots < k_n,$$

so

$$1 = (1+x)(\alpha_{k_1}x^{k_1} + \cdots + \alpha_{k_n}x^{k_n})$$
$$= \alpha_{k_1}x^{k_1} + \cdots + \alpha_{k_n}x^{k_n} + \alpha_{k_1}x^{k_1+1} + \cdots + \alpha_{k_n}x^{k_n+1}.$$

Multiplying both sides of the above equation by x^{-k_1}, we get

$$x^{-k_1} = \alpha_{k_1} + \alpha_{k_2}x^{k_2-k_1} + \cdots + \alpha_{k_n}x^{k_n-k_1}$$
$$+ \alpha_{k_1}x + \alpha_{k_2}x^{k_2-k_1+1} + \cdots + \alpha_{k_n}x^{k_n-k_1+1} \in R = \mathbb{R}[x].$$

Hence $-k_1 \ge 0$, so $k_1 \le 0$. From

$$1 = \alpha_{k_1}x^{k_1} + \cdots + \alpha_{k_n}x^{k_n} + \alpha_{k_1}x^{k_1+1} + \cdots + \alpha_{k_n}x^{k_n+1},$$

we must have $k_1 = 0$. Then the term x^{k_n+1} has the exponent $k_n + 1 > 0$, which is a contradiction. This completes the proof. $\square$

If instead of considering extension of a lattice order, we want to extend a partial order or a total order from an integral domain to its quotient field, the situation becomes relatively easy. For a partially ordered ring R with the positive cone R^+ and any ring S containing R, we say that the partial order on R can be extended to S if S is a partially ordered ring with the positive cone P such that $R^+ = R \cap P$. We may also say that (R, R^+) can be embedded into (S, P) for this situation. It is clear that for any ring S containing R, S becomes a partially ordered ring with the same positive cone $P = R^+$, and $R^+ = R \cap P$, that is, a partial order on a partially ordered ring R can be extended to any ring containing R. However clearly this extension is not interesting.

The partial order $\ge$ of a partially ordered ring R is called *division-closed* if $ab > 0$ and one of $a, b > 0$, then so is the other, for any $a, b \in R$. Clearly a total order is division-closed. For a partially ordered ring R with a division closed partial order, we will just call R as division-closed.

Lemma 4.14. *Let R be a partially ordered ring.*

(1) Suppose that R is division-closed. If R is unital and $R^+ \ne \{0\}$, then identity element $1 > 0$ and inverse of each positive invertible element is positive.

(2) If R is a partially ordered division ring with $R^+ \neq \{0\}$, then R is division-closed if and only if the inverse of each nonzero positive element is positive.

(3) If R is a lattice-ordered division ring, then R is division-closed if and only if R is a totally ordered division ring.

Proof. (1) Take $0 < a \in R$. Then $a = 1a > 0$ and $a > 0$ implies that $1 > 0$. Suppose that $0 < u \in R$ and u is invertible. Then $uu^{-1} = 1 > 0$ and $u > 0$ implies that $u^{-1} > 0$.

(2) Suppose that the inverse of each nonzero positive element is positive. If $ab > 0$ and $a > 0$, for $a, b \in R$, then $b = a^{-1}(ab) > 0$ since $a^{-1} > 0$. Similarly, $ba > 0$ and $a > 0$ implies that $b > 0$.

(3) If R is a lattice-ordered division ring and division-closed, then each $u > 0$ is a d-element by (2) and Theorem 1.20(2), that is, R is a d-ring. Then by Theorem 1.28(4), R is totally ordered. ☐

Let's look at some examples of partially ordered rings that are division-closed.

Example 4.3.

(1) (Exercise 1.43) Let $R = \mathbb{R} \times \mathbb{R}$ be the direct sum of two copies of $\mathbb{R}$. Define the positive cone $P = \{(a, b) \mid b > 0\} \cup \{(0, 0)\}$. Then R is a commutative partially ordered ring. If $(a, b)(x, y) = (ax, by) > 0$ and $(a, b) > 0$, then $by > 0$ and $b > 0$. Thus $y > 0$, so $(x, y) > 0$. Hence R is division-closed.

(2) Let $R = \mathbb{R}[x]$ be the polynomial ring over $\mathbb{R}$. For $f(x) = a_n x^n + \cdots + a_1 x + a_0 \in R$ with leading coefficient $a_n \neq 0$ and $n \geq 0$. Define the positive cone
$$P = \{f(x) \mid n = 4k, a_n > 0 \text{ or } n = 4k + 2, a_n < 0\} \cup \{0\}.$$
Then R is a partially ordered integral domain that is division-closed (Exercise 79). We note that for a nonzero polynomial $f(x)$, if $f(x)$ has an even degree, then $(f(x))^2 > 0$ and if $f(x)$ has an odd degree, then $(f(x))^2 < 0$.

Let R be a partially ordered integral domain and Q be its quotient field. Define the subset P of Q as follows. If $R^+ = \{0\}$, $P = \{0\}$. If $R^+ \neq \{0\}$, then
$$P = \{q \in Q \mid \text{ there exist } a, b \in R, a \geq 0, b > 0 \text{ such that } q = ab^{-1}\}.$$

Theorem 4.21. *Let R be a partially ordered integral domain and Q be its quotient field, and P be defined as above.*

(1) (Q, P) *is a partially ordered field which is division-closed and* $R^+ \subseteq R \cap P$.

(2) (R, R^+) *can be embedded into* (Q, P) *if and only if* (R, R^+) *is division-closed.*

Proof. (1) It is clear that $P + P \subseteq P$, $PP \subseteq P$, and $P \cap -P = \{0\}$. Thus (Q, P) is a partially ordered field. For $0 < a \in R$, then $a = a^2 a^{-1}$ implies that $a \in P$. Thus $R^+ \subseteq R \cap P$. Let $p, q \in Q$ with $pq > 0$ and $p > 0$. Then $qp = ab^{-1}$, $p = a_1 b_1^{-1}$ with $a > 0, b > 0, a_1 > 0$, and $b_1 > 0$ in R. Hence $q = (ab_1)(a_1 b)^{-1}$ with $ab_1 > 0$ and $a_1 b > 0$. Therefore $q > 0$ in Q and Q is division-closed.

(2) If $R^+ = \{0\}$, then $P = \{0\}$ and $R^+ = R \cap P$. Suppose that $R^+ \neq \{0\}$ and R is division-closed. If $a \in R \cap P$. Then $a = xy^{-1}$ for some $x, y \in R$ with $x > 0$, $y > 0$, so $ay = x > 0$ and $y > 0$ implies that $a > 0$ in R. Thus $R^+ = R \cap P$. Conversely suppose that $R^+ = R \cap P$. If $ab > 0$ and $a > 0$ for some $a, b \in R$. Then $b \in P$, and hence $b \in R^+ = R \cap P$. Therefore R is division-closed. $\square$

Theorem 4.22. *Let R be a division-closed ℓ-ring.*

(1) If R is unital, then 1 is a weak unit, and hence R is an almost f-ring.
(2) If R is ℓ-reduced, then R is an ℓ-domain.

Proof. (1) We notice that the identity element 1 must be positive since $R^+ \neq \{0\}$. Suppose that $a > 0$ and $1 \wedge a = 0$. Then

$$(a^2 - a + 1)(a + 1) = a^3 - a^2 + a + a^2 - a + 1 = a^3 + 1 > 0,$$

and $a + 1 > 0$ implies that $a^2 - a + 1 > 0$. Thus $a < a^2 + 1$. Since $1 \wedge a = 0$,

$$a = (a^2 + 1) \wedge a \leq a^2 \wedge a + 1 \wedge a = a^2 \wedge a \leq a^2.$$

So $2a^2 > a$ and $a(2a - 1) > 0$. By division-closed property, we have $2a > 1$, which is a contradiction. Therefore for any $a \in R^+$, $1 \wedge a = 0$ implies $a = 0$, that is, 1 is a weak unit, and hence R is an almost f-ring by Corollary 4.7.

(2) Suppose that R is not an ℓ-domain. Then there exist $a > 0, b > 0$ such that $ab = 0$. Hence $a(a - b) = a^2 > 0$ and $a > 0$ implies that $a > b$, so $ab \geq b^2 = 0$, which is a contradiction. Thus R is an ℓ-domain. $\square$

As a direct consequence of Theorem 4.22, a unital division-closed lattice-ordered domain must be a totally ordered domain.

Consider complex field $\mathbb{C}$ with the positive cone $\mathbb{R}^+ = \{r \in \mathbb{R} \mid r \geq 0\}$. Clearly $(\mathbb{C}, \mathbb{R}^+)$ is a division-closed partially ordered field. We show that no division-closed partial order on $\mathbb{C}$ properly contains $\mathbb{R}^+$.

Lemma 4.15. *Suppose* $(\mathbb{C}, P)$ *is a partially ordered ring with the positive cone* P *that is division-closed and contains* $\mathbb{R}^+$. *Then* $P = \mathbb{R}^+$.

Proof. We show that for any $0 \neq z = x + iy \in P$, $x \geq 0$ in $\mathbb{R}$. In fact, $(\mathbb{C}, P)$ is division-closed implies that $z^{-1} = (x^2 + y^2)^{-1}(x - iy) \in P$, and hence $x - iy \in P$. So $2x \in P$. If $x < 0$ in $\mathbb{R}$, then since $R^+ \subseteq P$, we will have $-2x \in P \cap -P$, which is a contradiction. Thus we must have $x \geq 0$ in $\mathbb{R}$. Therefore, for each $z = x + iy \in P$, $x \geq 0$ in $\mathbb{R}$, so $P \subseteq \mathbb{R}$ (Exercise 86). Hence $P = \mathbb{R}^+$. □

Now let's consider extending total orders. Let R be a totally ordered integral domain and Q be its quotient field. Define the positive cone P as follows:

$$P = \{q \in Q \mid q = \frac{a}{b} \text{ with } a, 0 \neq b \in R \text{ and } ab \in R^+\}.$$

It is easy to check that P is a well-defined total order to make Q into a totally ordered field and R becomes a totally ordered subring of Q. We leave the verification of these facts as an exercise (Exercise 40).

4.5   Matrix ℓ-algebras over totally ordered integral domains

In this section R denotes a totally ordered integral domain, that is, R is a unital commutative totally ordered domain, and F denotes the totally ordered quotient field of R. Then

$$F = \{\frac{a}{b} \mid a, b \in R, b \neq 0\} \text{ and } \frac{a}{b} \geq 0 \text{ if } ab \geq 0 \text{ in } R.$$

We establish connection between ℓ-algebras $M_n(R)$ over R and the ℓ-algebras $M_n(F)$ over F so that we are able to generalize results on matrix ℓ-algebras over totally ordered fields to matrix ℓ-algebras over totally ordered integral domains.

Suppose that $M_n(R)$ is an ℓ-algebra and an f-module over R. We first extend the lattice order on $M_n(R)$ to $M_n(F)$. Define

$$P = \{x \in M_n(F) \mid \alpha x \in M_n(R)^+ \text{ for some } 0 < \alpha \in R\},$$

where $M_n(R)^+ = \{z \in M_n(R) \mid z \geq 0\}$.

Theorem 4.23. *The P defined above is the positive cone of a lattice order on $M_n(F)$ to make it into an ℓ-algebra over F such that $M_2(R)^+ = M_2(R) \cap P$.*

Proof. We first notice that the definition of P is actually not depending on α. Suppose that $x \in P$. For any $0 < \beta \in R$, if $\beta x \in M_n(R)$, then $\beta x \in M_n(R)^+$. In fact, suppose that $\alpha x \in M_n(R)^+$ for some $0 < \alpha \in R$. Then $\alpha(\beta x) = \beta(\alpha x) \in M_n(R)^+$ implies that $\alpha(\beta x \wedge 0) = \alpha(\beta x) \wedge 0 = 0$ since $M_n(R)$ is an f-module over R. Thus $\beta x \wedge 0 = 0$ since R is a domain, that is, $\beta x \in M_n(R)^+$.

It is clear that $P + P \subseteq P$, $PP \subseteq P$, $P \cap -P = \{0\}$, and $F^+P \subseteq P$, so $M_n(F)$ becomes a partially ordered algebra over F with the positive cone P (Exercise 41). From Theorem 1.18 the partial order is defined by $x \leq y$ for any $x, y \in M_n(F)$ if $y - x \in P$. We show that the partial order "$\leq$" is a lattice order. Given $x \in M_n(F)$, there exists $0 < k \in R$ such that $kx \in M_n(R)$. Let $kx \vee 0 = y$ in $M_n(R)$. We show $x \vee 0 = k^{-1}y$ in $M_n(F)$. Since $k(k^{-1}y) = y \in M_n(R)^+$, $(k^{-1})y \geq 0$ in $M_n(F)$, and since $k(k^{-1}y - x) = y - kx \in M_n(R)^+$, $k^{-1}y \geq x$ in $M_n(F)$. So $k^{-1}y$ is an upper bound for x and 0 in $M_n(F)$. Let $z \in M_n(F)$ and $z \geq x, 0$. Then there exist $0 < k_1 \in R$ and $0 < k_2 \in R$ such that $k_1 z \in M_n(R)^+$ and $k_2(z - x) \in M_n(R)^+$. So

$$k_2(z - x) \in M_n(R)^+ \Rightarrow kk_1k_2(z - x) \in M_n(R)^+$$
$$\Rightarrow kk_2(k_1z) - k_1k_2(kx) \in M_n(R)^+$$
$$\Rightarrow kk_2(k_1z) \geq k_1k_2(kx) \text{ in } M_n(R).$$

Also $kk_2(k_1z) \in M_n(R)^+$. Hence $kk_2(k_1z)$ is an upper bound of $k_1k_2(kx), 0$ in $M_n(R)$. From $kx \vee 0 = y$ and that $M_n(R)$ is an f-module over R, we have $k_1k_2(kx) \vee 0 = k_1k_2y$. Thus $kk_2(k_1z) \geq k_1k_2y$ in $M_n(R)$, that is, $k_1k_2(kz - y) \in M_n(R)^+$. Thus $kz - y \in P$, so $z - k^{-1}y \in P$. Hence $z \geq k^{-1}y$ in $M_n(F)$. Therefore $k^{-1}y$ is the least upper bound of x and 0 in $M_n(F)$, so $M_n(F)$ is an ℓ-algebra over F.

For any $f, g \in M_n(R)$, $f \geq g$ in $M_n(R)$ if and only if $f \geq g$ in $M_n(F)$, so $M_n(R)^+ = M_n(R) \cap P$. $\square$

In the following, the ℓ-algebra $M_n(F)$ defined above is called the *order extension* of the given ℓ-algebra $M_n(R)$. We collect some basic relations between $M_n(R)$ and its order extension $M_n(F)$ in the following result and leave the proof as an exercise to the reader (Exercise 42).

Lemma 4.16. *Let $M_n(R)$ be an ℓ-algebra and f-module over R, and let $M_n(F)$ be its order extension.*

(1) $x \in M_n(R)$ is basic in $M_n(R)$ if and only if x is basic in $M_n(F)$.

(2) A set $S \subseteq M_n(R)$ is disjoint in $M_n(R)$ if and only if S is disjoint in $M_n(F)$.

(3) $x \in M_n(R)$ *is an* f-*element* (*d*-*element*) *in* $M_n(R)$ *if and only if* x *is an* f-*element* (*d*-*element*) *in* $M_n(F)$.

We need a well-known result in general ring theory which states that every automorphism of the matrix algebra over a field is inner. This result is generally stated as a consequence of Skolem-Noether theorem (see [Jacobson (1980)]). We present a nice direct proof due to P. Semrl [Semrl (2005)].

Theorem 4.24. *Let* K *be a field. If* φ *is an automorphism of matrix algebra* $M_n(K)$, *then there exists an invertible matrix* $f \in M_n(K)$ *such that* $\varphi(x) = fxf^{-1}$ *for every* $x \in M_n(K)$.

Proof. Let K^n denote the n-dimensional column space over K. For any vector $w \in K^n$, w^t denotes the transpose of w. Choose and fix $u, v \in K^n$ with $u \neq 0, v \neq 0$. Then $0 \neq uv^t \in M_n(K)$, and hence $\varphi(uv^t) \neq 0$ implies that $\varphi(uv^t)z \neq 0$ for some $z \in K^n$. Define $f : K^n \to K^n$ by $f(w) = \varphi(wv^t)z$, $w \in K^n$. Clearly the linearity of f follows from the linearity of φ. Hence we may identify f as a matrix in $M_n(K)$ with $fw = f(w)$ for any $w \in K^n$. For any $x \in M_n(K)$ and $w \in K^n$ we have

$$(fx)w = f(xw) = \varphi((xw)v^t)z = \varphi(x(wv^t))z = \varphi(x)\varphi(wv^t)z = \varphi(x)fw,$$

so $fx = \varphi(x)f$. For $w \in K^n$, since $fu = \varphi(uv^t)z \neq 0$ and φ is subjective, there exists $y \in M_n(K)$ such that $w = \varphi(y)(fu) = f(yu)$, that is, f is surjective. Therefore f is invertible and $\varphi(x) = fxf^{-1}$ for every $x \in M_n(K)$. $\qquad\square$

For a square matrix a, det$a$ denotes the *determinant* of a.

Theorem 4.25. *Given an* ℓ-*algebra* $M_n(R)$ *which is an* f-*module over* R, *let* $M_n(F)$ *be its order extension. Then* $M_n(R)$ *is* ℓ-*isomorphic to the* ℓ-*algebra* $M_n(R)$ *with the entrywise order if and only if the following two conditions are satisfied.*

(1) $M_n(F)$ *is* ℓ-*isomorphic to the* ℓ-*algebra* $M_n(F)$ *with the entrywise order,*

(2) $M_n(R)$ *has a basis that spans* $M_n(R)$ *as a module over* R.

Proof. "$\Rightarrow$" Suppose that φ is an ℓ-isomorphism from the ℓ-algebra $M_n(R)$ with the entrywise order to $M_n(R)$. Then clearly $S = \{\varphi(e_{ij}) \mid 1 \leq i, j \leq n\}$ satisfies condition (2). Since S is a disjoint set of basic elements in $M_n(R)$, S is also a disjoint set of basic elements in $M_n(F)$ by Lemma

4.16, so S is also a basis for $M_n(F)$, and hence S is a vector space basis of the vector space $M_n(F)$ over F. Define the mapping from $M_n(F)$ with the entrywise order to $M_n(F)$ by

$$\sum_{1 \leq i,j \leq n} q_{ij} e_{ij} \to \sum_{1 \leq i,j \leq n} q_{ij} \varphi(e_{ij}), \quad q_{ij} \in F.$$

Then it is clear that $M_n(F)$ is ℓ-isomorphic to $M_n(F)$ with the entrywise order, so (1) is also true.

"$\Leftarrow$" Suppose that conditions (1) and (2) are true. We show that $M_n(R)$ is ℓ-isomorphic to the ℓ-algebra $M_n(R)$ with the entrywise order. Recall that $e_{ij}, 1 \leq i,j \leq n$, denote standard matrix units. Since $\{e_{ij} \mid 1 \leq i,j \leq n\}$ is a basis for the ℓ-algebra $M_n(F)$ with the entrywise order, and a vector space basis of $M_n(F)$ over F, by (1) and Theorem 4.24 there exists an invertible matrix $h \in M_n(F)$ such that $\{he_{ij}h^{-1} \mid 1 \leq i,j \leq n\}$ is a basis for $M_n(F)$ over F. By (2) $M_n(R)$ has a basis $S = \{b_{ij} \mid 1 \leq i,j \leq n\}$ and S spans $M_n(R)$ over R. Since S is also a disjoint set of basic elements in $M_n(F)$ by Lemma 4.16(2), each element in S is a scalar product of a positive scalar in F and an element in $\{he_{ij}h^{-1} \mid 1 \leq i,j \leq n\}$. Thus without loss of generality, we may just assume that

$$b_{ij} = t_{ij}(he_{ij}h^{-1}), \quad \text{where } 0 < t_{ij} \in F, \ 1 \leq i,j \leq n.$$

Thus $b_{ij}b_{rs} = 0$ if $j \neq r$, and $b_{ij}b_{js} = t_{ij}t_{js}t_{is}^{-1}b_{is}$. Therefore $t_{ij}t_{js}t_{is}^{-1} \in R$, $1 \leq i,j,s \leq n$ since S spans $M_n(R)$ over R (Exercise 43).

We claim that $\prod_{1 \leq i,j \leq n} t_{ij}$ is a positive unit of R. Form an $n^2 \times n^2$ matrix B in the following fashion. For each b_{ij}, form a column vector with n^2 elements by arranging the second column in b_{ij} under the first column, the third column under the second column, and so forth. Then use the resulting column vector to form the $((i-1)n + j)^{th}$ column in B. Since $\{b_{ij}\}$ spans $M_n(R)$ over R, each e_{rs} $(1 \leq r,s \leq n)$ in $M_n(R)$ can be written as a linear combination of the b_{ij} and hence the identity matrix of $M_{n^2}(R)$ can be written as a product BC for some $C \in M_{n^2}(R)$. Thus $\det B \in R$ is a unit of R. Now since $b_{ij} = t_{ij}he_{ij}h^{-1}$, we can also create B by applying a similar process to $t_{ij}he_{ij}h^{-1}$. For this construction, let $h^{-1} = (r_{ij})$ and define column vectors $v_{ij}^1, v_{ij}^2, \ldots, v_{ij}^n$, each with n coordinates, by letting the kth component in v_{ij}^k be r_{ij} and the other components in v_{ij}^k be zero. For each $i = 1, \ldots, n$, let f_i be the $n \times n$ matrix

$$f_i = \left(t_{i1} \begin{pmatrix} r_{11} \\ \vdots \\ r_{1n} \end{pmatrix} \cdots t_{in} \begin{pmatrix} r_{n1} \\ \vdots \\ r_{nn} \end{pmatrix} \right);$$

let A and F be the $n^2 \times n^2$ matrices

$$A = \begin{pmatrix} f_1 & & \\ & \ddots & \\ & & f_n \end{pmatrix} \quad \text{and} \quad F = \begin{pmatrix} h & & \\ & \ddots & \\ & & h \end{pmatrix};$$

and let J be the $n^2 \times n^2$ matrix

$$J = \begin{pmatrix} t_{11}v_{11}^1 & \cdots & t_{1n}v_{n1}^1 & \cdots & t_{n1}v_{11}^n & \cdots & t_{nn}v_{n1}^n \\ t_{11}v_{12}^1 & \cdots & t_{1n}v_{n2}^1 & \cdots & t_{n1}v_{12}^n & \cdots & t_{nn}v_{n2}^n \\ \vdots & & \vdots & & \vdots & & \vdots \\ t_{11}v_{1n}^1 & \cdots & t_{1n}v_{nn}^1 & \cdots & t_{n1}v_{1n}^n & \cdots & t_{nn}v_{nn}^n \end{pmatrix}.$$

We leave it to the reader to check that (Exercise 44)

$$B = \begin{pmatrix} t_{11}hv_{11}^1 & \cdots & t_{1n}hv_{n1}^1 & \cdots & t_{n1}hv_{11}^n & \cdots & t_{nn}hv_{n1}^n \\ t_{11}hv_{12}^1 & \cdots & t_{1n}hv_{n2}^1 & \cdots & t_{n1}hv_{12}^n & \cdots & t_{nn}hv_{n2}^n \\ \vdots & & \vdots & & \vdots & & \vdots \\ t_{11}hv_{1n}^1 & \cdots & t_{1n}hv_{nn}^1 & \cdots & t_{n1}hv_{1n}^n & \cdots & t_{nn}hv_{nn}^n \end{pmatrix}$$

$$= FJ.$$

Also a series of elementary row operations converts J to A. Then $\det(B) = \det(FJ) = \det(F)(\pm\det(A))$. However, for each $1 \le i \le n$, $\det(f_i) = t_{i1}\cdots t_{in}\det(h^{-1})$ and hence $\det(A) = (\prod_{1 \le i,j \le n} t_{ij})\det(h^{-1})^n$. So, since $\det(F) = \det(h)^n$,

$$\det(B) = \det(h)^n \left(\prod_{1 \le i,j \le n} t_{ij} \right) \det(h^{-1})^n = \prod_{1 \le i,j \le n} t_{ij},$$

and hence $\prod_{1 \le i,j \le n} t_{ij} = \gamma \in R$ is a positive unit. Then we have

$$\prod_{1 \le i,j,s \le n} t_{ij}t_{js}(t_{is})^{-1} = \prod_{i=1}^{n} \prod_{j=1}^{n} \prod_{s=1}^{n} \frac{t_{ij}t_{js}}{t_{is}}$$

$$= \prod_{i=1}^{n} \left[\left(\frac{t_{i1}t_{11}}{t_{i1}} \cdots \frac{t_{i1}t_{1n}}{t_{in}} \right) \cdots \left(\frac{t_{in}t_{n1}}{t_{i1}} \cdots \frac{t_{in}t_{nn}}{t_{in}} \right) \right]$$

$$= \prod_{i=1}^{n} \left(\prod_{1 \le u,v \le n} t_{uv} \right) = \gamma^n.$$

Therefore, since each $0 < t_{ij}t_{js}(t_{is})^{-1} \in R$ and γ^n is a unit in R, each $t_{ij}t_{js}(t_{is})^{-1}$ must be a positive unit in R.

For simplicity, let $t_{ij}t_{js}(t_{is})^{-1} = v_{ijs}$, $1 \leq i,j,s \leq n$. We show that there exist positive units α_{ij} in R, $1 \leq i,j \leq n$, such that $\alpha_{ij}\alpha_{js}(\alpha_{is})^{-1} = v_{ijs}$. To this end, define

$$\alpha_{ij} = \begin{cases} t_{ii}(= v_{iis}), & \text{if } i = j = 1, ..., n \\ 1, & \text{if } i = 1 \text{ and } j = 2, ..., n \\ v_{1ij}, & \text{if } 2 \leq i < j \leq n \\ \alpha_{ji}^{-1}t_{ii}v_{iji}, & \text{if } 1 \leq j < i \leq n. \end{cases}$$

It is clear that each α_{ij} defined above is a positive unit in R. All we need to do is to check that

$$(*) \quad \alpha_{ij}\alpha_{js}\alpha_{is}^{-1} = v_{ijs}, \quad \text{for } 1 \leq i,j,s \leq n.$$

We first note that if $j = i$ or $j = s$, then $\alpha_{ij}\alpha_{js}\alpha_{is}^{-1} = \alpha_{jj} = t_{jj}$, so $(*)$ is true. Let's, for instance, check the case $1 \leq s < i < j \leq n$.

If $s = 1$, then

$$\begin{aligned} \alpha_{ij}\alpha_{js}\alpha_{is}^{-1} &= v_{1ij}\alpha_{sj}^{-1}t_{jj}v_{jsj}(\alpha_{si}^{-1}t_{ii}v_{isi})^{-1} \\ &= (t_{1i}t_{ij}t_{1j}^{-1})t_{jj}(t_{js}t_{sj}t_{jj}^{-1})t_{ii}^{-1}(t_{is}t_{si}t_{ii}^{-1})^{-1} \\ &= t_{ij}t_{js}t_{is}^{-1} \\ &= v_{ijs}. \end{aligned}$$

If $s \geq 2$, then

$$\begin{aligned} \alpha_{ij}\alpha_{js}\alpha_{is}^{-1} &= v_{1ij}\alpha_{sj}^{-1}t_{jj}v_{jsj}(\alpha_{si}^{-1}t_{ii}v_{isi})^{-1} \\ &= (t_{1i}t_{ij}t_{1j}^{-1})(t_{1s}t_{sj}t_{1j}^{-1})^{-1}t_{jj}(t_{js}t_{sj}t_{jj}^{-1})(t_{1s}t_{si}t_{1i}^{-1}) \\ &\quad t_{ii}^{-1}(t_{is}t_{si}t_{ii}^{-1})^{-1} \\ &= t_{ij}t_{js}t_{is}^{-1} \\ &= v_{ijs}. \end{aligned}$$

The verification of other possible values of i, j, and s is similar. We omit the detail and leave them to the reader.

Now define $\varphi : M_n(R) \to M_n(R)$ by

$$\sum_{1 \leq i,j \leq n} \beta_{ij}b_{ij} \to \sum_{1 \leq i,j \leq n} \beta_{ij}(\alpha_{ij}e_{ij}), \quad \beta_{ij} \in R.$$

Since $\varphi(b_{ij}b_{rs}) = 0$ if $j \neq r$, and

$$\varphi(b_{ij}b_{js}) = \varphi(v_{ijs}b_{is}) = v_{ijs}(\alpha_{is}e_{is}) = \alpha_{ij}\alpha_{js}e_{is} = \varphi(b_{ij})\varphi(b_{js}),$$

φ is an ℓ-isomorphism from $M_n(R)$ to the ℓ-algebra $M_n(R)$ with the entry-wise order. This completes the proof. $\quad\square$

Here is a brief history on research of matrix ℓ-rings. It seems that matrix ring over totally ordered field with the entrywise order first appeared in [Birkhoff and Pierce (1956)]. In 1966, E. Weinberg studied $M_2(\mathbb{Q})$. He claimed that he found all the lattice orders of $M_2(\mathbb{Q})$ to make it into an ℓ-ring and only for the entrywise order (up to ℓ-isomorphism), the identity matrix is positive [Weinberg (1966)]. E. Weinberg conjectured that for any ℓ-ring $M_n(\mathbb{Q})$ $(n \geq 2)$, if the identity matrix is positive, then it is ℓ-isomorphic to the $M_n(\mathbb{Q})$ with the entrywise order. This is so-called Weinberg's conjecture. In 2000, S. Steinberg found and corrected a mistake in E. Weinberg's proof on lattice orders of $M_2(\mathbb{Q})$ and showed that the proof is true for 2×2 matrix algebra over any totally ordered field. In 2002, Weinberg's conjecture was solved by P. Wojciechowski and present author not only for $\mathbb{Q}$ but also for any totally ordered subfield of $\mathbb{R}$ [Ma, Wojciechowski (2002)]. Then in 2007, the result was proved to be true for ℓ-ring $M_n(\mathbb{Z})$, where $\mathbb{Z}$ is the totally ordered ring of integers. These results and their proofs are presented in [Steinberg (2010)]. In 2013, the result was further proved to be true for any greatest common divisor domain which is a totally ordered subring of $\mathbb{R}$ [Li, Bai and Qiu (2013)]. Using Theorem 4.25, we are able to show that Weinberg's conjecture is true for certain totally ordered integral domains.

For some totally ordered integral domains, the condition (1) in Theorem 4.25 implies the condition (2). An integral domain R is called a *greatest common divisor* (GCD) domain if for any $a, b \in R$, a and b have a greatest common divisor, denoted by $\gcd(a, b)$. We review a few definitions and properties on GCD domains. An element d in an arbitrary integral domain is called a *greatest common divisor* (gcd) of two elements a, b if $d|a$ and $d|b$, and for all element e if $e|a$ and $e|b$, then $e|d$. A GCD domain is an integral domain in which any two elements have at least one gcd. We note that if $a = b = 0$, then 0 is the gcd. Also two elements a and b may have more than one gcd. In fact, if d is a gcd of a, b, then for any unit u, du is also a gcd of a, b, and if d, d' both are gcd of a, b, then there exists a unit v such that $d = d'v$. We use $\gcd(a, b)$ to denote any greatest common divisor of a and b. The following result collects some basic properties of GCD domains that will be used later. The verification of them is left to the reader (Exercise 45).

Lemma 4.17. *Let R be a GCD domain and $a, b, c \in R$.*

(1) $\gcd(ab, ac) = a(\gcd(b, c))$.

(2) If gcd(a, b) = d, then gcd($\frac{a}{d}, \frac{b}{d}$) = 1.

(3) If gcd(a, b) = 1, gcd(a, c) = 1, then gcd(a, bc) = 1.

(4) If a|bc and gcd(a, b) = 1, then a|c.

Theorem 4.26. *Let R be an ℓ-simple totally ordered greatest common divisor domain. Suppose that $M_n(R)$ is an ℓ-algebra and an f-module over R. If its order extension $M_n(F)$ is ℓ-isomorphic to $M_n(F)$ with the entrywise order, then $M_n(R)$ is ℓ-isomorphic to the ℓ-algebra $M_n(R)$ over R with the entrywise order.*

Proof. As in the proof of Theorem 4.25, since $M_n(F)$ is ℓ-isomorphic to the ℓ-algebra $M_n(F)$ with the entrywise order, there exists an invertible matrix $h \in M_n(F)$ such that $T = \{he_{ij}h^{-1} \mid 1 \le i, j \le n\}$ is a basis for $M_n(F)$ and also a vector space basis of $M_n(F)$ over F. By the definition of the order on its order extension $M_n(F)$, there exist $0 < \alpha_{ij} \in R$ such that $b_{ij} = \alpha_{ij}(he_{ij}h^{-1}) \in M_n(R)^+$. Then the set $\{b_{ij} \mid 1 \le i, j \le n\}$ is a maximal disjoint set of basic elements in $M_n(R)$.

We show that $M_n(R)$ is Archimedean over R. Suppose that $x, y \in M_n(R)^+$, and $\alpha x \le y$ for all $\alpha \in R^+$. We claim that $x = 0$. Take $0 < \frac{a}{b} \in F$. We may assume that $a, b \in R^+$. The ℓ-ideal generated by b is equal to R, and hence $a \le \alpha b$ for some $0 < \alpha \in R$. Thus $\frac{a}{b}x \le \alpha x \le y$, so for any $0 < f \in F$, $fx \le y$. It follows from Theorem 1.31 that $M_n(F)$ is Archimedean over F since it is well known in general ring theory that $M_n(F)$ is simple, and hence $x = 0$. Therefore $M_n(R)$ is Archimedean over R, then by Theorem 1.17 we have the direct sum

$$M_n(R) = \sum_{1 \le i, j \le n} b_{ij}^{\perp\perp}.$$

Since R is a greatest common divisor domain, we may assume that $0 < \beta_{ij} \in R$ is the greatest common divisor of the entries in each matrix b_{ij}. Let $a_{ij} = \frac{1}{\beta_{ij}}b_{ij}$, $1 \le i, j \le n$. We show each $b_{ij}^{\perp\perp} = Ra_{ij}$, $1 \le i, j \le n$. Clearly $Ra_{ij} \subseteq b_{ij}^{\perp\perp}$. On the other hand, we know that $\{a_{ij} \mid 1 \le i, j \le n\}$ is a basis in $M_n(F)$ that also spans $M_n(F)$ as a vector space over F since $M_n(F)$ is an n^2-dimensional over F and the disjoint set $\{a_{ij} \mid 1 \le i, j \le n\}$ is linearly independent over F by Theorem 1.13. Let $0 \ne x \in b_{ij}^{\perp\perp}$. Then in $M_n(F)$, $x = q_{ij}a_{ij}$ for some $0 \ne q_{ij} \in F$. Since x and a_{ij} are both in $M_n(R)$, and the greatest common divisor of the entries in matrix a_{ij} is a unit in R, we must have $q_{ij} \in R$, so $b_{ij}^{\perp\perp} = Ra_{ij}$ for each $i, j = 1, \ldots, n$. Then $M_n(R) = \sum_{1 \le i, j \le n} Ra_{ij}$, so $\{a_{ij} \mid 1 \le i, j \le n\}$ is a basis for $M_n(R)$ that spans $M_n(R)$ as a module over R.

Therefore, condition (2) in Theorem 4.25 is satisfied, and hence $M_n(R)$ is ℓ-isomorphic to the ℓ-algebra $M_n(R)$ with the entrywise order.   □

Corollary 4.8. *Let R be a totally ordered GCD domain that is a subring of $\mathbb{R}$. If $M_n(R)$ is an ℓ-algebra over R such that identity matrix is positive, then $M_n(R)$ is ℓ-isomorphic to the ℓ-algebra $M_n(R)$ with the entrywise order.*

Proof.   Let $F \subseteq \mathbb{R}$ be the totally ordered quotient field of R and $M_n(F)$ be the order extension of $M_n(R)$. Then the identity matrix is also positive in $M_n(F)$, so $M_n(F)$ is ℓ-isomorphic to the ℓ-algebra over F with the entrywise order [Ma, Wojciechowski (2002)]. By Theorem 4.26, we just need to show that R is ℓ-simple and $M_n(R)$ is an f-module over R. Since R is Archimedean, R is ℓ-simple. For any $0 < \alpha \in R$, if $x \wedge y = 0$ for some $x, y \in M_n(R)$, then $\alpha x \wedge y \leq n x \wedge y$ for some positive integer n implies that $\alpha x \wedge y = 0$. Thus $M_n(R)$ is an f-module over R.   □

An integral domain is called a *local domain* if it contains a unique maximal ideal. For examples of local domains, we refer the reader to [Lam (2001)]. We show that the result in Corollary 4.8 is true for matrix ℓ-algebras over local domains. First we review a few definitions and results from general ring theory whose proofs are omitted.

Let R be a unital ring and M be a left R-module. A subset X of M is called *linearly independent* provided that for any distinct elements $x_1, \cdots, x_n \in X$ and $r_1, \cdots, r_n \in R$,

$$r_1 x_1 + \cdots + r_n x_n = 0 \Rightarrow r_i = 0, \ i = 1, \cdots, n.$$

A nonempty subset X of left R-module M is called a *module basis* of M over R if X is linearly independent and each element in M is a linear combination of elements in X, that is, for any $a \in M$, $a = s_1 u_1 + \cdots + s_t u_t$, where $u_i \in M$ and $s_i \in R$. A left R-module M is called a *free R-module* if it contains a nonempty module basis. Generally two module bases of a free R-module may have different cardinality. However if R is commutative, then any two module bases of a free R-module have the same cardinality [Hungerford (1974)]. In this case the cardinal number of any module basis of a free R-module is called the *rank* or *dimension*.

An R-module P over a unital ring is called *projective* if it is a direct summand of a free R-module, that is, there is a free R-module F and an R-module M such that $F \cong P \oplus M$. For a unital ring R, each free R-module over R is projective, however a projective R-module may not be a free R-module [Hungerford (1974)]. I. Kaplansky proved that a projective module over a unital local ring R (even R is not commutative) is free [Kaplansky (1958)].

Theorem 4.27. *Let R be an ℓ-simple totally ordered local domain. Suppose that $M_n(R)$ is an ℓ-algebra and an f-module over R. If its order extension $M_n(F)$ is ℓ-isomorphic to $M_n(F)$ with the entrywise order, then $M_n(R)$ is ℓ-isomorphic to the ℓ-algebra $M_n(R)$ over R with the entrywise order.*

Proof. Similar to the proof of Theorem 4.26, we have the direct sum
$$M_n(R) = \sum_{1 \le i,j \le n} b_{ij}^{\perp\perp},$$
where $\{b_{ij} \mid 1 \le i,j \le n\}$ is a basis of $M_n(R)$. Since $M_n(R)$ is a free R-module, each R-module $b_{ij}^{\perp\perp}$ is projective, so that R is local implies that each $b_{ij}^{\perp\perp}$ is a free R-module. Since $M_n(R)$ has rank n^2 over R, each of its n^2 summands must have rank 1, and hence each $b_{ij}^{\perp\perp} = Rs_{ij}$ for some $0 < s_{ij} \in M_n(R)$. Hence $\{s_{ij} \mid 1 \le i,j \le n\}$ is a basis which spans $M_n(R)$ as an R-module. Therefore condition (2) in Theorem 4.25 is satisfied, and hence $M_n(R)$ is ℓ-isomorphic to the ℓ-algebra $M_n(R)$ over R with the entrywise order. □

The proof of the following result is similar to that of Corollary 4.9, and hence is omitted.

Corollary 4.9. *Let R be a totally ordered local domain that is a subring of $\mathbb{R}$. If $M_n(R)$ is an ℓ-algebra over R such that identity matrix is positive, then $M_n(R)$ is ℓ-isomorphic to the ℓ-algebra $M_n(R)$ with the entrywise order.*

For the remainder of this section, we determine lattice orders on $M_2(R)$ to make it into an ℓ-algebra over R, where R is a GCD domain. We first consider lattice orders on $M_2(F)$, where F is a totally ordered field, by using the idea of *invariant cones*. We need some preparations on invariant cones first.

Let $F^2 = F \oplus F$ be the 2-dimensional vector space over F. Each vector in F^2 is written as a column vector. A *cone* in F^2 is the positive cone of a partially ordered vector space F^2 over F. Let P be the positive cone of an ℓ-algebra $M_2(F)$ over F. A cone O in F^2 is said to be a *P-invariant cone* if for every $f \in P$, $fO \subseteq O$, where $fO = \{fv \mid v \in O\}$. If $O \ne \{0\}$, then O is called a *nontrivial P-invariant cone*. As an example, for the coordinatewise order on F^2, the cone $F^+ + F^+$ is $M_2(F^+)$-invariant, where $M_2(F^+)$ is the positive cone of the entrywise order on $M_2(F)$.

For a subset K of F^2, define

$$\operatorname{cone}_F(K) = \{\sum \alpha_i v_i \mid \alpha_i \in F^+, v_i \in K\},$$

where the sum is certainly a finite sum. It is easily verified that $\operatorname{cone}_F(K)$ is closed under the addition of F^2 and positive scalar multiplication (Exercise 46). If a cone $O = \operatorname{cone}_F(K)$ for some finite subset K of F^2 and K is a minimal finite set generating the cone, then vectors in K are called the *edges* of O. That K is a minimal set generating the cone means that any proper subset of K cannot generate the cone.

Theorem 4.28. *Let F be a totally ordered field and $M_2(F)$ be an ℓ-algebra over F with the positive cone P. Then $M_2(F)$ is ℓ-isomorphic to an ℓ-algebra $M_2(F)$ with the positive cone $P_1 \subseteq M_2(F^+)$.*

Proof. Since $M_2(F)$ is Archimedean and finite-dimensional over F, $M_2(F)$ is a direct sum of totally ordered subspaces over F by Corollary 1.3. $M_2(F)$ must be a direct sum of four totally ordered subspaces, that is, $M_2(F) = T_1 \oplus T_2 \oplus T_3 \oplus T_4$, where each T_i is a totally ordered subspace over F, so each T_i is 1-dimensional over F. We may assume that $T_i = Ff_i$ for some $0 < f_i \in T_i$. Then $T_i^+ = F^+ f_i$, $i = 1, 2, 3, 4$. We omit the proof of this fact and refer the reader to [Steinberg (2010)].

We divide the proof of Theorem 4.28 into several lemmas.

Lemma 4.18. *There is a nontrivial P-invariant cone in F^2.*

Proof. Let $\mathcal{M} = \{N \subseteq F^2 \mid N \text{ is a null space of some nonzero } f \in P\}$. Take $N \in \mathcal{M}$ with largest dimension and $u \notin N$. Define

$$O = \{gu \mid g \in P \text{ and } gN = 0\}.$$

Then $O \neq \{0\}$ since $fu \neq 0$, $O + O \subseteq O$ and $F^+ O \subseteq O$. If $v \in O \cap -O$, then $v = fu = -gu$ for some $f, g \in P$ and $fN = gN = \{0\}$. Thus $(f + g)u = 0$ and $(f + g)N = \{0\}$ implies that $f + g = 0$, and hence $f = g = 0$. Therefore $v = 0$ and $O \cap -O = \{0\}$. Hence O is a cone of F^2 and it is clear that O is a P-invariant cone. $\qquad\square$

Lemma 4.19. *Each nontrivial P-invariant cone of F^2 contains two linearly independent vectors over F.*

Proof. Let O be a P-invariant cone of F^2. Consider the subspace M spanned by O. Then $fM \subseteq M$ for each $f \in P$. Then since each matrix in $M_2(F)$ is a difference of two matrices in P, $gM \subseteq M$ for each matrix $g \in M_2(F)$. Hence $M = F^2$ (Exercise 47), so O contains two linearly independent vectors since O spans M. $\qquad\square$

Lemma 4.20. *Suppose that $O = cone_F(K)$ is a cone with a minimal finite set K, that is, for any proper subset K_1 of K, $O \neq cone_F(K_1)$. If $0 < w < k$ for some $w \in F^2$, $k \in K$, then $w = \alpha k$ for some $\alpha \in F^+$.*

Proof. Suppose that $K = \{k, k_1, \cdots, k_n\}$. Then $w \in O$ implies that $w = \alpha k + \alpha_1 k_1 + \cdots + \alpha_n k_n$ and $k - w \in O$ implies that $k - w = \beta k + \beta_1 k_1 + \cdots + \beta_n k_n$ with $\alpha, \beta, \alpha_i, \beta_i \in F^+$, so

$$k = (\alpha + \beta)k + (\alpha_1 + \beta_1)k_1 + \cdots + (\alpha_n + \beta_n)k_n.$$

Since K is minimal, $\alpha + \beta = 1$ and $\alpha_i + \beta_i = 0$, $i = 1, \cdots, n$, and hence $\alpha_i = \beta_i = 0$, $i = 1, \cdots, n$. Hence $w = \alpha k$ with $0 < \alpha \in F$. $\square$

Lemma 4.21. *Let O be a nontrivial P-invariant cone. For any $0 \neq v \in O$, $Pv \subseteq O$ is a nontrivial P-invariant cone. Moreover $Pv = cone_F(\{k_1, k_2\})$ is a lattice order of F^2, where k_1, k_2 are disjoint basic elements.*

Proof. Since $Pv \subseteq O$, $Pv \cap -Pv \subseteq O \cap -O = \{0\}$, and hence Pv is a P-invariant cone of F^2. Since

$$M_2(F) = Ff_1 \oplus Ff_2 \oplus Ff_3 \oplus Ff_4,$$

$$Pv = F^+(f_1 v) + F^+(f_2 v) + F^+(f_3 v) + F^+(f_4 v).$$

Let $k_i = f_i v$, $i = 1, \cdots, 4$. Then each $k_i \in Pv$ and $Pv = cone_F(K)$ with $K = \{k_1, k_2, k_3, k_4\}$. Certainly some k_j may be zero. Since Pv is a nontrivial P-invariant cone, Pv contains two linearly independent vectors by Lemma 4.19, so among nonzero vectors in K, there are at least two of them that are linearly independent since if any two different nonzero vectors in K are linearly dependent, then it is not possible for Pv to contain two linearly independent vectors. We may assume k_1 and k_2 are linearly independent over F. Suppose that $k_1, k_2 \in K' \subseteq K$ and K' is minimal with the property that $Pv = cone_F(K')$. We claim that $K' = \{k_1, k_2\}$.

Suppose that, for instance, $k_3 \neq 0$ and $k_3 \in K'$. Since k_1, k_2, k_3 are linearly dependent, $k_3 = \gamma_1 k_1 + \gamma_2 k_2$ for some $\gamma_1, \gamma_2 \in F$. We claim that $\gamma_1 = 0$ or $\gamma_2 = 0$. The key to show this is using Lemma 4.20. Suppose that $\gamma_1 > 0$ and $\gamma_2 > 0$. By Lemma 4.20, $\gamma_1 k_1 = \alpha_1 k_3$ and $\gamma_2 k_2 = \alpha_2 k_3$ for some $\alpha_1, \alpha_2 \in F^+$ since $\gamma_1 k_1 < k_3$ and $\gamma_2 k_2 < k_3$, which is a contradiction. Certainly γ_1, γ_2 cannot be both negative. Suppose that $\gamma_1 > 0$ and $\gamma_2 < 0$. Then $k_3 + (-\gamma_2)k_2 = \gamma_1 k_1$, and similarly since k_1 is an edge of Pv, $k_3 = \beta_1(\gamma_1 k_1)$ and $-\gamma_2 k_2 = \beta_2(\gamma_1 k_1)$, which is a contradiction again. Similarly $\gamma_1 < 0$ and $\gamma_2 > 0$ are not possible. Thus we must have $\gamma_1 = 0$ or $\gamma_2 = 0$, and hence $k_3 = \gamma_2 k_2$ or $k_3 = \gamma_1 k_1$, which contradicts with the minilarity of K'. Therefore $K' = \{k_1, k_2\}$ and $Pv = \operatorname{cone}_F(\{k_1, k_2\})$.

Now we show that Pv is actually a lattice order in F^2. To this end, we show that for any $\alpha, \beta \in F$, $\alpha k_1 + \beta k_2 \geq 0$ if and only if $\alpha, \beta \in F^+$. Certainly if $\alpha, \beta \in F^+$, then $\alpha k_1 + \beta k_2 \geq 0$. Conversely suppose that $\alpha k_1 + \beta k_2 \geq 0$. Then clearly α, β cannot be both less than zero. Assume that $\alpha > 0$ and $\beta < 0$. Then we have $\alpha k_1 \geq -\beta k_2 > 0$ and by Lemma 4.20, $-\beta k_2 = \gamma \alpha k_1$ for some $0 < \gamma \in F$, which contradicts with the fact that k_1, k_2 are linearly independent over F. Similarly that $\alpha < 0$ and $\beta > 0$ is impossible. Therefore we must have $\alpha, \beta \in F^+$. Therefore Pv is a lattice order and F^2 is a vector lattice over F with the positive cone Pv. $\square$

We are ready to complete the proof of Theorem 4.28. Let O be a nontrivial P-invariant cone and $0 \neq v \in O$. Then $Pv = \operatorname{cone}_F(\{k_1, k_2\})$ by Lemma 4.21. Define matrix $h = (k_1, k_2) \in M_2(F)$. h is invertible since k_1, k_2 are linearly independent, and hence h defines the inner isomorphism $x \rightarrow h^{-1}xh$ from $M_2(F)$ to $M_2(F)$. Let $P_1 = h^{-1}Ph$ and $O_1 = h^{-1}(Pv)$. Then P_1 is a lattice order on $M_2(F)$ and O_1 is a P_1-invariant cone. Since $Pv = F^+k_1 + F^+k_2$,

$$O_1 = h^{-1}(Pv) = h^{-1}(F^+k_1 + F^+k_2) = h^{-1}h(F^+)^2 = (F^+)^2,$$

and hence for any $f \in P_1$, $fO_1 \subseteq O_1$ implies that $P_1 \subseteq M_2(F^+)$. This completes the proof of Theorem 4.28. $\square$

The above nice idea of using P-invariant cones to connect ℓ-algebras $M_n(F)$ with vector lattices F^n was due to P. Wojciechowski when we spent some pleasant time working on Weinberg's conjecture around year 2000. It provides us a useful method in studying ℓ-rings. We are going to use this method again next section.

In the following we give a more concrete description of lattice orders on $M_2(F)$ to make it into an ℓ-algebra over F. By Theorem 4.28, an ℓ-algebra $M_2(F)$ is ℓ-isomorphic to the ℓ-algebra with the positive cone that is contained in $M_2(F^+)$. Thus we just need to consider $M_2(F)$ with the positive cone $P \subseteq M_2(F^+)$. Working on $P \subseteq M_2(F^+)$ will simplify calculation, for instance, if $0 \neq f \in P$ is nilpotent, since each entry of f is in F^+, then either $f = ae_{12}$ or $f = be_{21}$ for some $0 < a, 0 < b \in F$.

As we mentioned before, $M_2(F)$ is a direct sum of four totally ordered subspaces over F, and hence

$$M_2(F) = Ff_1 + Ff_2 + Ff_3 + Ff_4,$$

with the positive cone

$$P = F^+ f_1 + F^+ f_2 + F^+ f_3 + F^+ f_4,$$

where $f_i \in M_2(F^+)$, $i = 1, 2, 3, 4$.

Since f_1, f_2, f_3, f_4 are linearly independent, they contain at most two nilpotent elements. We consider the number of nilpotent elements among f_1, f_2, f_3, f_4.

(I) There are two nilpotents in $\{f_1, f_2, f_3, f_4\}$. In this case $(M_2(F), P)$ is ℓ-isomorphic to the ℓ-algebra $M_2(F)$ with the entrywise order.

Suppose $f_1 = ae_{12}$ and $f_2 = be_{21}$ with $0 < a, b \in F$. Multiplying f_1, f_2 by $\frac{1}{a}, \frac{1}{b}$ respectively, we may assume that $f_1 = e_{12}$ and $f_2 = e_{21}$. Then $f_1 f_2 = e_{11}$ and $f_2 f_1 = e_{22}$ are both in P, so we may assume that $e_{11} = cf_3$ and $e_{22} = df_4$ for some $0 < c, 0 < d \in F$. Thus we may replace f_3, f_4 by e_{11}, e_{22}. Therefore $P = F^+ e_{11} + F^+ e_{12} + F^+ e_{21} + F^+ e_{22} = M_2(F^+)$.

(II) There is one nilpotent element in $\{f_1, f_2, f_3, f_4\}$. We may assume that $f_1 = e_{12}$. Suppose that

$$f_2 = \begin{pmatrix} a_2 & b_2 \\ c_2 & d_2 \end{pmatrix}, \quad f_3 = \begin{pmatrix} a_3 & b_3 \\ c_3 & d_3 \end{pmatrix}, \quad f_4 = \begin{pmatrix} a_4 & b_4 \\ c_4 & d_4 \end{pmatrix}.$$

Since f_1, f_2, f_3, f_4 are linearly independent, one of c_2, c_3, c_4 is not zero. We may assume that $c_2 > 0$. Then

$$f_1 f_2 = \begin{pmatrix} c_2 & d_2 \\ 0 & 0 \end{pmatrix}, \quad f_2 f_1 = \begin{pmatrix} 0 & a_2 \\ 0 & c_2 \end{pmatrix},$$

imply that one of f_3, f_4 has zero second row and one of f_3, f_4 has zero first column. Since $c_2 > 0$, we may assume that

$$f_3 = \begin{pmatrix} 1 & a \\ 0 & 0 \end{pmatrix}, \quad f_4 = \begin{pmatrix} 0 & b \\ 0 & 1 \end{pmatrix}.$$

Then $1 = -(a+b)f_1 + f_3 + f_4$ with $a + b > 0$.

By Cayley-Hamilton equation or a direct calculation, for any $f \in M_2(F)$, $f^2 = (\mathrm{tr} f)f - (\det f)1$, where 1 is the identity matrix and $\mathrm{tr} f$ is the trace of f. Thus

$$f_2^2 = (\mathrm{tr} f_2)f_2 - (\det f_2)1$$
$$= (\mathrm{tr} f_2)f_2 - (\det f_2)(-(a+b)f_1 + f_3 + f_4).$$

Then $f_2^2 \geq 0$ implies that $(\det f_2)(a+b) \geq 0$ and $-(\det f_2) \geq 0$, so $\det f_2 = 0$, and hence $f_2^2 = (\mathrm{tr} f_2)f_2$. Since f_1 is the only nilpotent element, $\mathrm{tr} f_2 \neq 0$, so we may assume that f_2 is idempotent by changing f_2 to $(\mathrm{tr} f_2)^{-1} f_2$. That is, we may assume that $a_2 + d_2 = 1$.

Since

$$(f_2 + f_3)^2 = (\mathrm{tr}(f_2 + f_3))(f_2 + f_3) - \det(f_2 + f_3)1$$
$$= (\mathrm{tr}(f_2 + f_3))(f_2 + f_3) - \det(f_2 + f_3)(-(a + b)f_1 + f_3 + f_4)$$
$$\geq 0,$$

we have $\det(f_2 + f_3) = 0$. Thus

$$(f_2 + f_3)^2 = (\mathrm{tr}(f_2 + f_3))(f_2 + f_3) = 2(f_2 + f_3),$$

and hence $f_2 f_3 + f_3 f_2 = f_2 + f_3$ from $f_2^2 = f_2$ and $f_3^2 = f_3$. Since

$$f_2 f_3 = \begin{pmatrix} a_2 & a_2 a \\ c_2 & c_2 a \end{pmatrix} = f_2 + \alpha f_1 + \beta f_4,$$

with $\alpha, \beta \in F^+$ (Exercise 87), we must have $\alpha f_1 + \beta f_4 = 0$. Therefore $f_2 f_3 = f_2$, $f_3 f_2 = f_3$. Similarly $f_4 f_2 = f_2$ and $f_2 f_4 = f_4$ (Exercise 88).

Since $\mathrm{tr}(f_1 + f_2) = 1$ and $\det(f_1 + f_2) = -c_2$, by Cayley-Hamilton equation

$$(f_1 + f_2)^2 = f_1 + f_2 - c_2(a + b)f_1 + c_2 f_3 + c_2 f_4$$
$$= f_1 f_2 + f_2 f_1 + f_2$$

and hence $1 = c_2(a + b)$ and $f_1 f_2 + f_2 f_1 = c_2 f_3 + c_2 f_4$ (Exercies 89). Multiplying the equation from the left by f_4, we get $f_2 f_1 = c_2 f_4$, and it follows that $f_1 f_2 = c_2 f_3$.

By changing f_1 to $(a + b)f_1$, we have $1 = -f_1 + f_2 + f_3$ and following multiplication table for $\{f_1, f_2, f_3, f_4\}$,

	f_1	f_2	f_3	f_4
f_1	0	f_3	0	f_1
f_2	f_4	f_2	f_2	f_4
f_3	f_1	f_3	f_3	f_1
f_4	0	f_2	0	f_4

and $P = F^+ f_1 + F^+ f_2 + F^+ f_3 + F^+ f_4$.

(III) $M_2(F)$ is ℓ-reduced.

Take $a, b \in F$ with $a > b > 0$ and define the following matrices.

$$f_1 = \begin{pmatrix} 1 & a \\ 0 & 0 \end{pmatrix}, f_2 = \begin{pmatrix} 0 & 0 \\ b^{-1} & 1 \end{pmatrix}, \ f_3 = \begin{pmatrix} 1 & b \\ 0 & 0 \end{pmatrix}, \ f_4 = \begin{pmatrix} 0 & 0 \\ a^{-1} & 1 \end{pmatrix}.$$

Then

$$1 = -\frac{b}{a - b}f_1 + \frac{a}{a - b}f_2 + \frac{a}{a - b}f_3 - \frac{b}{a - b}f_4$$

and we have the following multiplication table.

	f_1	f_2	f_3	f_4
f_1	f_1	$-\alpha^{-1}\beta f_3$	f_3	f_1
f_2	$-\alpha^{-1}\beta f_4$	f_2	f_2	f_4
f_3	f_1	f_3	f_3	$-\beta^{-1}\alpha f_1$
f_4	f_4	f_2	$-\beta^{-1}\alpha f_2$	f_4

where $\alpha = -b/(a-b)$ and $\beta = a/(a+b)$. Then $P = F^+ f_1 + F^+ f_2 + F^+ f_3 + F^+ f_4$. We omit the proof of this case and refer the reader to [Steinberg (2010)].

We notice that the identity matrix with respect to the lattice orders in (II) and (III) is not positive. The lattice orders in (II) and (III) could be obtained by using the method in Theorem 1.19(2) from the entrywise order $M_2(F^+)$. In fact, using the following matrices

$$f = \begin{pmatrix} 1 & 1 \\ 1 & 0 \end{pmatrix}, \quad g = \begin{pmatrix} 1 & 1 \\ a & b \end{pmatrix}, \quad a > b > 0,$$

the lattice order in (II) is ℓ-isomorphic to $f M_2(F^+)$ and the lattice order in (III) is ℓ-isomorphic to $g M_2(F^+)$. For instance, for the positive cone $f M_2(F^+)$, the following matrices are disjoint and a vector space basis over F.

$$h_1 = \begin{pmatrix} 0 & 1 \\ 0 & 0 \end{pmatrix}, h_2 = \begin{pmatrix} 1 & 0 \\ 1 & 0 \end{pmatrix}, \quad h_3 = \begin{pmatrix} 1 & 0 \\ 0 & 0 \end{pmatrix}, \quad h_4 = \begin{pmatrix} 0 & 1 \\ 0 & 1 \end{pmatrix}.$$

The multiplication table of $\{h_1, h_2, h_3, h_4\}$ is exactly the same as the table in (II), and hence the lattice order in (II) is ℓ-isomorphic to $f M_2(F^+)$. We leave the verification of these facts as an exercise (Exercise 48).

Now for an ℓ-simple totally ordered greatest common divisor domain R, suppose that $M_2(R)$ is an ℓ-algebra and an f-module over R. We describe lattice orders on $M_2(R)$ using the results on its order extension $M_2(F)$. By Theorem 4.23, the lattice order on $M_2(R)$ is extended to a lattice order on $M_2(F)$, where F is the totally ordered quotient field of R. By the above results, the lattice order on $M_2(F)$ is ℓ-isomorphic to $u M_2(F^+)$ for some invertible matrix $u \in M_2(F^+)$. Similar to the proof of Theorem 4.26,

$$M_2(R) = \sum_{1 \leq i,j \leq 2} R a_{ij},$$

where $a_{ij} = q_{ij}(hue_{ij}h^{-1})$ are disjoint with $0 < q_{ij} \in F$ and $h \in M_2(F^+)$ invertible. Let $u = (u_{ij})$. A direct calculation shows that for $1 \leq i, j \leq 2$,

$$a_{ij}a_{rs} = u_{jr}q_{ij}q_{rs}q_{is}^{-1}a_{is},$$

and hence each $u_{jr}q_{ij}q_{rs}q_{is}^{-1} \in R$. By a calculation similar to that in Theorem 4.25, we have $(\Pi_{1 \leq i,j \leq 2}q_{ij})(\det u)^2$ is a unit in R (Exercise 49).

We know that positive cone $uM_2(F^+)$ has three nonisomorphic cases and if u is identity matrix, then it is the entrywise order and by Corollary 4.8, $M_2(R)$ is ℓ-siomorphic to $M_2(R)$ with the entrywise order. We consider below the other two cases.

In the second case (II), $u = \begin{pmatrix} 1 & 1 \\ 1 & 0 \end{pmatrix}$. Since $\det u = -1$, $(\Pi_{1 \leq i,j \leq 2}q_{ij})$ is a unit in R, and since $a_{11}^2 = q_{11}a_{11}$, $a_{12}^2 = q_{12}a_{12}$, and $a_{21}^2 = q_{21}a_{21}$, we have $q_{11}, q_{12}, q_{21} \in R$. Suppose that

$$1 = k_{11}a_{11} + k_{12}a_{12} + k_{21}a_{21} + k_{22}a_{22},$$

for some $k_{11}, k_{12}, k_{21}, k_{22} \in R$, where 1 is the identity matrix. Then

$$1 = k_{11}q_{11}(ue_{11}) + k_{12}q_{12}(ue_{12}) + k_{21}q_{21}(ue_{21}) + k_{22}q_{22}(ue_{22}).$$

However we know that $1 = -ue_{22} + ue_{12} + ue_{21}$, and hence $k_{12}q_{12} = 1$ and $k_{21}q_{21} = 1$. Hence q_{12}, q_{21} are unit in R. Then since $q_{11}q_{12}q_{21}q_{22} \in R$ is a unit, $q_{11}q_{22} \in R$ is a unit. Define

$$c_{11} = a_{11}, \; c_{12} = q_{12}^{-1}a_{12}, \; c_{21} = q_{21}^{-1}a_{21}, \; c_{22} = (q_{11}q_{22})^{-1}a_{22}.$$

Since $q_{12}, q_{21}, q_{11}q_{22}$ are positive unit in R, $\{c_{11}, c_{12}, c_{21}, c_{22}\}$ is also a basis that spans $M_2(R)$ as an R-module. It is straightforward to verify the following multiplication table (Exercise 50).

	c_{11}	c_{12}	c_{21}	c_{22}
c_{11}	$q_{11}c_{11}$	$q_{11}c_{12}$	c_{11}	c_{12}
c_{12}	c_{11}	c_{12}	0	0
c_{21}	$q_{11}c_{21}$	$q_{11}c_{22}$	c_{21}	c_{22}
c_{22}	c_{21}	c_{22}	0	0

Define

$$w = \begin{pmatrix} q_{11} & 1 \\ 1 & 0 \end{pmatrix} \in M_2(R^+).$$

From $\det w = -1$, w is invertible in $M_2(R)$, and hence $wM_2(R^+)$ is the positive cone of a lattice order on $M_2(R)$ by Theorem 1.19(2). Define

$$h_{11} = we_{11}, \; h_{12} = we_{12}, \; h_{21} = we_{21}, \; h_{22} = we_{22}.$$

The multiplication table for $\{h_{11}, h_{12}, h_{21}, h_{22}\}$ is exactly the same as the table for $\{c_{11}, c_{12}, c_{21}, c_{22}\}$. Therefore the ℓ-algebra $M_2(R)$ is ℓ-isomorphic to the ℓ-algebra $M_2(R)$ with the positive cone $wM_2(R^+)$.

In the third case (III), $u = \begin{pmatrix} 1 & 1 \\ a & b \end{pmatrix}$ with $a, b \in F$ and $a > b > 0$. Since

$$a_{11}^2 = q_{11}a_{11}, \quad a_{12}^2 = aq_{12}a_{12}, \quad a_{21}^2 = q_{21}a_{21}, \quad a_{22}^2 = bq_{22}a_{22},$$

$q_{11}, aq_{12}, q_{21}, bq_{22} \in R$. Similar to case (II), for some $k_{11}, k_{12}, k_{21}, k_{22} \in R$

$$1 = k_{11}q_{11}(ue_{11}) + k_{12}q_{12}(ue_{12}) + k_{21}q_{21}(ue_{21}) + k_{22}q_{22}(ue_{22})$$

$$= -\frac{b}{a-b}(ue_{11}) + \frac{1}{a-b}(ue_{12}) + \frac{a}{a-b}(ue_{21}) - \frac{1}{a-b}(ue_{22}),$$

and hence

$$-\frac{b}{a-b} = k_{11}q_{11}, \quad \frac{1}{a-b} = k_{12}q_{12}, \quad \frac{a}{a-b} = k_{21}q_{21}, \quad -\frac{1}{a-b} = k_{22}q_{22}.$$

So if we set $m = \frac{a}{a-b}$, then $m \in R^+$, and $m - 1 = \frac{b}{a-b} \in R^+$,

$$k_{12}aq_{12} = k_{21}q_{21} = m \quad \text{and} \quad k_{22}bq_{22} = k_{11}q_{11} = 1 - m.$$

We know that $q_{11}q_{12}q_{21}a_{22}(a - b)^2$ is a unit in R, and it follows that

$$q_{11}(aq_{12})q_{21}(bq_{22}) = m(m-1)q_{11}q_{12}q_{21}a_{22}(a-b)^2.$$

Then $\gcd(m, m-1) = 1$ implies that $(aq_{12})q_{21} = mr$ and $q_{11}(bq_{22}) = (m-1)s$, where $r, s \in R^+$ are unit, and hence $q_{11}q_{22} = q_{12}q_{21}v$ and $v = r^{-1}s \in R^+$ is a unit. Define $d_{11} = a_{11}, d_{12} = a_{12}, d_{21} = a_{21}$, and $d_{22} = v^{-1}a_{22}$. Then $\{d_{11}, d_{12}, d_{21}, d_{22}\}$ is a disjoint set that spans $M_2(R)$ and has the following multiplication table.

	d_{11}	d_{12}	d_{21}	d_{22}
d_{11}	$q_{11}d_{11}$	$q_{11}d_{12}$	$q_{21}d_{11}$	$q_{21}d_{12}$
d_{12}	$aq_{12}d_{11}$	$aq_{12}d_{12}$	$v^{-1}(bq_{22})d_{11}$	$v^{-1}(bq_{22})d_{12}$
d_{21}	$q_{11}d_{21}$	$q_{11}d_{22}$	$q_{21}d_{21}$	$q_{21}d_{22}$
d_{22}	$aq_{12}d_{21}$	$aq_{12}d_{22}$	$v^{-1}(bq_{22})d_{21}$	$v^{-1}(bq_{22})d_{22}$

Define the matrix

$$y = \begin{pmatrix} q_{11} & q_{21} \\ aq_{12} & v^{-1}(bq_{22}) \end{pmatrix} \in M_2(R^+).$$

Since

$$\det y = v^{-1}q_{11}(bq_{22}) - q_{21}(aq_{12}) = v^{-1}(m-1)s - mr = -r$$

is a unit in R, y is invertible in $M_2(R)$ and $yM_2(R)$ defines the positive cone of a lattice ordered on $M_2(R)$ by Theorem 1.19(2). Let $m_{ij} = ye_{ij}$, $1 \le i, j \le 2$. It is easily verified that m_{ij} and d_{ij} have the same multiplication table, so the ℓ-algebra $M_2(R)$ is ℓ-isomorphic to $M_2(R)$ with the positive cone $yM_2(R^+)$.

Therefore we have described all the lattice orders on $M_2(R)$. An interesting fact is that for a totally ordered subfield F of $\mathbb{R}$, any ℓ-algebra $M_n(F)$ is ℓ-isomorphic to an ℓ-algebra $M_n(F)$ with the positive cone $fM_n(F^+)$, where $f \in M_n(F^+)$ is an invertible matrix. The reader is referred to [Steinberg (2010)] for more details. However it is still an open question if this fact is true for matrix ℓ-algebras over non-Archimedean totally ordered fields and totally ordered Archimedean GCD (UFD, PID) domains.

4.6 d-elements that are not positive

When we define d-elements in chapter 1, we assume that they are positive. In this section we consider d-elements that are not positive. Those elements arise when considering unital ℓ-rings in which $1 \not\geq 0$.

Let R be a unital ℓ-ring. An element $a \in R$ is called an f-element (d-element) if for any $x, y \in R$,

$$x \wedge y = 0 \Rightarrow ax \wedge y = xa \wedge y = 0 \ (ax \wedge ay = xa \wedge ya = 0).$$

We may call f-element and d-element defined in chapter 1 as *positive* f-element and d-element. Define

$$\overline{f}(R) = \{a \in R \mid a \text{ is an } f\text{-element of } R\},$$

and

$$\overline{d}(R) = \{a \in R \mid a \text{ is a } d\text{-element of } R\}.$$

So $f(R)^+ = \overline{f}(R) \cap R^+$ and $d(R) = \overline{d}(R) \cap R^+$. Clearly $1 \in \overline{f}(R) \subseteq \overline{d}(R)$. It is also clear that $\overline{f}(R)$ is closed under the addition and multiplication of R and $\overline{d}(R)$ is closed under the multiplication. If a is a d-element of R, then $aR^+ \subseteq R^+$ and $R^+a \subseteq R^+$. For a unital ℓ-ring, if $1 \geq 0$, then any d-element e is positive since $e \wedge 0 = e(1 \wedge 0) = e0 = 0$. Thus a unital ℓ-ring has a d-element that is not positive if and only if $1 \not\geq 0$. The following example shows that there are unital ℓ-rings with $1 \not\geq 0$ that contain positive f-element.

Example 4.4. Consider $R = M_3(\mathbb{Q})$ and matrix

$$f = \begin{pmatrix} 1 & 0 & 0 \\ 0 & 1 & 0 \\ 0 & 1 & 1 \end{pmatrix} \in M_3(\mathbb{Q}^+).$$

By Theorem 1.19, $P = fM_3(\mathbb{Q}^+)$ is the positive cone of a lattice order on $M_3(\mathbb{Q})$ to make it into an ℓ-ring. Since

$$f^{-1} = \begin{pmatrix} 1 & 0 & 0 \\ 0 & 1 & 0 \\ 0 & -1 & 1 \end{pmatrix} \notin M_3(\mathbb{Q}^+),$$

identity matrix 1 is not positive with respect to P. Now $e_{11} = fe_{11} \in P$ and it is straightforward to check that e_{11} is an f-element with respect to P (Exercise 51).

The ℓ-ring in Example 4.4 is not ℓ-reduced. For an ℓ-reduced unital ℓ-ring, the situation is different.

Theorem 4.29. *Let R be a unital ℓ-ring*

(1) If R is ℓ-reduced and 1 is not positive, then each nonzero d-element is not positive.

(2) Let $u \in R$ be an invertible element. Then u is a d-element if and only if $uR^+ \subseteq R^+, R^+u \subseteq R^+$, and $u^{-1}R^+ \subseteq R^+, R^+u^{-1} \subseteq R^+$.

(3) Let u be an invertible d-element. If $a \in R$ is a basic element, then au and ua are both basic elements.

Proof. (1) We first assume that R is an ℓ-domain. Suppose that $a > 0$ is a d-element of R. Then $a(1^-) = (-a) \vee 0 = 0$ implies that $1^- = 0$, so $1 > 0$, which is a contradiction. Then R has no nonzero positive d-element. Now suppose that R is ℓ-reduced. By Theorem 1.30 R is a subdirect product of ℓ-domains, that is, there are ℓ-ideals I_k such that $\cap I_k = \{0\}$ and each R/I_k is an ℓ-domain. Let a be a d-element of R and $\bar{x} = x + I_k, \bar{y} = y + I_k \in R/I_k$ with $\bar{x} \wedge \bar{y} = 0$ in R/I_k. Then $(x - z) \wedge (y - z) = 0$ for some $z \in I_k$ and $(ax - az) \wedge (ay - az) = 0$, so $\bar{a}\bar{x} \wedge \bar{a}\bar{y} = 0$ in R/I_k. Similarly $\bar{x}\bar{a} \wedge \bar{y}\bar{a} = 0$. Thus $\bar{a} = a + I_k$ is a d-element of R/I_k. Since 1 is not positive in R, there is at least one k such that $\bar{1}$ is not positive in R/I_k, and hence $\bar{a}$ is not positive in R/I_k by the above argument. It follows that a is not positive in R.

(2) Suppose that u is a d-element. Then $uR^+ \subseteq R^+$ and $R^+u \subseteq R^+$. For $x \in R^+$, $u(u^{-1}x \wedge 0) = x \wedge 0 = 0$ implies $u^{-1}x \wedge 0 = 0$, so $u^{-1}x \geq 0$, and hence $u^{-1}R^+ \subseteq R^+$. Similarly $R^+u^{-1} \subseteq R^+$.

Conversely, suppose that $uR^+ \subseteq R^+$ and $u^{-1}R^+ \subseteq R^+$. If $x \wedge y = 0$ for $x, y \in R$, then

$$0 \leq u^{-1}(ux \wedge uy) \leq u^{-1}ux \wedge u^{-1}uy = x \wedge y = 0,$$

and hence $ux \wedge uy = 0$. Similarly $xu \wedge yu = 0$. Hence u is a d-element.

(3) Clearly $au > 0$ and $ua > 0$. Let $0 \leq x, y \leq au$. Then $0 \leq xu^{-1}, yu^{-1} \leq a$ implies that xu^{-1} and yu^{-1} are comparable, so x, y are comparable. Thus au is basic. Similarly ua is also basic. $\square$

In Chapter 1, we give a general method to construct lattice orders with $1 \not> 0$ on an ℓ-unital ℓ-ring. R. Redfield discovered another method to produce lattice orders with $1 \not> 0$ by changing multiplication of ℓ-unital ℓ-rings.

Let R be an ℓ-unital ℓ-ring with the identity element 1 and u in the center of R. Define a new multiplication $*$ on R for any $x, y \in R$,

$$x * y = xyu^{-1}.$$

Then $(R, +, *)$ is a ring with u as an identity element (Exercise 52). Now suppose that $u \not> 0$ and $u^{-1} > 0$. If $x \geq 0$ and $y \geq 0$, then $x*y = xyu^{-1} \geq 0$. Thus $(R, +, *)$ is an ℓ-ring with the identity element $u \not> 0$.

The ℓ-ring $(R, +, *)$ may be obtained by using Theorem 1.19(2).

Theorem 4.30. *Let ℓ-ring $(R, +, *)$ be defined as above. Then there exists a lattice order on $(R, +, *)$ with the positive cone P such that $u \in P$ and $R^+ = 1 * P$ with $1 \in P$.*

Proof. Define $P = uR^+$. Clearly $P + P \subseteq P$, and $P \cap -P = \{0\}$. For ua, ub with $a, b \in R^+$,

$$(ua) * (ub) = (ua)(ub)u^{-1} = u(ab) \in uR^+.$$

Thus $P * P \subseteq P$. So P is a partial order on $(R, +, *)$. For $x \in R$, with respect to P, $x \vee 0 = u(u^{-1}x \vee 0)$, where $u^{-1}x \vee 0$ is the sup of $u^{-1}x, 0$ with respect to R^+. Therefore $(R, +, *)$ is an ℓ-ring with the positive cone P and $u = u1 \in P$. For any $a \in R^+$, $a = 1 * (ua)$, so $R^+ = 1 * P$. We also have $1 = uu^{-1} \in uR^+ = P$. $\square$

Theorem 4.31. *Let L be an ℓ-field with $1 \not> 0$ and an ℓ-algebra over a totally ordered field F. Suppose that L satisfies the following conditions.*

(1) There is a vector space basis B of L over F with $B \subseteq \bar{d}(L)$.
(2) L has a basic element a such that $a^{\perp\perp} = Fa$.

Then there exists a lattice order $\succeq$ on L to make it into an ℓ-field with $1 \succ 0$ such that $L^+ = aP$, where P is the positive cone of $\succeq$.

Proof. Define $P = \{x \in L \mid xL^+ \subseteq L^+\}$. It is straightforward to check that $P + P \subseteq P$, $PP \subseteq P$, and $P \cap -P = \{0\}$ (Exercise 53). For any $x, y \in L$, define $x \succeq y$ if $x - y \in P$. Then L is a partially ordered field with respect to $\succeq$.

We show that $\succeq$ is actually a lattice order. First of all, for any $u \in B$, by Theorem 4.29, ua is basic, and $(ua)^{\perp\perp} = F(ua)$ (Exercise 54). Next if $u, v \in B$ and $u \neq v$, then $ua \wedge va = 0$. In fact, since ua and va are both basic elements, if $ua \wedge va \neq 0$, then $(ua)^{\perp\perp} = (va)^{\perp\perp}$ by Theorem 1.14, and hence $F(ua) = F(va)$. Hence u and v are linearly dependent over F, which is a contradiction. Therefore $ua \wedge va = 0$. Suppose that $z \succeq 0$ and $z = \alpha_1 b_1 + \cdots + \alpha_n b_n$, where $b_1, \cdots, b_n \in B$ are distinct and $\alpha_1, \cdots, \alpha_n \in F$. We show that each $\alpha_i \in F^+$. Suppose that $\alpha_1 < 0, \cdots, \alpha_k < 0$ in F and $\alpha_{k+1} > 0, \cdots, \alpha_n > 0$, $1 \leq k < n$. Then

$$0 \preceq -\alpha_1 b_1 \preceq \alpha_{k+1} b_{k+1} + \cdots + \alpha_n b_n$$

implies that

$$0 \leq -\alpha_1 b_1 a \leq \alpha_{k+1} b_{k+1} a + \cdots + \alpha_n b_n a.$$

Since $ua \wedge va = 0$ for any $u, v \in B$ and $u \neq v$, and $F^+ 1 \subseteq \overline{f}(L)$,

$$
\begin{aligned}
0 &\leq -\alpha_1 b_1 a \\
&= -\alpha_1 b_1 a \wedge (\alpha_{k+1} b_{k+1} a + \cdots + \alpha_n b_n a) \\
&\leq (-\alpha_1 b_1 a \wedge \alpha_{k+1} b_{k+1} a) + \cdots + (-\alpha_1 b_1 a \wedge \alpha_n b_n a) \\
&= 0.
\end{aligned}
$$

Thus $-\alpha_1 b_1 a = 0$, which is a contradiction. Hence each $\alpha_i \geq 0$ in F.

Now for $x \in L$, if $x = \beta_1 c_1 + \cdots + \beta_m c_m$ for some $\beta_1, \cdots, \beta_m \in F$ and $c_1, \cdots, c_m \in B$ are distinct. The least upper bound of x and 0 with respect to $\succeq$ is

$$x \vee_{\succeq} 0 = \beta_1^+ c_1 + \cdots + \beta_m^+ c_m.$$

We leave the verification of this fact as an exercise (Exercise 55). Therefore $\succeq$ is a lattice order on L. Clearly $1 \in P$ and $\{au \mid u \in B\}$ is disjoint and a vector space basis of L over F. Therefore $L^+ = aP$. $\qquad\square$

Let's look at an example that Theorem 4.31 may apply.

Example 4.5. Consider the field

$$L = \mathbb{Q}[\sqrt{2}, \sqrt{3}] = \{\alpha + \beta\sqrt{2} + \gamma\sqrt{3} + \delta\sqrt{6} \mid \alpha, \beta, \gamma, \delta \in \mathbb{Q}\}.$$

With respect to the coordinatewise order, L is an ℓ-field in which identity element 1 is positive. Suppose now that L is an arbitrary ℓ-field with the positive cone L^+. Since L is finite-dimensional over $\mathbb{Q}$, L has basic elements. If $\sqrt{2}L^+ \subseteq L^+$, $\sqrt{3}L^+ \subseteq L^+$, and there is a basic element a such that $a^{\perp\perp} = \mathbb{Q}a$, then by Theorem 4.31, $L^+ = aP$, where P is the positive cone of the coordinatewise order.

There are rings and algebras that cannot be made into an ℓ-ring and ℓ-algebra. In the following, we use idea of P-invariant cones to show complex field $\mathbb{C}$ and division algebra $\mathbb{H}$ of real quaternions cannot be an ℓ-algebra over $\mathbb{R}$. We prove that only finite-dimensional ℓ-algebra over $\mathbb{R}$ is $\mathbb{R}$ itself. We first review a few definitions and results on n-dimensional Euclidean space $\mathbb{R}^n$.

Let S be a subset of $\mathbb{R}^n$. A *cover* of S is a collection $\{U_i \mid i \in I\}$ of sets in $\mathbb{R}^n$ such that

$$S \subseteq \bigcup_{i \in I} U_i.$$

A cover of S is called an *open cover* if each U_i is an open set and a *finite cover* if index set I is fine. A *subcover* of the cover $\{U_i \mid i \in I\}$ is a collection $\{U_j \mid j \in J\}$ with $J \subseteq I$ such that

$$S \subseteq \bigcup_{j \in J} U_j.$$

A subset S of $\mathbb{R}^n$ is called *compact* if every open cover of S has a finite subcover. It is well-known that a subset S of $\mathbb{R}^n$ is compact if and only if it is closed and bounded in $\mathbb{R}^n$. Let S be a compact set and $\{K_i \mid i \in I\}$ be a collection of closed subsets of S. As a direct consequence of compactness of S, if for each finite set of indices $i_1, \cdots, i_n$, $K_{i_1} \cap \cdots \cap K_{i_n} \neq \emptyset$, then

$$\bigcap_{i \in I} K_i \neq \emptyset.$$

For a subset B of $\mathbb{R}^n$, $\overline{B}$ denotes the *closure* of B, which is the intersection of all closed sets containing B, and hence $\overline{B}$ is the smallest closed subset containing B. We first prove a basic result which will be used later in the proof.

Lemma 4.22. *Suppose that N is a subspace of $\mathbb{R}^n$ over $\mathbb{R}$ which is totally ordered. If $\overline{N^+} \cap \overline{-N^+} = \{0\}$, then N is 1-dimensional over $\mathbb{R}$.*

Proof. Since N is a subspace of $\mathbb{R}^n$, N must be closed (Exercise 56). Since $N = N^+ \cup -N^+$,

$$N = \overline{N} \Rightarrow N^+ \cup -N^+ = \overline{N^+} \cup \overline{-N^+}.$$

Then $\overline{N^+} \cap \overline{-N^+} = \{0\}$ implies that $N^+ = \overline{N^+}$ and $-N^+ = \overline{-N^+}$, so $N^+, -N^+$ are closed. Therefore N must be 1-dimensional over $\mathbb{R}$ (Exercise 57). $\qquad \square$

Theorem 4.32. *Suppose that A is a finite-dimensional division ℓ-algebra over $\mathbb{R}$. Then A must be totally ordered.*

Proof. If A is 1-dimensional over $\mathbb{R}$, then $A = \mathbb{R}1$ is totally ordered. Suppose that $\dim_{\mathbb{R}} A = n \geq 2$. We use P to denote the positive cone of ℓ-algebra A. By Lemma 2.5, we may consider A as a subalgebra of $M_n(\mathbb{R})$. As before, $\mathbb{R}^n$ denotes the n-dimensional Euclidean column space over $\mathbb{R}$.

For each $0 \neq v \in \mathbb{R}^n$, Pv is a nontrivial P-invariant cone. It is clear that $Pv + Pv \subseteq Pv$, $\mathbb{R}^+(Pv) \subseteq Pv$, and Pv is P-invariant. We show that $Pv \cap -Pv = \{0\}$. Suppose that $u \in Pv \cap -Pv$. Then $u = fv = -gv$ for some $f, g \in P$, so $(f + g)v = 0$. If $f + g \neq 0$, then A is a division algebra implies that $v = 0$, which is a contradiction. Thus $f + g = 0$, and hence $f = g = 0$ and $u = 0$. Therefore Pv is a P-invariant cone.

Let M be the subspace spanned by Pv. Then $fM \subseteq M$ for each $f \in A$ since Pv is P-invariant. Let $g_1, \cdots, g_n$ be a vector space basis of A over $\mathbb{R}$ and $0 \neq w \in M$. Then $g_1 w, \cdots, g_n w \in M$ are linearly independent, so M is an n-dimensional subspace. It follows that Pv contains n linearly independent vectors since Pv spans M. Let $f_1 v, \cdots, f_n v \in Pv$ be linearly independent over $\mathbb{R}$, where $f_1, \cdots, f_n \in P$. Then $f_1, \cdots, f_n$ are linearly independent, and $\mathrm{cone}_{\mathbb{R}}(K_v) \subseteq Pv$, where $K_v = \{f_1 v, \cdots, f_n v\}$. We note that $\mathrm{cone}_{\mathbb{R}}(K_v)$ is a closed subset of $\mathbb{R}^n$ (Exercise 58).

Since A is finite-dimensional, by Corollary 1.3, A is a finite direct sum of maximal convex totally ordered subspaces of A over $\mathbb{R}$. We show that each direct summand is 1-dimensional. Let T be a direct summand in the direct sum of A. For some $0 \neq v \in \mathbb{R}^n$, Tv is a totally ordered subspace of $\mathbb{R}^n$ with the positive cone $T^+ v$. Since $T^+ v \subseteq Pv$ and $\overline{Pv} \cap \overline{-Pv} = \{0\}$ (Exercise 59), we have $\overline{T^+ v} \cap \overline{-T^+ v} = \{0\}$. Thus by Lemma 4.22, Tv is 1-dimensional. Take $0 \neq f \in T$. Then $fv \in Tv$ is a basis over $\mathbb{R}$. For any $g \in T$, $gv = \alpha(fv)$ implies that $(g - \alpha f)v = 0$, so $g - \alpha f = 0$ and $g = \alpha f$. Thus T is 1-dimensional. It follows that A is a direct sum of n direct summands, and hence A contains n disjoint elements $f_1, \cdots, f_n$. As a direct consequence of this fact, we have for any $0 \neq v \in \mathbb{R}^n$, $Pv = \mathrm{cone}_{\mathbb{R}}(f_1 v, \cdots, f_n v)$ is closed.

Consider partially ordered set $\mathcal{M} = \{Pv \mid 0 \neq v \in \mathbb{R}^n\}$ under set inclusion. We show that $\mathcal{M}$ has a minimal element by Zorn's Lemma. Let

$\{Pv_\alpha \mid \alpha \in \Gamma\}$ be a chain in $\mathcal{M}$ and $S = \{v \in \mathbb{R}^n \mid |v| = 1\}$ be the unit sphere, where $|v|$ denotes the length of the vector v. Then the collection $\{Pv_\alpha \cap S \mid \alpha \in \Gamma\}$ is a chain of closed sets of S and each $Pv_\alpha \cap S \neq \emptyset$. Since S is closed and bounded in $\mathbb{R}^n$, S is compact, and hence

$$\bigcap_{\alpha \in \Gamma} (Pv_\alpha \cap S) \neq \emptyset.$$

Take $v \in \cap_{\alpha \in \Gamma}(Pv_\alpha \cap S)$. Then $v \neq 0$ since $0 \notin S$. Pv is a P-invariant cone contained in each Pv_α, that is, Pv is a lower bound of the chain $\{Pv_\alpha \mid \alpha \in \Gamma\}$ in $\mathcal{M}$. Therefore by Zorn's Lemma, $\mathcal{M}$ has a minimal element.

Suppose $Pu \in \mathcal{M}$ is a minimal element for some $0 \neq u \in \mathbb{R}^n$. Take $0 < f \in P$. Then $0 \neq fu \in Pu$ implies that $P(fu) \subseteq Pu$, and hence $P(fu) = Pu$. It follows that there is a $g \in P$ such that $gfu = fu$, so $gf = f$ and $g = 1$. Hence $1 \in P$ and for any $0 < h \in P$, $P(hu) = Pu$ implies that $jh = 1$ for some $j \in P$. Therefore we have proved that $1 > 0$ and for each nonzero positive element in A, its inverse is also positive. Then by Theorem 1.20(2), A is a d-ring, and hence A is totally ordered by Theorems 1.27 and 1.28. $\square$

Let $\mathbb{H}$ be the 4-dimensional vector space over $\mathbb{R}$ with the vector space basis $\{1, i, j, k\}$ having the following multiplication table.

	1	i	j	k
1	1	i	j	k
i	i	-1	k	$-j$
j	j	$-k$	-1	i
k	k	j	$-i$	-1

Then $\mathbb{H}$ is a 4-dimensional algebra over $\mathbb{R}$. For an element $x = a + bi + cj + dk \in \mathbb{H}$, where $a, b, c, d \in \mathbb{R}$, define $\bar{x} = a - bi - cj - dk$. Then $x\bar{x} = \bar{x}x = a^2 + b^2 + c^2 + d^2 \in \mathbb{R}$. Thus if $x \neq 0$, then x has the inverse $(a^2 + b^2 + c^2 + d^2)^{-1}\bar{x}$. Therefore $\mathbb{H}$ is a division ring, and hence $\mathbb{H}$ is actually a division algebra over $\mathbb{R}$, which is called division algebra of *real quaternions*.

Frobenius's Theorem in general ring theory states that a finite-dimensional division algebra over $\mathbb{R}$ is isomorphic to $\mathbb{R}, \mathbb{C}$, or $\mathbb{H}$ [Lam (2001)]. Since $\mathbb{C}$ and $\mathbb{H}$ cannot be a totally ordered algebra over $\mathbb{R}$ because of $i^2 = -1$, they cannot be ℓ-algebra over $\mathbb{R}$ by Theorem 4.32, so $\mathbb{R}$ is the only finite-dimensional division ℓ-algebra over $\mathbb{R}$.

Complex field $\mathbb{C}$ cannot be an ℓ-algebra over $\mathbb{R}$ was first proved by G. Birkhoff and R. S. Pierce [Birkhoff and Pierce (1956)]. Then R. McHaffey noticed that $\mathbb{H}$ cannot be an ℓ-algebra over $\mathbb{R}$ [McHaffey (1962)]. Their proofs are much simpler than the proof presented in Theorem 4.32. However P-invariant cone method may be used to prove more examples that cannot be an ℓ-algebra, for instance, matrix algebras $M_n(\mathbb{C})$ and $M_n(\mathbb{H})$ cannot be made into an ℓ-algebra over $\mathbb{R}$ [Steinberg (2010)].

Let F be a totally ordered subfield of $\mathbb{R}$. Define

$$C_F = \{a + bi \mid a, b \in F\},$$

and

$$H_F = \{a + bi + cj + dk \mid a, b, c, d \in F\}.$$

Then C_F is called the complex field over F and H_F is called the division algebra of quaternions over F. By using the same argument as in Theorem 4.32 with some modification, it can be shown that C_F, H_F cannot be made into an ℓ-algebra over F.

Also by using Theorem 4.33 below, for any integral domain R which is a totally ordered subring of $\mathbb{R}$, complex numbers and quaternions over R cannot be an ℓ-ring. In particular, for $R = \mathbb{Z}$, it means that complex integers and quaternion integers cannot be an ℓ-ring.

In section 4.4, we have considered extending lattice order on a lattice-ordered integral domain with positive identity to its quotient field. The results can be generalized to lattice-ordered integral domains with $1 \not\geq 0$.

Theorem 4.33. *Let R be a lattice-ordered integral domain. If for any nonzero element a of R, $Ra \cap \overline{f}(R) \neq \{0\}$, then its quotient field F can be made into an ℓ-ring extension of R.*

Proof. For $q \in F$, we have $q = \frac{a}{b}$, $a, b \in R$ with $b \neq 0$. Since $b \neq 0$, $Rb \cap \overline{f}(R) \neq \{0\}$, so there is $c \in R$ such that $0 \neq cb = d \in \overline{f}(R)$. Thus each element $q \in F$ can be written as $q = \frac{a}{d}$ with $0 \neq d \in \overline{f}(R)$.

Define the positive cone P on F as follows:

$$P = \{q \in F \mid q = \frac{a}{d}, \ 0 \leq a \in R, 0 \neq d \in \overline{f}(R)\}.$$

If $\frac{a}{d} = \frac{c}{e}$ with $a, c \in R, 0 \leq a, 0 \neq d, 0 \neq e \in \overline{f}(R)$, then $ae = cd$. Since $a \geq 0$ and $e \in \overline{f}(R)$, $ae \geq 0$, and hence $cd = |cd| = |c|d$. Thus $c = |c| \geq 0$ and P is well-defined.

It is clear that $P + P \subseteq P$, $PP \subseteq P$, and $P \cap -P = \{0\}$. For $q = \frac{a}{b}$ with $a \in R$, $0 \neq b \in \overline{f}(R)$, $q \vee 0 = \frac{a \vee 0}{b}$. These proofs are similar to that given in section 4.4 and we leave it as an exercise. $\square$

We note that Theorem 4.33 is also true for left (or right) Ore domains and we omit the proof which is similar to the proof of Theorem 4.17.

4.7　Lattice-ordered triangular matrix algebras

In this section we study lattice-ordered triangular matrix algebras. We use $T_n(F)$ to denote the $n \times n$ $(n \geq 2)$ upper triangular matrix algebras over a totally ordered field F. We construct all the lattice orders on $T_2(F)$ to make it into an ℓ-algebra over F. In section 1, we first construct all the lattice orders in which the identity matrix is positive, and then in section 2 we show each lattice order on $T_2(F)$ in which the identity matrix is not positive can be obtained from a lattice order in which the identity matrix is positive by using Theorem 1.19(2). In section 3, some conditions are provided for $T_n(F)$ to be ℓ-isomorphic to the ℓ-algebra $T_n(F)$ with the entrywise order.

4.7.1　*Lattice orders on $T_2(F)$ with $1 > 0$*

In this section we describe all the lattice orders on $T_2(F)$ to make it into an ℓ-algebra over F with identity matrix $1 > 0$. First we construct three Archimedean lattice orders over F.

We use P_0 to denote the positive cone of the entrywise order on $T_2(F)$, that is, $P_0 = T_2(F^+)$.

Recall that $e_{11}, e_{22}, e_{12} \in T_2(F)$ denote the standard matrix units. It is clear that $\{1, e_{22}, e_{12}\}$ is a vector space basis of vector space $T_2(F)$ over F and we have the following multiplication table for $\{1, e_{22}, e_{12}\}$.

	1	e_{22}	e_{12}
1	1	e_{22}	e_{12}
e_{22}	e_{22}	e_{22}	0
e_{12}	e_{12}	e_{12}	0

By Theorem 1.19(1), $T_2(F)$ becomes an Archimedean ℓ-unital ℓ-algebra over F with the positive cone

$$P_1 = F^+1 + F^+e_{22} + F^+e_{12}.$$

Since $e_{12}^2 = 0$, $(T_2(F), P_1)$ is not ℓ-reduced.

Let $k = e_{12} + e_{22}$. It is also clear that $\{1, e_{22}, k\}$ is a vector space basis for vector space $T_2(F)$ over F and we have the following multiplication table.

	1	e_{22}	k
1	1	e_{22}	k
e_{22}	e_{22}	e_{22}	e_{22}
k	k	k	k

By Theorem 1.19(1) again, $T_2(F)$ becomes an Archimedean ℓ-unital ℓ-algebra over F with the positive cone

$$P_2 = F^+1 + F^+e_{22} + F^+k.$$

Clearly, $(T_2(F), P_2)$ is ℓ-reduced. We also notice that $P_2 \subsetneq P_1 \subsetneq P_0$. We show that an Archimedean ℓ-unital ℓ-algebra $T_2(F)$ is ℓ-isomorphic or anti-ℓ-isomorphic to the ℓ-algebra $T_2(F)$ with the positive cone P_1, P_2, or P_3.

We first state a lemma that will be used in proofs. Recall that $f(T_2(F)) = \{a \in T_2(F) \mid |a|$ is an f-element of $T_2(F)\}$.

Lemma 4.23. *Let $T_2(F)$ be an ℓ-algebra over F.*

(1) If $T_2(F)$ is ℓ-reduced, then $T_2(F)$ is an ℓ-domain. Moreover if $T_2(F)$ is ℓ-unital, then $f(T_2(F))$ is totally ordered.

(2) Suppose that $T_2(F)$ is ℓ-unital. If $f(T_2(F))$ is totally ordered and $f(T_2(F))^\perp$ contains nonzero positive nilpotent elements, then $f(T_2(F)) = F1$.

Proof. (1) Let $0 \le u, v \in T_2(F)$ with $uv = 0$. Then $(vu)^2 = 0$, and hence $vu = 0$ since $T_2(F)$ is ℓ-reduced. Thus $(vzu)^2 = (uzv)^2 = 0$, for any $z \in T_2(F)^+$. Hence $vzu = uzv = 0$, for any $z \in T_2(F)^+$ since $T_2(F)$ is ℓ-reduced. Therefore

$$vT_2(F)u = uT_2(F)v = \{0\}.$$

By a direct calculation, we have that u is nilpotent or v is nilpotent (Exercise 60), and hence $u = 0$ or $v = 0$. If $T_2(F)$ is ℓ-unital, then by Theorem 1.27, $f(T_2(F))$ is also a totally ordered domain.

(2) Let $0 < a \in f(T_2(F))^\perp$ with $a^2 = 0$. Then $a = \alpha e_{12}$ for some $0 \ne \alpha \in F$. We notice that $T_2(F)$ cannot be an f-ring by Theorem 1.22(3) since it contains idempotent elements which are not central. We claim that $f(T_2(F))$ cannot be two-dimensional over F. In fact, if $f(T_2(F))$ is two-dimensional, then $T_2(K) = f(T_2(F)) \oplus f(T_2(F))^\perp$ as a vector lattice and $f(T_2(F))^\perp = Fa$. Let $e_{11} = b + c$, where $b \in f(T_2(F))$ and $c \in f(T_2(F))^\perp = Fa$. Then $b = e_{11} - c$ is an idempotent element, so $b = 1$ or $b = 0$ by Theorem 1.22, which is a contradiction. Thus $f(T_2(F))$ cannot be two-dimensional over F, and hence $f(T_2(F)) = F1$. $\qquad\square$

We also notice that if $T_2(F)$ is an ℓ-unital ℓ-algebra over F and $a \ge 0$ is a nilpotent element in $T_2(F)$, then Fa is an ℓ-ideal of $T_2(F)$ (Exercise 61).

An *anti-isomorphism* φ between two rings R and S is a group isomorphism between underlying additive groups of R and S, and for any $a, b \in R$, $\varphi(ab) = \varphi(b)\varphi(a)$. For instance, $\varphi : T_2(F) \to T_2(F)$ defined by

$$\begin{pmatrix} x & y \\ 0 & z \end{pmatrix} \to \begin{pmatrix} z & y \\ 0 & x \end{pmatrix}$$

is an anti-isomorphism. An anti-ℓ-isomorphism between two ℓ-rings is a ring anti-isomorphism which preserves the lattice orders.

Theorem 4.34. *Let $T_2(F)$ be an Archimedean ℓ-unital ℓ-algebra over F. If $T_2(F)$ is not ℓ-reduced, then*

(1) $T_2(F)$ is ℓ-isomorphic to $(T_2(F), P_0)$ provided 1 is not a basic element;
(2) $T_2(F)$ is ℓ-isomorphic or anti-ℓ-isomorphic to $(T_2(F), P_1)$ provided 1 is a basic element.

Proof. Let

$$I = \left\{ \begin{pmatrix} 0 & x \\ 0 & 0 \end{pmatrix} : x \in F \right\}.$$

Since $T_2(F)$ is not ℓ-reduced, there exists $a > 0$ which is nilpotent, so $a \in I$, and hence $I = Fa$ and I is an ℓ-ideal of $T_2(F)$.

Since $f(T_2(F))$ is an Archimedean f-algebra over F with identity element, it contains no nilpotent element by Lemma 3.3(1), and hence $f(T_2(F))$ is a finite direct sum of unital totally ordered algebras over F by Corollary 4.5. Let $0 \leq b \in f(T_2(F))$. Then $a \wedge b$ is a positive nilpotent f-element implies $a \wedge b = 0$. Thus we have the direct sum $f(T_2(F)) \oplus Fa$ as vector lattices. We consider the following two cases.

(1) Suppose 1 is not basic in $T_2(F)$. Since 1 is not basic in $T_2(F)$, $f(T_2(F))$ is a finite direct sum of at least two totally ordered algebras, and since $T_2(F)$ is three-dimensional, $f(T_2(F))$ is a direct sum of exactly two totally ordered algebras. Thus $f(T_2(F))$ is two-dimensional and $T_2(K) = f(T_2(F)) \oplus Fa$ as a vector lattice.

Now let $1 = e + f$, where $e > 0, f > 0$, and $e \wedge f = 0$. Then we have $e^2 = e, f^2 = f$, and $ef = fe = 0$ since $0 \leq e, f \leq 1$ implies that e and f are f-elements. Thus $T_2(F) = Ff \oplus Fe \oplus Fa$ as a vector lattice. Without loss of generality, we may assume that

$$f = \begin{pmatrix} 1 & u \\ 0 & 0 \end{pmatrix}, \quad e = \begin{pmatrix} 0 & v \\ 0 & 1 \end{pmatrix}$$

with $v = -u \in F$ (Exercise 62). Also, suppose that

$$a = \begin{pmatrix} 0 & r \\ 0 & 0 \end{pmatrix}, \quad \text{where } 0 \neq r \in F,$$

and define

$$q = \begin{pmatrix} 1 & u \\ 0 & r \end{pmatrix}.$$

Consider the *inner automorphism* $i_q : T_2(F) \to T_2(F)$. Then

$$i_q(e_{11}) = q^{-1}e_{11}q = f, \ i_q(e_{22}) = q^{-1}e_{22}q = e, \ i_q(e_{12}) = q^{-1}e_{12}q = a.$$

Thus i_q defines an ℓ-isomorphism from ℓ-algebra $(T_2(F), P_0)$ to ℓ-algebra $T_2(F) = Ff \oplus Fe \oplus Fa$.

(2) Suppose 1 is basic in $T_2(F)$. Since 1 is basic, $f(T_2(F))$ is totally ordered since if $x, y \in f(T_2(F))$ with $x \wedge y = 0$, then $(1 \wedge x) \wedge (1 \wedge y) = 0$ implies $1 \wedge x = 0$ or $1 \wedge y = 0$, and hence $x = 0$ or $y = 0$. Hence

$$T_2(F) = f(T_2(F)) \oplus f(T_2(F))^\perp$$

by Lemma 4.7 since $T_2(F)$ is Archimedean over F. Thus $Fa \subseteq f(T_2(F))^\perp$. By Lemma 4.23, $f(T_2(F)) = F1$, and hence $f(T_2(F))^\perp$ is two-dimensional. Let $0 < a_1 \in f(T_2(F))^\perp \setminus Fa$. Since $T_2(F)$ is Archimedean over F, there exists $0 < a_2 \in Fa$ such that $a_2 \nleq a_1$. Let $a_1 \wedge a_2 = a_3$. Then $(a_1 - a_3) \wedge (a_2 - a_3) = 0$, and

$$0 < (a_1 - a_3) \in f(T_2(F))^\perp \setminus Fa, \ 0 < (a_2 - a_3) \in Fa.$$

Let $e_1 = a_1 - a_3$ and $f_1 = a_2 - a_3$. Then $0 < e_1 \in f(T_2(F))^\perp \setminus Fa$, $0 < f_1 \in Fa$, and $e_1 \wedge f_1 = 0$, so

$$f(T_2(F))^\perp = Fe_1 \oplus Ff_1,$$

as a vector lattice, and hence

$$T_2(F) = F1 \oplus Fe_1 \oplus Ff_1,$$

as a vector lattice. Now we determine e_1. Let

$$e_1 = \begin{pmatrix} x & y \\ 0 & z \end{pmatrix}, \quad \text{where } x, y, z \in F.$$

Since $e_1 f_1 = x f_1$ and $f_1 e_1 = z f_1$, $x \geq 0$ and $z \geq 0$. Since $\{1, e_1, f_1\}$ is linearly independent, $x \neq z$. Otherwise e_1 is a linear combination of 1 and f_1. Let

$$e_1^2 = \begin{pmatrix} x^2 & (x+z)y \\ 0 & z^2 \end{pmatrix} = \alpha + \beta e_1 + \gamma f_1,$$

for some $\alpha, \beta, \gamma \in F^+$. Then we have
$$x^2 - \beta x - \alpha = 0 \text{ and } z^2 - \beta z - \alpha = 0,$$
and hence $x + z = \beta$ and $xz = -\alpha$.

If x and z are both not zero, then one of them must be negative since $xz = -\alpha \leq 0$, which is a contradiction. Thus we have $x = 0$ or $z = 0$.

Suppose $x = 0$. Then $z > 0$ since e_1 is not nilpotent, $\alpha = 0$, and
$$e_1^2 = \begin{pmatrix} 0 & zy \\ 0 & z^2 \end{pmatrix} = ze_1.$$

Let
$$i = z^{-1}e_1 = \begin{pmatrix} 0 & z^{-1}y \\ 0 & 1 \end{pmatrix}.$$
Then $T_2(F) = F1 \oplus Fi \oplus Ff_1$ as a vector lattice. Now let
$$f_1 = \begin{pmatrix} 0 & r_1 \\ 0 & 0 \end{pmatrix}, \quad \text{where } 0 \neq r_1 \in F,$$
and define
$$q = \begin{pmatrix} 1 & -z^{-1}y \\ 0 & r_1 \end{pmatrix}.$$

Then
$$i_q(e_{22}) = q^{-1}e_{22}q = i, \quad i_q(e_{12}) = q^{-1}e_{12}q = f_1.$$
Thus i_q is an ℓ-isomorphism from ℓ-algebra $(T_2(F), P_1)$ to ℓ-algebra $T_2(F) = F1 \oplus Fi \oplus Ff_1$.

Suppose $z = 0$. Then $x > 0$, $\alpha = 0$, and
$$e_1^2 = \begin{pmatrix} x^2 & xy \\ 0 & 0 \end{pmatrix} = xe_1.$$

Let
$$j = x^{-1}e_1 = \begin{pmatrix} 1 & x^{-1}y \\ 0 & 0 \end{pmatrix}.$$
Then $T_2(F) = F1 \oplus Fj \oplus Ff_1$, a direct sum as vector lattices. Define
$$q = \begin{pmatrix} 1 & -x^{-1}y \\ 0 & r_1 \end{pmatrix}.$$

Then
$$\varphi i_q(e_{22}) = \varphi(q^{-1}e_{22}q) = j, \quad \varphi i_q(e_{12}) = \varphi(q^{-1}e_{12}q) = f_1,$$
where $\varphi : T_2(F) \to T_2(F)$ is defined by
$$\begin{pmatrix} x & y \\ 0 & z \end{pmatrix} \to \begin{pmatrix} z & y \\ 0 & x \end{pmatrix}.$$

Thus φi_q is an anti-ℓ-isomorphism from ℓ-algebra $(T_2(F), P_1)$ to ℓ-algebra $T_2(F) = F1 \oplus Fj \oplus Ff_1$ (Exercise 63). $\square$

Theorem 4.35. *Let $T_2(F)$ be an Archimedean ℓ-unital ℓ-algebra over F. If $T_2(F)$ is ℓ-reduced, then $T_2(F)$ is ℓ-isomorphic or anti-ℓ-isomorphic to $(T_2(F), P_2)$.*

Proof. Since $T_2(F)$ is ℓ-reduced, by Lemma 4.23, $f(T_2(F))$ is totally ordered, and hence 1 is basic. Since $f(T_2(F))$ is totally ordered, $T_2(F) = f(T_2(F)) \oplus f(T_2(F))^{\perp}$ as a vector lattice by Lemma 4.7.

Let $a_1 = (e_{12})^+$ and $b_1 = (e_{12})^-$. Since $T_2(F)$ is ℓ-reduced, $a_1 > 0$ and $b_1 > 0$. It follows from $a_1 \wedge b_1 = 0$ that $(a_1 \wedge 1) \wedge (b_1 \wedge 1) = 0$, and hence $a_1 \wedge 1 = 0$ or $b_1 \wedge 1 = 0$ since 1 is basic. In the following we suppose $a_1 \wedge 1 = 0$. A similar argument may be used to prove the case that $b_1 \wedge 1 = 0$ and we leave the verification of this fact as an exercise. Let

$$a_1 = \begin{pmatrix} x_1 & x_2 \\ 0 & x_3 \end{pmatrix}.$$

Then $a_1^2 = (x_1 + x_3)a_1 + (-x_1 x_3)1 \geq 0$ implies $x_1 + x_3 \geq 0$ and $-x_1 x_3 \geq 0$ since $a_1 \wedge 1 = 0$.

First we claim that b_1 is not an f-element. Suppose b_1 is an f-element. From

$$(a_1 - b_1)^2 = a_1^2 - a_1 b_1 - b_1 a_1 + b_1^2 = 0,$$

we have

$$(x_1 + x_3)a_1 + (-x_1 x_3)1 - a_1 b_1 - b_1 a_1 + b_1^2 = 0,$$

and hence

$$(-x_1 x_3)1 + b_1^2 = 0 \text{ and } (x_1 + x_3)a_1 - a_1 b_1 - b_1 a_1 = 0,$$

since $(-x_1 x_3)1 + b_1^2 \in f(T_2(F))$ and $(x_1 + x_3)a_1 - a_1 b_1 - b_1 a_1 \in f(T_2(F))^{\perp}$. It follows from $(-x_1 x_3)1 + b_1^2 = 0$ that $b_1^2 = 0$, and hence $b_1 = 0$, which is a contradiction. Thus b_1 is not an f-element.

Since $T_2(F)$ is Archimedean over F, there exists $0 < \alpha \in F$ such that $\alpha 1 \not\leq b_1$. Let $b_1 \wedge \alpha 1 = c$. Then $c < \alpha 1$, and $c < b_1$ since b_1 is not an f-element. Thus

$$(b_1 - c) \wedge (\alpha 1 - c) = 0, \text{ with } b_1 - c > 0 \text{ and } \alpha 1 - c > 0,$$

so $b_1 - c \in f(T_2(F))^{\perp}$ since $0 < (\alpha 1 - c) \in f(T_2(F))$. Let $d = b_1 - c$. Then $0 < a_1, d \in f(T_2(F))^{\perp}$ and $a_1 \wedge d = 0$ since $d \leq b_1$. Thus $T_2(F) = F1 \oplus Fa_1 \oplus Fd$ as a vector lattice.

Now we determine a_1 and d. Recall that $e_{12} = a_1 - b_1 = a_1 - d - c$.

Since $a_1^2 = (x_1 + x_3)a_1 + (-x_1x_3)1$, we have $-x_1x_3 \geq 0$, and hence $x_1 \leq 0$ or $x_3 \leq 0$. Suppose $x_1 \leq 0$. From $e_{12} \leq a_1$, we have $a_1e_{12} \leq a_1^2$, and hence

$$x_1a_1 - x_1d - x_1c \leq (x_1 + x_3)a_1 + (-x_1x_3)1,$$

so

$$-x_1d - x_1c \leq x_3a_1 + (-x_1x_3)1.$$

Since $\{1, a_1, d\}$ is a disjoint set, we have $-x_1d = 0$ and hence $x_1 = 0$. By a similar argument, if $x_3 \leq 0$ then $x_3 = 0$ (Exercise 64). Thus we have $x_1 = 0$ or $x_3 = 0$ but not both of them are zero since a_1 is not nilpotent.

Let

$$d = \begin{pmatrix} y_1 & y_2 \\ 0 & y_3 \end{pmatrix}, \text{ where } y_1, y_2, y_3 \in F.$$

Then $d^2 = (y_1 + y_3)d + (-y_1y_3)1 \geq 0$ implies that $(y_1 + y_3) \geq 0$ and $-y_1y_3 \geq 0$, so $y_1 \leq 0$ or $y_3 \leq 0$. Suppose $y_1 \leq 0$. From $-e_{12} \leq b_1 = d + c$, we have $-de_{12} \leq d^2 + dc$, and hence

$$-y_1(a_1 - d - c) \leq (y_1 + y_3)d + (-y_1y_3)1 + dc,$$

so

$$-y_1a_1 \leq y_3d + (-y_1y_3)1 + dc - y_1c.$$

Since c is an f-element, a_1 is disjoint with d, 1, dc, and c, so we have $-y_1a_1 = 0$, and hence $y_1 = 0$. Similarly, if $y_3 \leq 0$, then $y_3 = 0$. Therefore, we have $y_1 = 0$ or $y_3 = 0$ but not both of them are zero.

If $x_1 = 0$ and $y_3 = 0$, then $a_1d = 0$, which is a contradiction by Lemma 4.23. Similarly, x_3 and y_1 cannot be both zero. Thus we have the following two cases.

(i) $x_1 = 0$ and $y_1 = 0$. Let

$$u = x_3^{-1}a_1 = \begin{pmatrix} 0 & x_3^{-1}x_2 \\ 0 & 1 \end{pmatrix}, \quad v = y_3^{-1}d = \begin{pmatrix} 0 & y_3^{-1}y_2 \\ 0 & 1 \end{pmatrix}.$$

Then $T_2(F) = F1 \oplus Fu \oplus Fv$ as a vector lattice. Define

$$q = \begin{pmatrix} 1 & -x_3^{-1}x_2 \\ 0 & y_3^{-1}y_2 - x_3^{-1}x_2 \end{pmatrix}.$$

Then q is invertible, and

$$i_q(e_{22}) = q^{-1}e_{22}q = u, \quad i_q(k) = q^{-1}kq = v,$$

where $k = e_{12} + e_{22}$. Thus i_q is an ℓ-isomorphism from ℓ-algebra $(T_2(F), P_2)$ to ℓ-algebra $T_2(F) = F1 \oplus Fu \oplus Fv$.

(ii) $x_3 = 0$ and $y_3 = 0$. Now let

$$u = x_1^{-1}a_1 = \begin{pmatrix} 1 & x_1^{-1}x_2 \\ 0 & 0 \end{pmatrix}, \quad v = y_1^{-1}d = \begin{pmatrix} 1 & y_1^{-1}y_2 \\ 0 & 0 \end{pmatrix}.$$

Then we have $T_2(K) = K1 \oplus Ku \oplus Kv$ as a vector lattice. Define

$$q = \begin{pmatrix} 1 & -x_1^{-1}x_2 \\ 0 & y_1^{-1}y_2 - x_1^{-1}x_2 \end{pmatrix}.$$

Then

$$\varphi i_q(e_{22}) = \varphi(q^{-1}e_{22}q) = u, \quad \varphi i_q(k) = \varphi(q^{-1}kq) = v,$$

where φ is defined in Theorem 4.34. Therefore, φi_q is an anti-ℓ-isomorphism from ℓ-algebra $(T_2(F), P_2)$ to ℓ-algebra $T_2(F) = F1 \oplus Fu \oplus Fv$. $\square$

Finally we determine non-Archimedean lattice orders on $T_2(F)$ in which 1 is positive. $T_2(F)$ can be made into a vector lattice as follows:

$$T_2(F) = F1 \oplus (Fe_{22} \underset{\rightarrow}{\oplus} Fe_{12}),$$

where $Fe_{22} \underset{\rightarrow}{\oplus} Fe_{12}$ is the lexicographic order, that is, $\alpha e_{22} + \beta e_{12} \geq 0$ if and only if $\alpha > 0$ or $\alpha = 0$ and $\beta \geq 0$. We denote the positive cone of this lattice order on $T_2(F)$ by P_3. Then

$$P_3 = \{\alpha 1 + \beta e_{22} + \gamma e_{12} : \alpha \geq 0, \beta > 0, \text{ or } \alpha \geq 0, \beta = 0, \gamma \geq 0, \forall \alpha, \beta, \gamma \in F\}.$$

We leave the routine checking that P_3 is closed under the multiplication in $T_2(F)$ as an exercise (Exercise 66). Thus $(T_2(F), P_3)$ becomes an ℓ-unital ℓ-algebra which is not Archimedean over F.

Theorem 4.36. *Let $T_2(F)$ be an ℓ-unital ℓ-algebra which is not Archimedean over F. Then $T_2(F)$ is ℓ-isomorphic or anti-ℓ-isomorphic to $(T_2(F), P_3)$.*

Proof. Since $T_2(F)$ is non-Archimedean over F, $T_2(F)$ is not ℓ-reduced by Theorem 1.31. Let $a > 0$ and $a^2 = 0$.

We first claim that a cannot be an f-element. Suppose that a is an f-element. Then $a, 1 \in f(T_2(F))$ and the square of each element in $f(T_2(F))$ is positive implies for any $0 \leq \alpha \in F$, $0 \leq (1 - \alpha a)^2$, so $\alpha a < 1$ for each $\alpha \in F^+$. Hence a and 1 are linearly independent over F. Then that $T_2(F)$ cannot be an f-ring implies that $f(T_2(F))$ is two-dimensional and totally ordered. Thus $T_2(F) = f(T_2(F)) \oplus f(T_2(F))^\perp$ as a vector lattice and $f(T_2(F))^\perp$ is one-dimensional over F. Let $0 < b \in f(T_2(F))^\perp$. Then $f(T_2(F))^\perp = Fb$. Since a is an f-element, $ab, ba \in f(T_2(F))^\perp$, then we have $ab = \gamma b$ and $ba = \beta b$, for some $\gamma, \beta \in F^+$. On the other hand, $ab, ba \in Fa$ since Fa is an ℓ-ideal of $T_2(F)$. Then we have $b^2 = 0$ (Exercise 67), so $b \in Fa \subseteq f(T_2(F))$, which is a contradiction. Therefore a is not an f-element.

Since $a \wedge 1 \in Fa$ and a is not an f-element, $a \wedge 1 = 0$, so $a \in f(T_2(F))^\perp$. If $f(T_2(F))$ is not totally ordered, then there are $0 < u, v \in f(T_2(F))$ with $u \wedge v = 0$, and hence

$$T_2(F) = Fu \oplus Fv \oplus Fa,$$

as a direct sum of vector lattices. Thus $T_2(F)$ is Archimedean over F, which is a contradiction. Therefore, $(T_2(F))$ is totally ordered. By Lemma 4.23, $f(T_2(F)) = F1$, and hence by Lemma 4.7

$$T_2(F) = (f(T_2(F)) \oplus f(T_2(F))^\perp) \cup U_f,$$

where

$$U_f = \{w \in T_2(F) : |w| \geq \alpha 1, \forall \alpha \in F\}.$$

If $0 < w \in U_f$, then $\alpha 1 \leq w$ for all $\alpha \in F$, so $\alpha a \leq wa$ for all $\alpha \in F$, which is a contradiction since $wa \in Fa$. Thus $U_f = \emptyset$, and $T_2(F) = f(T_2(F)) \oplus f(T_2(F))^\perp$, so $f(T_2(F))^\perp$ is two-dimensional over F.

Next we claim that $f(T_2(F))^\perp$ is totally ordered. If $f(T_2(F))^\perp$ is not totally ordered, then there exist $0 < s, t \in f(T_2(F))^\perp$ such that $s \wedge t = 0$, so

$$f(T_2(F))^\perp = Fs \oplus Ft, \text{ and } T_2(F) = F1 \oplus Fs \oplus Ft,$$

as vector lattices. Thus, again, $T_2(F)$ is Archimedean over F, which is a contradiction. Therefore, $f(T_2(F))^\perp$ is totally ordered. Let $0 < c \in f(T_2(F))^\perp$ such that a and c are linearly independent over F. If $c \leq \alpha a$ for some $\alpha \in F$, then $c \in Fa$ since Fa is an ℓ-ideal, so a and c are linearly dependent, which is a contradiction. Thus for all $\alpha \in F$, we have $\alpha a < c$. Let

$$c = \begin{pmatrix} z_1 & z_2 \\ 0 & z_3 \end{pmatrix} \quad \text{and} \quad a = \begin{pmatrix} 0 & x \\ 0 & 0 \end{pmatrix}.$$

Then $ac = z_3 a \geq 0$ and $ca = z_1 a \geq 0$ implies that $z_1 \geq 0$ and $z_3 \geq 0$. Since

$$c^2 = (z_1 + z_3)c + (-z_1 z_3)1 \geq 0,$$

we have $-z_1 z_3 \geq 0$, and hence $z_1 z_3 = 0$, so $z_1 = 0$ or $z_3 = 0$.

Suppose $z_1 = 0$. Then $z_3 > 0$. Let

$$d = z_3^{-1}c = \begin{pmatrix} 0 & z_3^{-1}z_2 \\ 0 & 1 \end{pmatrix}.$$

Then $T_2(F) = F1 \oplus (Fd \underset{\rightarrow}{\oplus} Fa)$ as a vector lattice. Define

$$q = \begin{pmatrix} 1 & -z_3^{-1}z_2 \\ 0 & x \end{pmatrix}.$$

Then

$$i_q(e_{22}) = q^{-1}e_{22}q = d, \quad i_q(e_{12}) = q^{-1}e_{12}q = a.$$

Thus i_q is an ℓ-isomorphism from ℓ-algebra $(T_2(K), P_3)$ to ℓ-algebra $T_2(K) = K1 \oplus (Kd \underset{\rightarrow}{\oplus} Ka)$.

Suppose $z_3 = 0$. Then $z_1 > 0$. Let

$$e = z_1^{-1}c = \begin{pmatrix} 1 & z_1^{-1}z_2 \\ 0 & 0 \end{pmatrix}.$$

Then $T_2(F) = F1 \oplus (Fe \underset{\rightarrow}{\oplus} Fa)$ as a vector lattice. Define

$$q = \begin{pmatrix} 1 & -z_1^{-1}z_2 \\ 0 & x \end{pmatrix}.$$

Then

$$\varphi i_q(e_{22}) = \varphi(q^{-1}e_{22}q) = e, \quad \varphi i_q(e_{12}) = \varphi(q^{-1}e_{12}q) = a,$$

and hence φi_q is an anti-ℓ-isomorphism from ℓ-algebra $(T_2(K), P_3)$ to ℓ-algebra $T_2(F) = F1 \oplus (Fe \underset{\rightarrow}{\oplus} Fa)$. $\qquad\square$

4.7.2 Lattice orders on $T_2(F)$ with $1 \not> 0$

In this section, we suppose that $T_2(F)$ is an ℓ-algebra over F in which the identity matrix $1 \not> 0$. In this case each lattice order can be obtained from a lattice order with $1 > 0$ using Theorem 1.19. As in the last section, we consider two cases in which $T_2(F)$ is not ℓ-reduced and ℓ-reduced respectively.

Suppose that $T_2(F)$ is not ℓ-reduced. Let $w = w_1 e_{12}$ be a positive nilpotent element, where $0 \neq w_1 \in F$. Then $I = Fw$ is an ℓ-ideal of $T_2(F)$ (Exercise 61). Suppose $u = 1^+$, $v = 1^-$. Then $u > 0$, $v > 0$, $1 = u - v$ and $u \wedge v = 0$. Let

$$u = \begin{pmatrix} u_1 & u_2 \\ 0 & u_3 \end{pmatrix} \quad \text{and} \quad v = \begin{pmatrix} v_1 & u_2 \\ 0 & v_3 \end{pmatrix},$$

where $u_1 - v_1 = 1$ and $u_3 - v_3 = 1$.

We first notice that for any $x = \begin{pmatrix} x_1 & x_2 \\ 0 & x_3 \end{pmatrix} \in T_2(F)$, $xw = x_1 w$ and $wx = x_3 w$. So if $x \geq 0$, then $x_1, x_3 \in F^+$. This fact will often be used later.

Since

$$\begin{aligned}
v^2 &= (v_1 + v_3)v - (v_1 v_3)1 \\
&= (v_1 + v_3)v - v_1 v_3(u - v) \\
&= -(v_1 v_3)u + (v_1 + v_3 + v_1 v_3)v \\
&\geq 0
\end{aligned}$$

and $u \wedge v = 0$, we have $-(v_1 v_3) \geq 0$. Thus $v_1 v_3 = 0$, so either $v_1 = 0$ or $v_3 = 0$. Without loss of generality, we may assume that $v_1 = 0$. (If $v_3 = 0$, we may use the anti-isomorphism $\varphi : \begin{pmatrix} x & y \\ 0 & z \end{pmatrix} \to \begin{pmatrix} z & y \\ 0 & x \end{pmatrix}$ to reduce to the case that $v_1 = 0$.) Then we have

$$u = \begin{pmatrix} 1 & u_2 \\ 0 & u_3 \end{pmatrix} \quad \text{and} \quad v = \begin{pmatrix} 0 & u_2 \\ 0 & v_3 \end{pmatrix}.$$

Since $v_3 \geq 0$, we have $u_3 = 1 + v_3 \geq 1$ and hence u is invertible. The element v_3 may be zero. In the following, we consider $v_3 > 0$ and $v_3 = 0$, respectively.

Theorem 4.37. *Let $T_2(F)$ be an ℓ-algebra over F with $1 \not> 0$. Suppose that $T_2(F)$ is not ℓ-reduced and u, v, w are defined as above. If $v_3 > 0$, then ℓ-algebra $T_2(F)$ is ℓ-isomorphic to ℓ-algebras $(T_2(F), rP_1)$, or $(T_2(F), rP_3)$, where $r \in P_1$ or P_3 is invertible, respectively.*

Proof. We first claim that for any $0 < \alpha \in F$, $u \wedge \alpha w = 0$. Suppose that $u \wedge \alpha w = p$. Then $p = \beta w$ for some $\beta \in F$, $0 \leq \beta \leq \alpha$, since $I = Fw$ is an ℓ-ideal. Then $\beta w \leq u$ implies $\beta(wv) \leq (uv)$. Since $wv = v_3 w$ and $uv = u_3 v$, we have $(\beta v_3)w \leq u_3 v$, so $\beta w \leq \frac{u_3}{v_3} v$. Since $u \wedge v = 0$ implies that $u \wedge \frac{u_3}{v_3} v = 0$, we have $p = \beta w = 0$.

We consider the following two cases.

(1) $\alpha w \leq v$ for each $0 \leq \alpha \in F$.

Since $v_3 > 0$, clearly $\{u, v, w\}$ is linearly independent over F, so $\{u, v, w\}$ is a vector space basis for $T_2(F)$ over F. Thus for each $f \in T_2(F)$, $f = \alpha u + \beta v + \gamma w$, where $\alpha, \beta, \gamma \in F$. It is straightforward to check that $f \geq 0$ if and only if $\alpha \geq 0$, $\beta > 0$ or $\alpha \geq 0$, $\beta = 0$, $\gamma \geq 0$ (Exercise 68).

Let $x = u^{-1}(\frac{u_3}{v_3} v)$ and $y = u^{-1} w$. Then $y = w$ and $\{1, x, y\}$ is linearly independent. The multiplication table for $\{1, x, y\}$ is given below.

	1	x	y
1	1	x	y
x	x	x	0
y	y	y	0

Now we define a positive cone

$$P = \{\alpha + \beta x + \gamma y \mid \alpha \geq 0, \beta > 0, \text{ or } \alpha \geq 0, \beta = 0, \gamma \geq 0\}.$$

Then $(T_2(F), P)$ is an ℓ-algebra in which $1 > 0$. Clearly $T_2(F)^+ = uP$ and since $u = 1 + v = 1 + v_3 x$, $u \in P$.

Define the mapping $\phi : T_2(F) \to T_2(F)$ by

$$\phi(\alpha + \beta x + \gamma y) = \alpha + \beta e_{22} + \gamma e_{12}.$$

Then ϕ is an ℓ-isomorphism from ℓ-algebra $(T_2(K), P)$ to ℓ-algebra $(T_2(F), P_3)$. Let $r = \phi(u)$. We have that $r \in P_3$ and ℓ-algebra $T_2(K)$ is ℓ-isomorphic to ℓ-algebra $(T_2(F), rP_3)$.

(2) $\beta w \not\leq v$ for some $0 < \beta \in F$.

Let $g = v \wedge \beta w$. Then $g = \delta w$ for some $\delta \in F^+$ since $I = Fw$ is an ℓ-ideal. Since $\beta w \not\leq v$, $\beta > \delta$, so $\beta w - g = (\beta - \delta) w > 0$. Now $(v - g) \wedge (\beta - \delta) w = 0$ implies that $(v - g) \wedge w = 0$ since $T_2(F)$ is an f-module over F and $\beta - \delta > 0$. Let $v' = v - g$. Then $v' = \begin{pmatrix} 0 & v_2 \\ 0 & v_3 \end{pmatrix}$, where $v_2 = u_2 - \delta w_1$; since $v_3 > 0$, we have $v' > 0$, and it is clear that the set $\{u, v', w\}$ is disjoint. Then for each $f \in T_2(F)$, $f = \alpha u + \beta v' + \gamma w$, where $\alpha, \beta, \gamma \in K$, we have that $f \geq 0$ if and only if $\alpha \geq 0$, $\beta \geq 0$, and $\gamma \geq 0$ by Theorem 1.13(2).

Let $x' = u^{-1}\left(\frac{u_3}{v_3} v'\right)$ and $y = u^{-1} w$. The multiplication table for $\{1, x', y\}$ is given below (Exercise 69).

	1	x'	y
1	1	x'	y
x'	x'	x'	0
y	y	y	0

Now we define the positive cone $P = \{\alpha + \beta x' + \gamma y \mid \alpha, \beta, \gamma \in F^+\}$. Then $(T_2(F), P)$ is an ℓ-algebra in which $1 > 0$ and $T_2(F)^+ = uP$. Since

$$u = 1 + v' + \delta w = 1 + v_3 x' + (v_3 \delta + \delta) y$$

and $v_3, \delta \in F^+$, we have $u \in P$.

By the same mapping as in (1), ℓ-algebra $T_2(F)$ is ℓ-isomorphic to ℓ-algebra $(T_2(F), rP_1)$ with $r = \phi(u) \in P_1$. $\qquad\square$

Next we consider the case that $v_3 = 0$.

Theorem 4.38. *Let $T_2(F)$ be an ℓ-algebra over F with $1 \not> 0$. Suppose that $T_2(F)$ is not ℓ-reduced and u, v, w are defined as above. If $v_3 = 0$, then ℓ-algebra $T_2(F)$ is ℓ-isomorphic or anti-ℓ-isomorphic to ℓ-algebras $(T_2(F), rP_0)$, $(T_2(F), rP_1)$, or $(T_2(F), rP_3)$, where $r \in P_0$, P_1, or P_3, respectively, is an invertible matrix.*

Proof. Since $v_3 = 0$, $v \in I = Fw$. Consider the quotient ℓ-algebra $T_2(F)/I$. Then $\bar{1} = 1 + I = \bar{u} = u + I > 0$ in $T_2(F)/I$. Since $T_2(F)/I$ has other idempotent elements except $\bar{1}$ and 0, $T_2(F)/I$ cannot be totally ordered, and since $T_2(F)/I$ contains no nilpotent element, $T_2(F)/I$ is Archimedean over F by Corollary 1.3. Thus $T_2(F)/I$ is a direct sum of two totally ordered subspaces over F. We need to consider two cases.

(i) $\bar{1} = \bar{u}$ is a basic element in $T_2(F)/I$.

Let $d \in T_2(F)^+$ such that $\bar{d} > 0$ and $\bar{u} \wedge \bar{d} = 0$. Then $u \wedge d \in I$, so $u \wedge d = \epsilon w$ for some $\epsilon \in F^+$. Since $u \wedge v = 0$, $\epsilon w \wedge v = 0$. But ϵw and v are both in $I = Fw$ and $v \neq 0$, so $\epsilon w = 0$. Thus $u \wedge d = 0$. Clearly $\{u, d, w\}$ is linearly independent. Let $d = \begin{pmatrix} d_1 & d_2 \\ 0 & d_3 \end{pmatrix}$. Then $d_1 \geq 0$ and $d_3 \geq 0$ since $dw = d_1 w$ and $wd = d_3 w$. Since

$$d^2 = (d_1 + d_3)d - (d_1 d_3)u + (d_1 d_3)v \geq 0,$$

$u \wedge d = 0$, and $u \wedge v = 0$, we have $-(d_1 d_3) \geq 0$, so either $d_1 = 0$ or $d_3 = 0$. We may assume that $d_1 = 0$. If $d_3 = 0$, then, by using the anti-isomorphism φ, we may reduce to the above situation. Then $d_3 > 0$ since $\bar{d} > 0$. There are two different lattice orders in this case.

(i_a) $\alpha w \leq d$ for all $\alpha \in F^+$.

In this case $\alpha u + \beta d + \gamma w \geq 0$ if and only if $\alpha \geq 0$, $\beta > 0$, or $\alpha \geq 0$, $\beta = 0$, $\gamma \geq 0$. Let $x = u^{-1}(\frac{1}{d_3}d)$ and $y = u^{-1}w$. The set $\{1, x, y\}$ is linearly independent with the following multiplication table.

	1	x	y
1	1	x	y
x	x	x	0
y	y	y	0

Now we define the positive cone

$$P = \{\alpha + \beta x + \gamma y \mid \alpha \geq 0, \beta > 0 \text{ or } \alpha \geq 0, \beta = 0, \gamma \geq 0\}.$$

Then $(T_2(F), P)$ is an ℓ-algebra in which $1 > 0$. Clearly $T_2(F)^2 = uP$ and since $u = 1 + v = 1 + \theta w = 1 + \theta y$ for some $\theta \in F^+$, $u \in P$.

Define the mapping $\phi : T_2(F) \to T_2(F)$ by

$$\phi(\alpha + \beta x + \gamma y) = \alpha + \beta e_{22} + \gamma e_{12}.$$

Then ϕ is an ℓ-isomorphism from ℓ-algebra $(T_2(F), P)$ to ℓ-algebra $(T_2(F), P_3)$. Let $r = \phi(u)$. We have that $r \in P_3$ and ℓ-algebra $T_2(F)$ is ℓ-isomorphic to ℓ-algebra $(T_2(F), rP_3)$.

(i_b) $\beta w \nleq d$ for some $0 < \beta \in F$.

Let $d \wedge \beta w = \delta w$. Then $0 \leq \delta < \beta$, and $(d - \delta w) \wedge (\beta - \delta)w = 0$. Let $d' = d - \delta w$. Then $d' > 0$, and since $\beta - \delta > 0$, $d' \wedge w = 0$. Thus $\{u, d', w\}$ is disjoint, so $\alpha u + \beta d' + \gamma w \geq 0$ if and only if $\alpha, \beta, \gamma \in F^+$ by Theorem 1.13(2). Let $x' = u^{-1}(\frac{1}{d_3}d')$, $y = u^{-1}w$. The set $\{1, x', y\}$ is linearly independent with the following multiplication table.

	1	x'	y
1	1	x'	y
x'	x'	x'	0
y	y	y	0

Now if we define a positive cone $P = \{\alpha 1 + \beta x' + \gamma y \mid \alpha, \beta, \gamma \in F^+\}$. Then $1 \in P$, $T_2(F)^+ = uP$ with $u \in P$, and $(T_2(F), P)$ is ℓ-isomorphic to $(T_2(F), P_1)$.

(ii) $\bar{1} = \bar{u}$ is not basic in $T_2(F)/I$.

Then there exist $f', g' \in T_2(F)$ such that $\bar{u} = \bar{f}' + \bar{g}'$, $\bar{f}' > 0$, $\bar{g}' > 0$, and $\bar{f}' \wedge \bar{g}' = 0$. Since $\bar{f}' \wedge \bar{g}' = 0$, $f' \wedge g' \in I = Fw$. Let $f' \wedge g' = p$, and let $f = f' - p$, $g = g' - p$. Then $f > 0$, $g > 0$, and $f \wedge g = 0$. Since $\bar{u} = \bar{f}' + \bar{g}' = \bar{f} + \bar{g}$, $u = f + g + \delta w$ for some $\delta \in F$. Since $f \wedge g = 0$, $(f \wedge w) \wedge (g \wedge w) = 0$, so either $f \wedge w = 0$ or $g \wedge w = 0$ since $(f \wedge w)$ and $(g \wedge w)$ are both in $I = Fw$. Without loss of generality, we may assume that $f \wedge w = 0$.

If $\alpha w \leq g$ for all $\alpha \in F^+$, then $1 = u - v = f + g + \delta w - v \geq 0$ since $v \in I = Fw$, which is a contradiction. Thus there exists $0 < \beta \in F$ such that $\beta w \nleq g$. Let $g \wedge \beta w = q$, $h = g - q$. Then $h \wedge (\beta w - q) = 0$ and $h > 0$, $\beta w - q > 0$. So $h \wedge w = 0$ since $\beta w - q = \sigma w$ for some $0 < \sigma \in F$. Now $\{f, h, w\}$ is disjoint and also a vector space basis for $T_2(F)$ over F.

Let

$$f = \begin{pmatrix} f_1 & f_2 \\ 0 & f_3 \end{pmatrix} \text{ and } h = \begin{pmatrix} h_1 & h_2 \\ 0 & h_3 \end{pmatrix}.$$

Then

$$f^2 = (f_1 + f_3)f - (f_1 f_3)1$$
$$= (f_1 + f_3)f - (f_1 f_3)(u - v)$$
$$= (f_1 + f_3)f - (f_1 f_3)(f + h + \delta w + q - v)$$
$$= (f_1 + f_3 - f_1 f_3)f - (f_1 f_3)h - (f_1 f_3)(\delta w + q - v) \geq 0$$

implies that $-(f_1 f_3) \geq 0$. Since $f \geq 0$, we also have $f_1, f_3 \geq 0$. Thus $f_1 f_3 = 0$, so either $f_1 = 0$ or $f_3 = 0$. By a similar argument, we have either $h_1 = 0$ or $h_3 = 0$.

Since $u = f + h + \delta w + q \geq 0$ and $\delta w + q \in I = Fw$, $\delta w + q \geq 0$, so $\delta w + q = (\delta w + q) \wedge u = 0$ since $u \wedge v = 0$ and $\delta w + q, v \in I = Fw$. Thus $u = f + h$. Then $f_2 + h_2 = u_2$, and f_1, h_1 cannot be both zero. Also f_3, h_3 cannot be both zero. We may assume that $f_3 = h_1 = 0$ and leave the verification for $f_1 = h_3 = 0$ to the reader (Exercise 70). Then $f_1 = 1$ and $h_3 = 1$. Then $f^2 = f$, $h^2 = h$, $hf = 0$ and $fh = v$. Now let $x = u^{-1}f$ and $y = u^{-1}h$. Then $\{x, y, w\}$ is linearly independent with the following multiplication table.

	x	y	w
x	x	0	w
y	0	y	0
w	0	w	0

If we define the positive cone $P = \{\alpha x + \beta y + \gamma w \mid \alpha, \beta, \gamma \in F^+\}$, then $1 = x + y \in P$, $T_2(F)^+ = uP$, and $(T_2(F), P)$ is ℓ-isomorphic to $(T_2(F), P_0)$. Since $u = f + h = x + y + v$, $u \in P$. This completes the proof of (ii). $\square$

Now we suppose that $T_2(F)$ is an ℓ-algebra over F in which the identity matrix $1 \not> 0$ and $T_2(F)$ is ℓ-reduced. Then $T_2(F)$ is an ℓ-domain by Lemma 4.23. Lattice orders on $T_2(F)$ for this case are characterized in the following result.

Theorem 4.39. *Suppose that $T_2(F)$ is an ℓ-reduced ℓ-algebra over F in which $1 \not> 0$. Then $T_2(F)$ is ℓ-isomorphic or anti-ℓ-isomorphic to ℓ-algebra $(T_2(F), rP_2)$, where $r \in P_2$ is an invertible matrix.*

Proof. Let $u = 1^+$, $v = 1^-$. Then $u > 0$, $v > 0$, $1 = u - v$ and $u \wedge v = 0$. As before, assume that

$$u = \begin{pmatrix} u_1 & u_2 \\ 0 & u_3 \end{pmatrix} \quad \text{and} \quad v = \begin{pmatrix} v_1 & u_2 \\ 0 & v_3 \end{pmatrix},$$

where $u_1, v_1, u_2, u_3, v_3 \in F$, $u_1 - v_1 = 1$, and $u_3 - v_3 = 1$.

Since $T_2(F)$ is ℓ-reduced, $T_2(F)$ is Archimedean over F by Theorem 1.31(3) and hence $T_2(F)$ is a direct sum of totally ordered subspaces over F by Corollary 1.3. We claim that $T_2(F)$ cannot be a direct sum of two totally ordered subspaces over F.

Suppose that $T_2(F) = W_1 \oplus W_2$, where W_1 and W_2 are totally ordered subspaces over F. Then $T_2(F)$ has a basis with two elements, so if $s \wedge t = 0$ and $s > 0, t > 0$, then s and t will be basic elements. Since $u \wedge v = 0$, we may assume that $u \in W_1$ and $v \in W_2$. Let

$$x = \begin{pmatrix} 0 & 1 \\ 0 & 0 \end{pmatrix}^{+} \quad \text{and} \quad y = \begin{pmatrix} 0 & 1 \\ 0 & 0 \end{pmatrix}^{-}.$$

Then $x > 0$, $y > 0$ since $T_2(F)$ is ℓ-reduced, $\begin{pmatrix} 0 & 1 \\ 0 & 0 \end{pmatrix} = x - y$ and $x \wedge y = 0$. Let

$$x = \begin{pmatrix} x_1 & x_2 \\ 0 & x_3 \end{pmatrix} \quad \text{and} \quad y = \begin{pmatrix} x_1 & y_2 \\ 0 & x_3 \end{pmatrix},$$

where $x_1, x_2, y_2, x_3 \in F$ and $x_2 - y_2 = 1$. Since $x \wedge y = 0$, x and y are not in the same totally ordered direct summand of $T_2(F)$. We may assume that $x \in W_1$ and $y \in W_2$. Since

$$x^2 = (x_1 + x_3)x - (x_1 x_3)1 = (x_1 + x_3)x - (x_1 x_3)u + (x_1 x_3)v \geq 0,$$

and $u \wedge v = x \wedge v = 0$, we have $x_1 x_3 \geq 0$. Since

$$y^2 = (x_1 + x_3)y - (x_1 x_3)1 = (x_1 + x_3)y - (x_1 x_3)u + (x_1 x_3)v \geq 0,$$

and $u \wedge v = u \wedge y = 0$, we have $-(x_1 x_3) \geq 0$. Thus $x_1 x_3 = 0$. So either $x_1 = 0$ or $x_3 = 0$.

We first consider the case that $x_1 = 0$.

Since $x^2 = x_3 x$, $x_3 \geq 0$, and since $T_2(F)$ contains no nonzero positive nilpotent element, $x_3 \neq 0$. So $x_3 > 0$. Since

$$u^2 = (u_1 + u_3)u - (u_1 u_3)1 = (u_1 + u_3 - u_1 u_3)u + (u_1 u_3)v \geq 0,$$

and $u \wedge v = 0$, we have $u_1 u_3 \geq 0$. Since $xv = v_3 x \geq 0$, $v_3 \geq 0$, and hence $u_3 = 1 + v_3 \geq 1$. Thus $u_1 \geq 0$. If $u_1 = 0$, then $v_1 = -1$. So $vy = x_3 u - y \geq 0$ and $u \wedge y = 0$ implies that $y = 0$, which is a contradiction. Thus $u_1 > 0$.

Let $0 < \alpha \in F$. If $u \leq \alpha x$, then $u^2 \leq \alpha x u$. Since

$$u^2 = (u_1 + u_3 - u_1 u_3)u + (u_1 u_3)v \quad \text{and} \quad xu = u_3 x,$$

we have

$$(u_1 + u_3 - u_1 u_3)u + (u_1 u_3)v \leq (\alpha u_3)x.$$

Since $u \wedge v = x \wedge v = 0$, we have $u_1 u_3 \leq 0$, which is a contradiction since $u_3 \geq 1$ and $u_1 > 0$. Therefore $u \not\leq \alpha x$. Since αx and u are both in W_1, which is totally ordered, $\alpha x \leq u$ for any $\alpha \in F^+$, which contradicts with the fact that $T_2(F)$ is Archimedean over F.

By a similar argument, the case $x_3 = 0$ will also cause a contradiction. Therefore, $T_2(F)$ cannot be a direct sum of two totally ordered subspaces over F.

Then $T_2(F) = W_1 \oplus W_2 \oplus W_3$, where each W_i is a totally ordered subspace over F, and $T_2(F)$ has a basis with three elements. Since $u \wedge v = 0$, u is a sum of at most two disjoint basic elements. Similarly v is a sum of at most two disjoint basic elements. We consider the following cases.

(I) $u = h + g$, where h and g are basic elements and $h \wedge g = 0$. Then $\{h, g, v\}$ is disjoint and a vector space basis over F. Let

$$h = \begin{pmatrix} h_1 & h_2 \\ 0 & h_3 \end{pmatrix} \text{ and } g = \begin{pmatrix} g_1 & g_2 \\ 0 & g_3 \end{pmatrix},$$

where $h_i, g_i \in F$, $i = 1, 2, 3$. Then

$$\begin{aligned} h^2 &= (h_1 + h_3)h - (h_1 h_3)1 \\ &= (h_1 + h_3)h - (h_1 h_3)(h + g - v) \\ &= (h_1 + h_3 - h_1 h_3)h - (h_1 h_3)g + (h_1 h_3)v \geq 0 \end{aligned}$$

implies that $-h_1 h_3 \geq 0$ and $h_1 h_3 \geq 0$. Thus $h_1 h_3 = 0$, so either $h_1 = 0$ or $h_3 = 0$. By a similar argument, we have either $g_1 = 0$ or $g_3 = 0$.

If $h_3 = g_1 = 0$, then $gh = 0$, which contradicts with the fact that $T_2(F)$ is an ℓ-domain. Similarly it is not possible that $h_1 = g_3 = 0$.

If $h_1 = g_1 = 0$, then since $u = h + g$, we have $u_1 = 0$, so $v_1 = -1$. Thus

$$v^2 = \begin{pmatrix} 1 & -u_2 + u_2 v_3 \\ 0 & v_3^2 \end{pmatrix} = \begin{pmatrix} 1 & -u_2 + u_2 v_3 \\ 0 & (u_3 - 1)v_3 \end{pmatrix} = -v + v_3 u \geq 0$$

implies that $v = 0$, which is a contradiction.

If $h_3 = g_3 = 0$, then $u_3 = 0$, so $v_3 = -1$. Thus $v^2 = -v + v_1 u \geq 0$, and hence $v = 0$, which is a contradiction.

Therefore, u cannot be a sum of two disjoint basic elements.

(II) $v = i + j$, where i and j are basic elements and $i \wedge j = 0$. Then $\{u, i, j\}$ is disjoint. Let

$$i = \begin{pmatrix} i_1 & i_2 \\ 0 & i_3 \end{pmatrix} \text{ and } j = \begin{pmatrix} j_1 & j_2 \\ 0 & j_3 \end{pmatrix},$$

where $i_k, j_k \in F$, $k = 1, 2, 3$. Then, by an argument similar to that in (I), we have $i_1 i_3 = 0$ and $j_1 j_3 = 0$. Since $T_2(F)$ is an ℓ-domain, $i_1 = j_3 = 0$ and $i_3 = j_1 = 0$ cannot happen.

Suppose that $i_1 = j_1 = 0$. Then $i_3 > 0$ and $j_3 > 0$. Since $v = i + j$, $v_1 = 0$, so $u_1 = 1$. Since $iu = u_3 i$, $u_3 > 0$. Thus u is invertible. Let $s = u^{-1}(\frac{u_3}{i_3} i)$ and $t = u^{-1}(\frac{u_3}{j_3} j)$. Then $\{1, s, t\}$ is linearly independent with the following multiplication table:

	1	s	t
1	1	s	t
s	s	s	s
t	t	t	t

Now we define the positive cone $P = \{\alpha + \beta s + \gamma t \mid \alpha, \beta, \gamma \in F^+\}$. Then $(T_2(F), P)$ is an ℓ-algebra in which $1 > 0$, $T_2(F)^+ = uP$, and $(T_2(F), P)$ is ℓ-isomorphic to $(T_2(F), P_2)$. Since $u = 1 + i_3 s + j_3 t$, $u \in P$.

In the case that $i_3 = j_3 = 0$, by a similar argument as above, there is a lattice order with the positive cone P on $T_2(F)$ such that $1 \in P$, $T_2(F)^+ = uP$ for some invertible matrix $u \in P$, and $(T_2(F), P)$ is anti-ℓ-isomorphic to $(T_2(F), P_2)$.

(III) u and v are both basic elements.

Then there is a basic element z such that $\{u, v, z\}$ is disjoint and a vector space basis of $T_2(F)$ over F. Let $z = \begin{pmatrix} z_1 & z_2 \\ 0 & z_3 \end{pmatrix}$. Then

$$z^2 = (z_1 + z_3)z - (z_1 z_3)1 = (z_1 + z_3)z - (z_1 z_3)u + (z_1 z_3)v \geq 0$$

implies that $-(z_1 z_3) \geq 0$ and $(z_1 z_3) \geq 0$. Thus $z_1 z_3 = 0$, so either $z_1 = 0$ or $z_3 = 0$.

We first consider the case $z_1 = 0$.

Since $u^2 = (u_1 + u_3 - u_1 u_3)u + (u_1 u_3)v \geq 0$, we have $u_1 u_3 \geq 0$. Since $zv = v_3 z$, $v_3 \geq 0$, and since $T_2(F)$ is an ℓ-domain, $v_3 > 0$. So $u_3 = v_3 + 1 > 1$ and $u_1 \geq 0$. If $u_1 = 0$, then $v_1 = -1$, so $v^2 = -v + v_3 u \geq 0$, which is a contradiction. Thus $u_1 > 0$. Similarly from $v^2 = (v_1 + v_3 + v_1 v_3)v - (v_1 v_3)u \geq 0$, we have $-(v_1 v_3) \geq 0$. Thus $v_1 \leq 0$. We claim that v_1 cannot be less than 0.

Suppose that $v_1 < 0$. Consider uz and vz. Since $uz - vz = (u - v)z = z$, we have that

$$uz = \alpha u + \beta v + \gamma_1 z \quad \text{and} \quad vz = \alpha u + \beta v + \gamma_2 z,$$

where $\alpha, \beta, \gamma_1, \gamma_2 \in F^+$ and $\gamma_1 - \gamma_2 = 1$. Since $z_1 = 0$, $z^2 = z_3 z$. Now multiplying the above equation for uz by z from the right, we get

$$z_3(uz) = \alpha(uz) + \beta(vz) + (\gamma_1 z_3)z.$$

Further substitutions of uz and vz result in:

$$(z_3\alpha)u + (z_3\beta)v + (z_3\gamma_1)z = (\alpha^2)u + (\alpha\beta)v + (\alpha\gamma_1)z + (\beta\alpha)u + (\beta^2)v + (\beta\gamma_2)z + (\gamma_1 z_3)z.$$

Comparing the coefficients of z, we have $z_3\gamma_1 = \alpha\gamma_1 + \beta\gamma_2 + \gamma_1 z_3$, so $\alpha\gamma_1 + \beta\gamma_2 = 0$. Thus $\alpha\gamma_1 = \beta\gamma_2 = 0$ since $\alpha, \beta, \gamma_1, \gamma_2 \in F^+$.

If $\alpha \neq 0$ and $\beta \neq 0$, then $\gamma_1 = \gamma_2 = 0$, which contradicts with the fact that $\gamma_1 - \gamma_2 = 1$. If $\alpha = 0$ and $\beta \neq 0$, then $\gamma_2 = 0$ and $\gamma_1 = 1$, so $uz = \beta v + z$, that is,

$$\begin{pmatrix} 0 & u_1 z_2 + u_2 z_3 \\ 0 & u_3 z_3 \end{pmatrix} = \begin{pmatrix} \beta v_1 & \beta u_2 \\ 0 & \beta v_3 \end{pmatrix} + \begin{pmatrix} 0 & z_2 \\ 0 & z_3 \end{pmatrix},$$

which is a contradiction since $\beta v_1 \neq 0$. Similarly the situation that $\alpha \neq 0$ and $\beta = 0$ cannot happen.

Finally we consider the case $\alpha = \beta = 0$. Then $uz = \gamma_1 z$ and $vz = \gamma_2 z$ implies

$$u_1 z_2 + u_2 z_3 = \gamma_1 z_2, \quad u_3 z_3 = \gamma_1 z_3, \quad v_1 z_2 + u_2 z_3 = \gamma_2 z_2, \quad v_3 z_3 = \gamma_2 z_3.$$

Thus $\gamma_1 = u_3$, $\gamma_2 = v_3$, and $(u_3 - u_1)z_2 = u_2 z_3 = (v_3 - v_1)z_2$. It is now straightforward to check that

$$(-v_1 z_2)u + (u_1 z_2)v + (-u_2)z = 0.$$

Thus $v_1 z_2 = 0$, $u_2 = 0$, so $z_2 = 0$ since $v_1 \neq 0$, but then $\{u, v, z\}$ will be linearly dependent, which is a contradiction.

Thus we must have $v_1 = 0$, and hence $u_1 = 1$. Let $s = u^{-1}(\frac{u_3}{v_3})v$ and $t = u^{-1}(\frac{u_3}{z_3}z)$. Then $\{1, s, t\}$ is linearly independent with the following multiplication table.

	1	s	t
1	1	s	t
s	s	s	s
t	t	t	t

Now if we define $P = \{\alpha + \beta s + \gamma t \mid \alpha, \beta, \gamma \in F^+\}$, then $(T_2(F), P)$ is an ℓ-algebra in which $1 > 0$ and $T_2(F)^+ = uP$. Since $u = 1 + v_3 s$, we have $u \in P$. By using the similar mapping as before, $(T_2(F), P)$ is ℓ-isomorphic to ℓ-algebra $(T_2(F), P_2)$.

If $z_3 = 0$, then by a similar argument, $T_2(F)^+ = uP$ for a lattice order on $T_2(F)$ with the positive cone P and an invertible matrix $u \in P$ such that $1 \in P$, and $(T_2(F), P)$ is anti-ℓ-isomorphic to the ℓ-algebra $(T_2(F), P_2)$. We leave the verification of this fact to the reader. This completes the proof of the theorem. □

Thus we have proved that if $T_2(F)$ is ℓ-algebra in which $1 \not> 0$, then there exists an ℓ-algebra $T_2(F)$ with the positive cone P such that $1 \in P$ and $(T_2(F), T_2(F)^+)$ is ℓ-isomorphic or anti-ℓ-isomorphic to $(T_2(F), uP)$, where $u \in P$ is invertible. However we don't know if this fact is true or not for $T_n(F)$ with $n > 2$.

4.7.3 Some lattice orders on $T_n(F)$ with $n \geq 3$

Although we have successfully described all the lattice orders on triangular matrix algebra $T_2(F)$ over a totally ordered field F. It seems very hard to do this for $T_n(F)$ when $n \geq 3$. By using *Mathematica*, Mike Bradley found over one hundred lattice orders on $M_3(F)$ to make it into an Archimedean ℓ-algebra over F in which $1 > 0$. Actually lattice orders P_1, P_2 on $T_2(F)$ in section 4.7.1 were first found by Mike using *Mathematica*. Since he couldn't produce more lattice orders when using different inputs, we were convinced that there are only three Archimedean lattice orders with $1 > 0$ on $T_2(F)$, and figured out a proof as shown in the last two sections.

We would like to present some lattice orders on $T_n(F)$ with $n \geq 3$ that are not the entrywise order on $T_n(F)$.

Example 4.6. For a positive integer $k = 1, \cdots, n-1$, define the positive cone on $T_n(F)$ as follows.

$$P_k = \{(a_{ij}) \mid a_{ij} \geq 0, \text{ if } 1 \leq i < j \leq n, \text{ and}$$
$$a_{11} \geq a_{nn}, \cdots, a_{kk} \geq a_{nn}, a_{k+1,k+1} \geq 0, \cdots, a_{nn} \geq 0\}.$$

We leave it to the reader to verify that $(T_n(F), P_k)$ is an ℓ-algebra over F with the following disjoint set (Exercise 71).

$$\{e_{ij}, 1 \leq i < j \leq n\} \cup \{e_{11}, \cdots, e_{n-1,n-1}, e_{11} + \cdots + e_{kk} + e_{nn}\}.$$

In $(T_n(F), P_k)$, $1 = e_{k+1,k+1} + \cdots + e_{n-1,n-1} + (e_{11} + \cdots + e_{kk} + e_{nn})$ is a sum of $n - k$ basic elements, $k = 1, \cdots, n - 1$.

We give a characterization of ℓ-algebra $T_3(F)$ with the entrywise order.

Theorem 4.40. *Let R be an ℓ-algebra over a totally ordered field F. R is ℓ-isomorphic to ℓ-algebra $T_3(F)$ with the entrywise order if and only if the following conditions are satisfied.*

(1) $\dim_F R = 6$ *and R is Archimedean over F,*
(2) 1 is a sum of 3 disjoint basic elements,
(3) R contains a nilpotent ℓ-ideal I with $I^2 \neq \{0\}$,
(4) if e is a basic element and an idempotent element, then eRe contains no nilpotent element.

Proof. Since R is finite dimensional and Archimedean over F, R is a finite direct sum of totally ordered subspaces of R over F by Theorem 1.17. Then each strictly positive element is a sum of disjoint basic elements. From $I^2 \neq \{0\}$, there exist two basic elements $x_1, x_2 \in I$ such that $x_1 x_2 \neq 0$. Let $a_{12} = x_1, a_{23} = x_2$, and $a_{13} = x_1 x_2$. Suppose that $1 = a + b + c$, where $\{a, b, c\}$ are disjoint basic elements. Then a, b, c are idempotent f-elements with $ab = ba = ac = ca = bc = cb = 0$.

From $1 = a + b + c$, $a_{12} = aa_{12} + ba_{12} + ca_{12}$. Since a_{12} is basic, any two of $aa_{12}, ba_{12}, ca_{12}$ are comparable. Suppose that $aa_{12} \neq 0$. If $aa_{12} \leq ba_{12}$, then $aa_{12} = a^2 a_{12} \leq aba_{12} = 0$, which is a contradiction. Thus $ba_{12} \leq aa_{12}$, and hence $ba_{12} = 0$. Similarly $ca_{12} = 0$, so $a_{12} = aa_{12}$. Let $a_{11} = a$. Similarly $a_{12}a = a_{12}$ or $a_{12}a = 0$. In the first case, $a_{12} = a_{11}a_{12}a_{11} \in a_{11}Ra_{11}$ implies that a_{12} is not nilpotent, which is a contradiction with the fact that a_{12} is in the nilpotent ℓ-ideal I. Thus $a_{12}a_{11} = 0$. Suppose that $a_{12}b = a_{12}$. Let $a_{22} = b$. Since $a_{12}a_{23} \neq 0$, we must have $a_{22}a_{23} = a_{23}$. Then by condition (4) again, $a_{23}a_{11} = a_{23}a_{22} = 0$ since a_{23}, a_{13} are nilpotent, so $a_{23}c = a_{23}$. Let $a_{33} = c$. Then for the elements in set $\{a_{ij} \mid 1 \leq i \leq j \leq 3\}$, $a_{ij}a_{rs} = \delta_{jr}a_{is}$, where δ_{jr} is Kronecker delta.

Then it is straightforward to check that $\{a_{ij} \mid 1 \leq i \leq j \leq 3\}$ is a linearly independent set, so it is a vector space basis of R over F since $\dim_F R = 6$. Moreover for a matrix $f \in T_3(F)$ with

$$f = \sum_{1 \leq i \leq j \leq 3} \alpha_{ij} a_{ij}, \ \alpha_{ij} \in F,$$

$f \geq 0$ if and only if each $\alpha_{ij} \geq 0$. Therefore R is ℓ-isomorphic to ℓ-algebra $T_3(F)$ with the entrywise order. $\qquad\qquad\square$

Theorem 4.40 is actually true for any positive integer $n \geq 3$ after some modifications. We state the result below and omit the proof.

Theorem 4.41. *Let R be an ℓ-algebra over a totally ordered field F. R is ℓ-isomorphic to ℓ-algebra $T_n(F)$ ($n \geq 3$) with the entrywise order if and only if the following conditions are satisfied.*

(1) $\dim_F R = \frac{n(n+1)}{2}$ and R is Archimedean over F,
(2) 1 is a sum of n disjoint basic elements,
(3) R contains a nilpotent ℓ-ideal I with $I^{n-1} \neq \{0\}$,
(4) if e is a basic element and an idempotent element, then eRe contains no nilpotent element.

At the end of this section, we provide a couple of examples to show that some conditions in Theorem 4.40 are necessary.

Example 4.7. This example shows condition (4) in Theorem 4.40 cannot be omitted. Let $S = \{a, b, c, d, d^2, e\}$ with the following multiplication table.

	a	b	c	d	d^2	e
a	a	0	0	d	d^2	0
b	0	b	0	0	0	e
c	0	0	c	0	0	0
d	d	0	0	d^2	0	0
d^2	d^2	0	0	0	0	0
e	0	0	e	0	0	0

It is straightforward to check that $S \cup \{0\}$ satisfies the associative law, so $S \cup \{0\}$ becomes a semigroup with zero (Exercise 80). Form the semigroup ℓ-algebra $F[S]$ with the coordinatewise order. Then $\text{Dim}_F F[S] = 6$, and the identity element is $1 = a + b + c$. Let

$$J = \{\alpha d + \beta d^2 + \gamma e \mid \alpha, \beta, \gamma \in F\}.$$

Then J is an ℓ-ideal of $F[S]$ such that $J^3 = 0$ and $J^2 \neq 0$ since $d^2 \neq 0$. But $aF[S]a = Fa + Fd + Fd^2$ contains nilpotent elements, since d^2 is a nilpotent element. Thus condition (4) in Theorem 4.40 is not satisfied. Therefore, $F[S]$ is not ℓ-isomorphic to the ℓ-algebra $T_3(F)$ with the entrywise order.

Example 4.8. This example presents an ℓ-algebra which does not satisfy condition (3) in Theorem 4.40. Let $S = \{a, b, c, d, e, f\}$ with the following multiplication table.

	a	b	c	d	e	f
a	a	0	0	d	e	0
b	0	b	0	0	0	f
c	0	0	c	0	0	0
d	0	0	d	0	0	0
e	0	0	e	0	0	0
f	f	0	0	0	0	0

Similarly it is straightforward to check that $S \cup \{0\}$ satisfies the associative law. Then $S \cup \{0\}$ becomes a semigroup with zero (Exercise 80). In the semigroup ℓ-algebra $F[S]$ with the coordinatewise order, $\mathrm{Dim}_F F[S] = 6$ and the identity element is $1 = a + b + c$. From the multiplication table, it is clear that for each $x \in \{a, b, c\}$, $xF[S]x = Fx$. Thus condition (4) in Theorem 4.40 is satisfied. Let $J = Fd + Fe + Ff$. Then J is an ℓ-ideal of $F[S]$ from the table. Clearly $J^2 = 0$. Let I be an ℓ-ideal of $F[S]$ with $I^m = 0$ for some $m \geq 1$. Since a, b, c are idempotent elements, they are not in I, so $I \subseteq J$. Thus $I^2 = 0$ for any ℓ-ideal I, so condition (3) in Theorem 4.40 is not satisfied. Clearly $F[S]$ is not ℓ-isomorphic to the ℓ-algebra $T_3(F)$ with the entrywise order.

Exercises

(1) Let T be an ℓ-unital ℓ-ring and $M_n(T)$ be the ℓ-ring with the entrywise order. Prove that for each $i = 1, \cdots, n$, $e_{ii}M_n(T)$ is a right ℓ-ideal and $e_{ii}M_n(T) \cong e_{jj}M_n(T)$ as right ℓ-modules over $M_n(T)$.

(2) Prove $a_i R = I_i$ in the proof of $(3) \Rightarrow (2)$ in Theorem 4.1.

(3) Show that $a_{ij}a_{k\ell} = \delta_{jk}a_{i\ell}$ in the proof of $(3) \Rightarrow (2)$ in Theorem 4.1.

(4) Prove $\beta_{ij} = \alpha_{ij}$, $1 \leq i, j \leq n$, in the proof of $(2) \Rightarrow (1)$ in Theorem 4.1.

(5) Prove that the φ defined in the proof of $(2) \Rightarrow (1)$ in Theorem 4.1 is one-to-one, onto, and for any $a, b \in R$, $\varphi(a + b) = \varphi(a) + \varphi(b)$.

(6) Let R be a unital ring and $e \in R$ be an idempotent. Prove that $\mathrm{End}_R(eR, eR)$ is a ring and $eRe \cong \mathrm{End}_R(eR, eR)$ as rings.

(7) Prove that for an ℓ-unital ℓ-ring with $e \in R^+$, the ring $\mathrm{End}_R(eR, eR)$ in problem (6) is a partially ordered ring with respect to the partial order defined by $\theta \geq 0$ if $\theta(eR^+) \subseteq eR^+$.

(8) Prove Theorem 4.2(2).

(9) Prove that for any $x \in R$, $(xg_t) \vee 0 = (x \vee 0)g_t$ in the proof of $(3) \Rightarrow (1)$ of Theorem 4.3.

(10) Verify that a, b, f defined in the proof of Lemma 4.3 satisfy $f^n = 0$ and $af^{n-1} + fb = 1$.

(11) Prove that in Lemma 4.3 $R = g_1 R + \cdots + g_n R$ is a direct sum and any two summands are isomorphic right R-module.

(12) Prove that $f^{m+n} = 0$ and $af^m + f^n b = 1$ in the proof of $(1) \Rightarrow (2)$ of Theorem 4.4.

(13) Prove that $\sum_{1 \leq i \leq k, 1 \leq j \leq n} e^i M e^j$ in Theorem 4.7 is a direct sum.

(14) Prove that in Theorem 4.7, each $a, eae^{n-1}, \cdots, e^{k-1}ae^{n-k+1}$ is in the sum for 1.

(15) Prove that $c_{ij} = e^i a e^{n-j}$, $1 \leq i, j \leq k$ in Theorem 4.7 are $k \times k$ matrix units.

(16) Prove that in Theorem 4.8, $H \cap J = \{0\}$.

(17) Prove that in Theorem 4.8, if x is in the centralizer of matrix units $\{c_{ij} \mid 1 \leq i, j \leq k\}$, then $x \in F + Fe^k + \cdots + Fe^{(\ell-1)k}$, where $\ell = n/k$.

(18) Suppose that T is a unital ring and $e = e_{12} + e_{23} + \cdots + e_{n-1,n} \in M_n(T)$. Prove the following.

(a) For $k = 1, \cdots, n-1$, $e^k = e_{1,k+1} + e_{2,k+2} + \cdots + e_{n-k,n} + \cdots + e_{n,k}$.
(b) If $x \in M_n(T)$ with $ex = xe$, then

$$x_{1,k} = x_{2,k+1} = \cdots = x_{n-k,n-1} = \cdots = x_{n,k-1}.$$

(19) Prove that in Theorem 4.9 for $i = 1, \cdots, n-1$, e^i is not in the center of R.

(20) Let R be a left Ore domain and Q be its classical left quotient ring. Prove that matrices in $M_n(R)$ that are linearly independent over R are also linearly independent over Q.

(21) Suppose that R is a totally ordered domain and $M_n(R)$ be a left f-module over R. Prove that a disjoint subset of M must be linearly independent over R.

(22) For a ring B, prove that $K = \{n \in \mathbb{Z} \mid n \text{ has an } n\text{-fier in } B\}$ is an ideal of $\mathbb{Z}$.

(23) Prove that in Theorem 4.10(2), for any $a \in R$, if $a \in U_f$, then one of $a^+, a^- \in U_f$, but not both of them.

(24) Prove that $I(k, x)$ defined before Lemma 4.9 is an ideal of $\overline{R}$.

(25) Prove that φ in Theorem 4.10 is an isomorphism between two additive groups of R_1 and A. Thus R_1 can be made into an ℓ-group such that R_1 and A are ℓ-isomorphic ℓ-groups.

(26) Prove that in Theorem 4.10 if $ne + ae \in f(R)$, then $\overline{(n,a)} \in f(R_1)$.

(27) Let R be an ℓ-ring and I be an ℓ-ideal of R. Prove that for each $a \in f(R)$, $a + I \in f(R/I)$.

(28) Let R be an ℓ-ring and I be an ℓ-ideal. Prove that $\ell(I)$ and $r(I)$ are ℓ-ideal. Moreover if ℓ-$N(R) = \{0\}$, then $\ell(I) = r(I)$.

(29) Prove that A' and B' in Lemma 4.11 are ℓ-annihilator ℓ-ideal.

(30) Let $R \neq \{0\}$ be an ℓ-ring. Prove that R is an ℓ-domain if and only if R is ℓ-prime and ℓ-reduced.

(31) Prove that Re in Theorem 4.12(4) is an ℓ-prime ℓ-ring.

(32) Prove that in the proof of Theorem 4.14, $ye(x \wedge e) = 0$, $x_1e_1 = e_1x_1 = 0$.

(33) Prove that in Lemma 4.13, $x_1e_1 \wedge y = 0$ implies that $yx_1e_1 = 0$ and $yex_1 = 0$.

(34) An element e in a ring R is called *regular* if for any $a \in R$, $ae = 0$ or $ea = 0$ implies that $a = 0$. Let R be a partially ordered ring. R is called *regular division-closed* if $ab > 0$, and one of a, b is a positive regular element, then another is positive. Prove that a d-ring is regular division-closed.

(35) Prove that in Theorem 4.16 the definition of the multiplication on Q is well-defined and Q becomes a ring.

(36) Show that $\varphi : R \to Q$ in Theorem 4.16 is a ring homomorphism.

(37) Show that P defined in Theorem 4.17 is a partial order on Q.

(38) Prove that in the skew polynomial ring $F[x; \sigma]$, for any left polynomials f, g with $f \neq 0$, there exist unique left polynomials q, r such that $g = qf + r$ with $r = 0$ or $\deg r < \deg f$.

(39) Show that $R = F[x; \sigma]$ is an ℓ-ring with respect to the order defined in Example 4.2 and $f(R) = F[x^2; \sigma]$.

(40) Let R be a totally ordered integral domain and Q be its quotient field. For $\frac{a}{b} \in Q$, define $\frac{a}{b} \geq 0$ if $ab \geq 0$ in R. Prove that F is a totally ordered field and $R^+ = R \cap Q^+$.

(41) Prove that the P defined in Theorem 4.23 is a partial order on $M_n(F)$.

(42) Prove Lemma 4.16.

(43) In Theorem 4.25, $b_{ij}b_{js} = t_{ij}t_{js}t_{is}^{-1}b_{is}$. Prove that $t_{ij}t_{js}t_{is}^{-1} \in R$.

(44) Check $B = FJ$ in Theorem 4.25.

(45) Prove Lemma 4.17.

(46) Let F be a totally ordered field and $K \subseteq F^n$. Prove that $\text{cone}_F(K) = \{\sum \alpha_i v_i \mid \alpha_i \in F^+, v_i \in K\}$ is closed under the addition of F^n and positive scalar multiplication.

(47) Let F be a field and M be a nonzero subspace of F^n. Prove that if for any $g \in M_n(F)$, $gM \subseteq M$, then $M = F^n$.

(48) For the following matrices

$$h_1 = \begin{pmatrix} 0 & 1 \\ 0 & 0 \end{pmatrix}, h_2 = \begin{pmatrix} 1 & 0 \\ 1 & 0 \end{pmatrix}, \quad h_3 = \begin{pmatrix} 1 & 0 \\ 0 & 0 \end{pmatrix}, \quad h_4 = \begin{pmatrix} 0 & 1 \\ 0 & 1 \end{pmatrix},$$

construct the multiplication table of $\{h_1, h_2, h_3, h_4\}$.

(49) Prove that $(\Pi_{1 \leq i,j \leq 2} q_{ij})(\det u)^2$ is a unit in R on p.168.

(50) Verify the multiplication table for $\{c_{11}, c_{12}, c_{21}, c_{22}\}$ on p.168.

(51) Show that f in Example 4.4 is a positive f-element.

(52) For a unital ring R and an element u in its center, define a new multiplication $*$ for any $x, y \in R$, $x * y = xyu^{-1}$. Show that $(R, +, *)$ is a ring and if R is division ring then $(R, +, *)$ is also a division ring.

(53) Prove that P defined in Theorem 4.31 is the positive cone of a partial order on L.

(54) Verify that in Theorem 4.31, $(ua)^{\perp\perp} = F(ua)$.

(55) In Theorem 4.31, prove that if $x = \beta_1 c_1 + \cdots + \beta_m c_m$ for some $\beta_1, \cdots, \beta_m \in F$ and $c_1, \cdots, c_m \in B$ are distinct. Then

$$x \vee_{\succeq} 0 = \beta_1^+ c_1 + \cdots + \beta_m^+ c_m.$$

(56) Suppose that T is a subspace of $\mathbb{R}^n$. Prove that T is a closed set.

(57) Suppose that T is a totally ordered subspace of $\mathbb{R}^n$ such that T^+ and $-T^+$ are closed sets. Then T must be 1-dimensional.

(58) For n linearly independent vectors $v_1, \cdots, v_n$ of $\mathbb{R}^n$, prove that $\text{cone}_{\mathbb{R}}(v_1, \cdots, v_n)$ is a closed set.

(59) Prove that in the proof of Theorem 4.32, $\overline{Pv} \cap \overline{-Pv} = \{0\}$.

(60) Suppose that $T_2(K)$ is the 2×2 upper triangular matrix algebra over a field K. Prove that if $uT_2(K)v = vT_2(K)u = \{0\}$ for some $u, v \in T_2(K)$, then $u^2 = 0$ or $v^2 = 0$.

(61) Suppose that $T_2(F)$ is an ℓ-unital ℓ-algebra over a totally ordered field F and $0 < a \in T_2(F)$ is a nilpotent element. Prove that Fa is an ℓ-ideal.

(62) Suppose that $f, e \in T_2(F)$ are idempotent elements with $1 = f + e$ and $ef = fe = 0$. Prove that

$$f = \begin{pmatrix} 1 & u \\ 0 & 0 \end{pmatrix}, \quad e = \begin{pmatrix} 0 & v \\ 0 & 1 \end{pmatrix},$$

and $v = -u \in F$.

(63) Prove that φi_q in Theorem 4.34 is an anti-ℓ-isomorphism.

(64) Verify that in Theorem 4.35, if $x_3 \leq 0$, then $x_3 = 0$.

(65) Verify that in Theorem 4.35, if $y_3 \leq 0$, then $y_3 = 0$.

(66) Check that $(T_2(F), P_3)$ in Theorem 4.36 is an non-Archimedean ℓ-unital ℓ-algebra over F.

(67) Prove that in Theorem 4.36, $b^2 = 0$.

(68) Prove that in Theorem 4.37(1), for $f = \alpha u + \beta v + \gamma w$, $f \geq 0$ if and only if $\alpha \geq 0$, $\beta > 0$ or $\alpha \geq 0$, $\beta = 0$, $\gamma \geq 0$.

(69) Verify the multiplication table for $\{1, x', y\}$ in Theorem 4.37(2).

(70) Prove that Theorem 4.38 is true when $f_1 = h_3 = 0$ in (ii).

(71) Verify that P_k in Example 4.6 is a lattice order on $T_n(F)$.

(72) Prove that the direct sum of two totally ordered domains is regular division-closed, but not division-closed.

(73) Consider the field $L = \mathbb{Q}[\sqrt{2}]$.

    (a) Describe all the lattice orders on L to make it into an ℓ-field in which $1 > 0$.

    (b) Prove that each lattice order on L in which $1 \not> 0$ can be obtained from a lattice ordered with $1 > 0$ by using Theorem 1.19.

(74) Consider group algebra $R = \mathbb{R}[G]$, where G is a cyclic group of order 2.

    (a) Describe all the lattice orders on R to make it into an ℓ-algebra over $\mathbb{R}$ in which $1 > 0$.

    (b) Prove that each lattice order on R in which $1 \not> 0$ can be obtained from a lattice ordered with $1 > 0$ by using Theorem 1.19.

(75) Let R be an ℓ-prime ℓ-ring with squares positive. Prove that for any $a \in R$, $r(a) \cap \ell(a) \neq \{0\}$ if and only if $|a|^2 = 0$.

(76) An ℓ-ring is called a *left d-ring* if for any $a \in R^+$, $x \wedge y = 0$ implies that $ax \wedge ay = 0$ for any $x, y \in R$. Prove that if R is an ℓ-prime left d-ring with squares positive, then R is an ℓ-domain.

(77) Let R be an ℓ-ring and $0 < e \in f(R)$. Prove that if $r(e) = \{0\}$ and e is a weak unit, then $(xex)^- = 0$ for any $x \in R$ and $aea = 0$ for any $a \in R^+, a^2 = 0$.

(78) Let A be a unital finite-dimensional Archimedean ℓ-algebra over a totally ordered field F. Suppose that A contains a positive element e with order $n \geq 2$ and $\dim_F i(e) = 1$, where $i(e) = \{a \in R \mid ae = ea = a\}$. Prove that if 1 is a basic element of A, then A is ℓ-siomorphic to the group ℓ-algebra $F[G]$ of a cyclic group G.

(79) Prove that the $R = \mathbb{R}[x]$ defined in Example 4.3(2) is a partially ordered ring.

(80) Prove that $S \cup \{0\}$ in Examples 4.7 and 4.8 is a semigroup with zero, that is, the multiplication on $S \cup \{0\}$ is associative.

(81) Find an ℓ-ring satisfying the polynomial constraint

$$f(x, y) = -(xy + yx) + (x^{2n} + y^{2n}) \geq 0, \text{ for a fixed } n \geq 1.$$

(82) Prove that a lattice-ordered division ring satisfying $x^{2n} \geq 0$ for some positive integer n must be a totally ordered field.

(83) Prove that an ℓ-semiprime ℓ-ring with squares positive and an f-superunit can be embedded in a unital ℓ-semiprime ℓ-ring with squares positive.

(84) Let R be a unital f-ring which is division-closed. Prove that $R/\ell\text{-}N(R)$ is a totally ordered domain.

(85) Let R be an ℓ-unital ℓ-ring. If I is an ℓ-ideal of R with $I \cap f(R) = \{0\}$, then I is contained in each maximal ℓ-ideal of R.

(86) Let $\mathbb{C}$ be a partially ordered field with the positive cone P. Prove that if for any $z \in P$, the real part of z is in $\mathbb{R}^+$, then $P \subseteq \mathbb{R}$.

(87) Prove that $f_2 f_3 = f_2 + \alpha f_1 + \beta f_4$ with $\alpha, \beta \in F^+$ on page 166.

(88) Show that $f_4 f_2 = f_2$ and $f_2 f_4 = f_4$ on page 166.

(89) Prove that $1 = c_2(a + b)$ and $f_1 f_2 + f_2 f_1 = c_2 f_3 + c_2 f_4$ on page 166.

(90) This problem is actually a conjecture. For a totally ordered field F and the $n \times n$ $(n \geq 3)$ upper triangular matrix algebra $T_n(F)$ over F, we conjecture that each lattice order on $T_n(F)$ in which $1 \not\geq 0$ can be obtained by using Theorem 1.19(2) from a lattice order on $T_n(F)$ in which $1 > 0$.

Chapter 5

ℓ-ideals of ℓ-unital lattice-ordered rings

In this chapter we always assume that R is an ℓ-unital ℓ-ring. We study properties of ℓ-ideals of R.

5.1 Maximal ℓ-ideals

For an ℓ-unital ℓ-ring R, let $\text{Max}_\ell(R)$ denote the set of all maximal ℓ-ideals of R. For a subset X of R, define

$$s(X) = \{M \in \text{Max}_\ell(R) \mid X \nsubseteq M\},$$

and

$$h(X) = \{M \in \text{Max}_\ell(R) \mid X \subseteq M\}.$$

If $X = \{x\}$, we will write $s(x)$ and $h(x)$ instead of $s(\{x\})$ and $h(\{x\})$. It is clear that $s(X) = s(\langle X \rangle)$ and $h(X) = h(\langle X \rangle)$, where $\langle X \rangle$ denote the ℓ-ideal generated by X. The sets $s(X), X \subseteq R$, form open sets of a topology as shown in the following result.

Theorem 5.1. *Let R be an ℓ-unital ℓ-ring.*

(1) $s(0) = \emptyset$, $s(1) = \text{Max}_\ell(R)$.
(2) $s(I) \cap s(J) = s(I \cap J)$, for any ℓ-ideals I and J.
(3) $\cup_\alpha s(I_\alpha) = s(\cup_\alpha I_\alpha)$, for any family $\{I_\alpha\}$ of ℓ-ideals of R.

Proof. (1) and (3) are clearly true. Let M be a maximal ℓ-ideal. Then M is ℓ-prime. If $I \cap J \subseteq M$, then $IJ \subseteq M$, and hence $I \subseteq M$ or $J \subseteq M$. Thus (2) is true. $\qquad\square$

By Theorem 5.1, the sets in $\{s(X) \mid X \text{ is a subset of } R\}$ constitute the open sets of a topology on $\text{Max}_\ell(R)$ which is called the *hull-kernel topology*.

We always endow $\text{Max}_\ell(R)$ with this topology. A *topological space basis* for a *topological space* is a collection of open sets such that each open set is a union of open sets in the collection. Clearly $s(a), a \in R^+$ form a topological space basis for the open sets (Exercise 1). A subset of a topological space is *closed* if its complement is open. Each $h(X)$ is closed in $\text{Max}_\ell(R)$ since $h(X) = \text{Max}_\ell(R) \setminus s(X)$. Recall that the closure of a subset $\mathcal{K}$ of $\text{Max}_\ell(R)$ is the smallest closed set containing $\mathcal{K}$.

Theorem 5.2. *Let R be an ℓ-unital ℓ-ring.*

(1) $\text{Max}_\ell(R)$ is compact.
(2) The closure of a subset $\mathcal{K}$ of $\text{Max}_\ell(R)$ is $h(\cap\{M \mid M \in \mathcal{K}\})$.

Proof. (1) Let $\text{Max}_\ell(R) \subseteq \cup_\alpha s(I_\alpha)$ for some ℓ-ideals I_α of R. Then $h(\sum_\alpha I_\alpha) = \emptyset$. Thus $\sum_\alpha I_\alpha = R$ implies $1 \in I_{\alpha_1} + \cdots + I_{\alpha_n}$, for some $I_{\alpha_1}, \cdots, I_{\alpha_n}$, and hence $\text{Max}_\ell(R) = \cup_{k=1}^n s(I_{\alpha_k})$. Therefore $\text{Max}_\ell(R)$ is compact.

(2) It is clear that $\mathcal{K} \subseteq h(\cap\{M \mid M \in \mathcal{K}\})$ and $h(\cap\{M \mid M \in \mathcal{K}\})$ is closed. Suppose that $\mathcal{K} \subseteq \mathcal{J}$ and $\mathcal{J}$ is closed. Then $\mathcal{J} = \text{Max}_\ell(R) \setminus s(I)$ for some ℓ-ideal I of R. For any $M \in \mathcal{K}$, $\mathcal{K} \subseteq \mathcal{J}$ implies that $I \subseteq M$, and hence $I \subseteq \cap\{M \mid M \in \mathcal{K}\}$. Therefore $h(\cap\{M \mid M \in \mathcal{K}\}) \subseteq \mathcal{J}$, that is, $h(\cap\{M \mid M \in \mathcal{K}\})$ is the smallest closed set containing $\mathcal{K}$. $\square$

For an ℓ-unital ℓ-reduced ℓ-ring R, we show that $\text{Max}_\ell(R)$ and $\text{Max}_\ell(f(R))$ are homeomorphic.

Lemma 5.1. *Let R be an ℓ-unital ℓ-reduced ℓ-ring.*

(1) If P be an ℓ-prime ℓ-ideal of R, then for any $x, y \in f(R)$, $xy \in P$ or $x \wedge y \in P$ implies $x \in P$ or $y \in P$.
(2) If $R = I + J$ for some left (right) ℓ-ideals I, J of R, then

    (a) $1 = x + y$ for some $0 \le x \in f(R) \cap I$, $0 \le y \in f(R) \cap J$, and
    (b) $1 = a + b + c$ for some $0 \le a \in f(R) \cap I$, $0 \le b \in f(R) \cap J$,
      $c \in f(R) \cap I \cap J$, and $ab = 0$.

Proof. (1) By Lemma 4.10, $f(R/P)$ is a totally ordered domain. For $x \in f(R)$, $\overline{x} = x + P \in f(R/P)$. Thus if $x, y \in f(R)$ with $xy \in P$ or $x \wedge y \in P$, then $(\overline{x})(\overline{y}) = 0$ or $\overline{x} \wedge \overline{y} = 0$ in R/P, and hence $\overline{x} = 0$ or $\overline{y} = 0$, that is, $x \in P$ or $y \in P$.

(2) Let $1 = w + z$ for some $w \in I$ and $z \in J$. Since $1 > 0$,

$$1 = |1| = |w + z| \le |w| + |z|,$$

and hence $1 = x+y$ for some $0 \le x \le |w|$ and $0 \le y \le |z|$. Thus $x \in f(R) \cap I$ and $y \in f(R) \cap J$.

Let $t = x \wedge y$ and $a = x - t$, $b = y - t$. We have $a \wedge b = 0$, so $ab = 0$ since a, b are f-element. Hence $1 = a + b + c$ with $a \in f(R) \cap I$, $b \in f(R) \cap J$, $c = 2t \in f(R) \cap I \cap J$. □

For a left ℓ-ideal I of R, define
$$(I : R) = \{a \in R \mid |a||x| \in I, \text{ for all } x \in R\}.$$

Lemma 5.2. *Let R be an ℓ-unital ℓ-ring and I be a left ℓ-ideal of R.*

(1) $(I : R)$ is the maximal ℓ-ideal contained in I.

(2) If I is a maximal left ℓ-ideal, then $(I : R)$ is an ℓ-prime ℓ-ideal.

Proof. (1) It is clear that $(I : R)$ is an ℓ-ideal. If $a \in (I : R)$, then $|a| = |a|1 \in I$, so $(I : R) \subseteq I$. Let J be an ℓ-ideal of R with $J \subseteq I$. Then clearly $J \subseteq (I : R)$.

(2) Suppose that I is a left maximal ℓ-ideal and H, K are ℓ-ideals of R such that $HK \subseteq (I : R)$. Assume that $K \not\subseteq (I : R)$, then $K \not\subseteq I$ by (1), and hence $R = K + I$. Thus $H = HR = HK + HI \subseteq I$, so $H \subseteq (I : R)$ by (1) again. Therefore $(I : R)$ is an ℓ-prime ℓ-ideal of R. □

Lemma 5.3. *Let R be a unital f-ring. Then maximal ℓ-ideals and maximal left (right) ℓ-ideals of R coincide.*

Proof. First assume that R is ℓ-simple. Then R is a totally ordered domain by Theorem 1.27, so R has no left and right ℓ-ideal except R and $\{0\}$ by Lemma 4.6.

Now let R be a unital f-ring and M be a maximal ℓ-ideal. Then $M \subseteq L$ for some maximal left ℓ-ideal L of R. Since R/M is ℓ-simple, by the above argument, $L/M = \{0\}$ in R/M, so $M = L$, that is, M is a maximal left ℓ-ideal of R. Let I be a maximal left ℓ-ideal. Then $(I : R)$ is an ℓ-prime ℓ-ideal contained in I by Lemma 5.2. By Theorem 1.27, $R/(I : R)$ is a totally ordered domain, so $I/(I : R)$ is an ℓ-ideal of $R/(I : R)$ by Lemma 4.6. Therefore I is an ℓ-ideal of R, so it must be a maximal ℓ-ideal. □

The above result is not true for general ℓ-rings. For example, in the matrix ℓ-algebra $M_n(\mathbb{R})$ $(n \ge 2)$ with the entrywise order, $M_n(\mathbb{R})$ is simple, but it contains more than one maximal left (right) ℓ-ideals.

Two topological spaces are called *homeomorphic* if there is a one-to-one and onto function between them that sends open (closed) sets to open (closed) sets.

Theorem 5.3. *Let R be an ℓ-unital ℓ-reduced ℓ-ring. Then $Max_\ell(R)$ and $Max_\ell(f(R))$ are homeomorphic topological spaces.*

Proof. Let M be a maximal ℓ-ideal of R. If $xy \in M \cap f(R)$, then $x \in M \cap f(R)$ or $y \in M \cap f(R)$ by Lemma 5.1. Thus $M \cap f(R)$ is an ℓ-prime ℓ-ideal of $f(R)$. By Lemma 4.10, the ℓ-ideals in an f-ring that contain an ℓ-prime ℓ-ideal form a chain, so there exists a unique maximal ℓ-ideal of $f(R)$ that contains $M \cap f(R)$. We denote this unique maximal ℓ-ideal of $f(R)$ by M_f and define $\varphi : Max_\ell(R) \to Max_\ell(f(R))$ by $\varphi(M) = M_f$.

Let $M, N \in Max_\ell(R)$ and $\varphi(M) = \varphi(N)$. If $M \neq N$, then $R = M + N$, and hence, by Lemma 5.1, there exist $0 \leq i \in M \cap f(R)$ and $0 \leq j \in N \cap f(R)$ such that $1 = i + j$, so $1 = i + j \in \varphi(M) = \varphi(N)$, which is a contradiction. Thus $\varphi(M) = \varphi(N)$ implies $M = N$, so φ is one-to-one.

Now let I be a maximal ℓ-ideal of $f(R)$. Then I is ℓ-prime in $f(R)$, so by Theorem 1.27, $f(R)^+ \setminus I$ is closed under multiplication, and hence it is an m-system. By Theorem 1.26, there is an ℓ-prime ℓ-ideal K of R such that $(f(R)^+ \setminus I) \cap K = \emptyset$. Then $(K \cap f(R)) \subseteq I$. Let M be a maximal ℓ-ideal of R containing K. Then $M \cap f(R)$ and I must be comparable since $K \cap f(R)$ is ℓ-prime in $f(R)$, and hence $M \cap f(R) \subseteq I$. Thus $I = M_f$, so φ is onto.

Let $\mathcal{K} = \{M_\alpha \mid \alpha \in \Gamma\}$ be a closed set in $Max_\ell(R)$. We show that $\varphi(\mathcal{K}) = \{(M_\alpha)_f \mid \alpha \in \Gamma\}$ is closed in $Max_\ell(f(R))$. Let I be a maximal ℓ-ideal of $f(R)$ such that

$$I \supseteq \cap_{\alpha \in \Gamma}(M_\alpha)_f \supseteq \cap_{\alpha \in \Gamma}(M_\alpha \cap f(R)).$$

If $I \notin \varphi(\mathcal{K})$, then $M_\alpha \cap f(R) \not\subseteq I$ for each $\alpha \in \Gamma$, and hence $(M_\alpha \cap f(R)) + I = f(R)$ for each $\alpha \in \Gamma$ since I is maximal. Thus by Lemma 5.1 for each $\alpha \in \Gamma$, there exist $x_\alpha, y_\alpha \in f(R)$,

$$0 \leq x_\alpha \in (M_\alpha \cap f(R)) \setminus I \text{ and } 0 \leq y_\alpha \in I \setminus (M_\alpha \cap f(R))$$

such that $1 = x_\alpha + y_\alpha$. Let $x_\alpha \wedge y_\alpha = z_\alpha$. Then $0 \leq z_\alpha \in (M_\alpha \cap f(R)) \cap I$ for each $\alpha \in \Gamma$. Since $y_\alpha \notin M_\alpha$ and $z_\alpha \in M_\alpha$, $y_\alpha - z_\alpha \notin M_\alpha$, for each $\alpha \in \Gamma$. Then $\{s(y_\alpha - z_\alpha) \mid \alpha \in \Gamma\}$ is an open cover for $\mathcal{K}$, and hence a finite subcover $s(y_i - z_i), i = 1, \ldots, n$, can be extracted out of this cover because $Max_\ell(R)$ is compact and $\mathcal{K}$ is closed. Then for every $M_\alpha \in \mathcal{K}$, there exists $y_j - z_j \notin M_\alpha$, for some $1 \leq j \leq n$. Since $(x_j - z_j) \wedge (y_j - z_j) = 0$ implies that $(x_j - z_j) \in M_\alpha$ by Lemma 5.1, $\wedge_{i=1}^n (x_i - z_i) \in M_\alpha$ for each $\alpha \in \Gamma$. Thus

$$\wedge_{i=1}^n (x_i - z_i) \in \cap_{\alpha \in \Gamma}(M_\alpha \cap f(R)) \subseteq \cap_{\alpha \in \Gamma}(M_\alpha)_f \subseteq I,$$

so $x_k - z_k \in I$ for some $1 \le k \le n$ by Theorem 1.27, and hence $x_k \in I$, which is a contradiction. Therefore, $I \in \varphi(\mathcal{K})$. We have shown that $\varphi(\mathcal{K})$ is equal to its closure $h(\cap_{\alpha \in \Gamma}(M_\alpha)_f)$, so it is closed.

Now let $\mathcal{J} = \{I_\alpha \mid \alpha \in \Gamma\} \subseteq \mathrm{Max}_\ell(f(R))$ be closed. We verify that $\varphi^{-1}(\mathcal{J})$ is closed in $\mathrm{Max}_\ell(R)$. Let $M \in \mathrm{Max}_\ell(R)$ be in the closure of $\varphi^{-1}(\mathcal{J})$. If $\cap_{\alpha \in \Gamma} I_\alpha \not\subseteq M_f$. Then $(\cap_{\alpha \in \Gamma} I_\alpha) + M_f = f(R)$, so $1 = x + y$, where $0 \le x \in (\cap_{\alpha \in \Gamma} I_\alpha) \setminus M_f$ and $0 \le y \in M_f \setminus (\cap_{\alpha \in \Gamma} I_\alpha)$. Since $x \in I_\alpha$ for each $\alpha \in \Gamma$, we have $y \notin I_\alpha$ for each $\alpha \in \Gamma$. Let $x \wedge y = z$. Then $(x - z) \wedge (y - z) = 0$. Let $I_\alpha = (M_\alpha)_f \supseteq M_\alpha \cap f(R)$, where M_α is a maximal ℓ-ideal of R for each $\alpha \in \Gamma$. Then $(y - z) \notin I_\alpha$ for each $\alpha \in \Gamma$ implies $(y - z) \notin M_\alpha$ for each $\alpha \in \Gamma$ since $y - z$ is an f-element, so $(x - z) \in M_\alpha$ for each $\alpha \in \Gamma$ by Lemma 5.1. However $\cap_{\alpha \in \Gamma} M_\alpha \subseteq M$ since $\varphi^{-1}(\mathcal{J}) = \{M_\alpha \mid \alpha \in \Gamma\}$, and hence $x - z \in M \cap f(R) \subseteq M_f$. It follows that $x \in M_f$, which is a contradiction. Hence $\cap_{\alpha \in \Gamma} I_\alpha \subseteq M_f$ and $M_f \in \mathcal{J}$ since $\mathcal{J}$ is closed. Then it follows that $M \in \varphi^{-1}(\mathcal{J})$. Therefore $\varphi^{-1}(\mathcal{J})$ is closed in $\mathrm{Max}_\ell(R)$. □

Corollary 5.1. *Suppose that R is an ℓ-unital ℓ-reduced ℓ-ring.*

(1) Each ℓ-prime ℓ-ideal of R is contained in a unique maximal ℓ-ideal.

(2) Every maximal ℓ-ideal of R is contained in a unique maximal left (right) ℓ-ideal of A, and each maximal left (right) ℓ-ideal of A contains a maximal ℓ-ideal of A.

Proof. (1) Let I be an ℓ-prime ℓ-ideal of R. Then I is contained in a maximal ℓ-ideal of R. Suppose that M and N are maximal ℓ-ideals both containing I. Then $I \cap f(R)$ is contained in $M \cap f(R)$ and $N \cap f(R)$. Let M_f and N_f be defined as in Theorem 5.3. Then M_f and N_f are comparable since they both contain $I \cap f(R)$ which is ℓ-prime in $f(R)$, and hence $M_f = N_f$. Therefore, $M = N$ by Theorem 5.3. Thus I is contained in a unique maximal ℓ-ideal.

(2) Let M be a maximal ℓ-ideal of R. Then M is contained in a maximal left ℓ-ideal by Zorn's lemma. Now suppose that M is contained in two different maximal left ℓ-ideals L_1 and L_2. Then we have $M \cap f(R) \subseteq L_1 \cap f(R)$ and $L_2 \cap f(R)$. Since $L_1 \cap f(R)$ and $L_2 \cap f(R)$ are left ℓ-ideals of $f(R)$, they are contained in some maximal left ℓ-ideals of $f(R)$, and since every maximal left (right) ℓ-ideal in $f(R)$ is a maximal ℓ-ideal by Lemma 5.3, there exist maximal ℓ-ideals I_1 and I_2 of $f(R)$ such that $L_1 \cap f(R) \subseteq I_1$ and $L_2 \cap f(R) \subseteq I_2$. Since $M \cap f(R)$ is contained in I_1 and I_2, $I_1 = I_2$. Now $R = L_1 + L_2$ implies $1 = i + j$, where $0 \le i \in L_1$ and $0 \le j \in L_2$

by Lemma 5.1. Since i and j are f-elements, $i \in L_1 \cap f(R) \subseteq I_1$ and $j \in L_2 \cap f(R) \subseteq I_2$, and hence $1 = i+j \in I_1 = I_2$, which is a contradiction. Thus each maximal ℓ-ideal M of R is contained in a unique maximal left ℓ-ideal.

Now let L be a maximal left ℓ-ideal of R. Then, by the same argument as above, $L \cap f(R) \subseteq M_f$ for some maximal ℓ-ideal M of R, where M_f is defined as in Theorem 5.3. If $M \nsubseteq L$, then $R = M+L$, and hence $1 = i+j$, where $0 \leq i \in M$ and $0 \leq j \in L$, so $1 \in M_f$, which is a contradiction. Thus $M \subseteq L$. □

For a unital f-ring, by Lemma 5.3, maximal ℓ-ideals and maximal left (right) ℓ-ideals coincide. By Corollary 5.1, in an ℓ-unital ℓ-reduced ℓ-ring, maximal ℓ-ideals and maximal left (right) ℓ-ideals are in one-to-one correspondence. However maximal ℓ-ideals and maximal left (right) ℓ-ideals are generally different in ℓ-unital ℓ-reduced ℓ-rings. We provide an example using *differential polynomial rings*.

Example 5.1. Let R be a ring and δ be a derivation on R. Define $R[x; \delta]$ to be the set consisting of all left polynomials $f(x) = \sum a_i x^i$. With coordinatewise addition, $R[x; \delta]$ becomes a group. Introduce the multiplication by repeatedly using $xa = ax + \delta(a)$ for $a \in R$. Then $R[x; \delta]$ is a ring (Exercise 3), called a differential polynomial ring. If R is a domain, then so is $R[x; \delta]$.

For an ℓ-ring R and a positive derivation δ on R. If we order $R[x; \delta]$ coordinatewisely, then $R[x; \delta]$ becomes an ℓ-ring. For instance, let $R = \mathbb{R}[y]$ be the polynomial ring in y with the total order in which a polynomial is positive if the coefficient of highest power is positive. Then R is a totally ordered domain. Take δ as the usual derivative on R. Then δ is a positive derivation on R. In the following, we assume $R = \mathbb{R}[y]$ and δ defined above and show that $R[x; \delta]$ is ℓ-simple, however it contains nonzero maximal left ℓ-ideal.

Let I be a nonzero ℓ-ideal of $R[x; \delta]$. Take a nonzero positive left polynomial $f(x) = a_n x^n + \cdots + a_1 x + a_0 \in I$ with the smallest degree n. We claim that $n = 0$. Suppose that $n \geq 1$. Since R is ℓ-simple, there is $b \in R^+$ such that $1 < ba_n$ (Exercise 4). Thus $x^n \leq bf(x)$ implies that $x^n \in I$. Then

$$x^n y = x^{n-1}(yx + 1) = x^{n-2}(yx^2 + 2x) = \cdots = yx^n + nx^{n-1} \in I$$

implies that $nx^{n-1} \in I$, so $x^{n-1} \in I$, which is a contradiction with the fact that $f(x)$ has the smallest degree in I. Hence we must have $n = 0$, so

$0 \neq a_0 \in I$ implies that the identity element $1 \in I$. Therefore $I = R$, that is, R is ℓ-simple.

It is clear that x is a right d-element of $R[x; \delta]$ in the sense that if $u \wedge v = 0$ for some $u, v \in R[x; \delta]$, then $ux \wedge vx = 0$. Thus it is straightforward to check that $R[x; \delta]x$ is a maximal left ℓ-ideal of $R[x; \delta]$ (Exercise 5). We also note that x is not a d-element since $1 \wedge yx = 0$, however

$$x \wedge x(yx) = x \wedge (yx + 1)x = x \wedge (yx^2 + x) = x.$$

A topological space is called *Hausdorff* if for any two distinct points x, y, there exist disjoint open sets containing x and y respectively. Let X be a compact Hausdorff space and $C(X)$ be the ring of real-valued continuous functions on X. With respect to the coordinatewise order, $C(X)$ is an f-ring (Exercise 2). In 1947, I. Kaplansky proved that if $C(X)$ and $C(Y)$ are isomorphic as lattices for two compact Hausdorff spaces X and Y, then X and Y are homeomorphic topological spaces [Kaplansky (1947)]. In 1968, H. Subramanian extended Kaplansky's argument to f-rings and proved that if two unital commutative ℓ-semisimple f-rings A and B are isomorphic as lattices, then $\text{Max}_\ell(A)$ and $\text{Max}_\ell(B)$ are homeomorphic [Subramanian (1968)]. Actually H. Subramanian's proof works for f-rings that are not commutative. In the following we present H. Subramanian's proof for f-rings and then consider how to generalize it to more general ℓ-rings.

We need some preparations to carry out the proof. Let L be a lattice. A *lattice-prime ideal* P of L is a nonempty proper subset of L satisfying the following properties.

(1) for all $a, b \in P$, $a \vee b \in P$,
(2) $b \leq a, a \in P$ and $b \in L \Rightarrow b \in P$,
(3) for all $a, b \in L, a \wedge b \in P \Rightarrow a \in P$ or $b \in P$.

Let A be a unital ℓ-semisimple f-ring. Recall that ℓ-semisimple means that the intersection of all maximal ℓ-ideals of A is zero. Then A must be reduced by Theorem 1.27(1). For a lattice-prime ideal P and a maximal ℓ-ideal M of A, we say that P is associated with M if for any $x \in P$, $y \in A$, $(y - x)^- \notin M \Rightarrow y \in P$.

Theorem 5.4. *Let R and S be unital ℓ-semisimple f-rings. If they are isomorphic as lattices, then $\text{Max}_\ell(R)$ and $\text{Max}_\ell(S)$ are homeomorphic.*

Proof. We achieve the proof by a series of steps.

(I) *Each lattice-prime ideal is associated with exactly one maximal ℓ-ideal.*

Suppose that P is a lattice-prime ideal. If P is not associated with any maximal ℓ-ideal, then for each maximal ℓ-ideal M_α there exist x_α, y_α such that $x_\alpha \in P$, $y_\alpha \notin P$ and yet $(y_\alpha - x_\alpha)^- \notin M_\alpha$. Define

$$s_\alpha = s((y_\alpha - x_\alpha)^-) = \{M \in \mathrm{Max}_\ell(R) \mid (y_\alpha - x_\alpha)^- \notin M\}.$$

Then $\{s_\alpha\}$ is an open cover for $\mathrm{Max}_\ell(R)$. Since $\mathrm{Max}_\ell(R)$ is compact, there exists a finite subcover $\{s_i\}$, $i = 1, \cdots, n$ for some positive integer n. Let $x = \vee_{i=1}^{n} x_i, y = \wedge_{i=1}^{n} y_i$. Since

$$(y - x)^- = (x - y) \vee 0 \geq (x_i - y_i) \vee 0 = (y_i - x_i)^-,$$

for each $i = 1, \cdots, n$, $(y-x)^- \notin M_\alpha$ for each maximal ℓ-ideal M_α, and hence $(y - x)^+ \in M_\alpha$ since $(y - x)^+ (y - x)^- = 0$. Thus A is ℓ-semisimple implies that $(y - x)^+ = 0$, so $y \leq x = \vee_{i=1}^{n} x_i \in P$. Then $y = \wedge_{i=1}^{n} y_i \in P$ and P is prime implies that $y_j \in P$ for some $1 \leq j \leq n$, which is a contradiction. Therefore P must be associated with at least one maximal ℓ-ideal.

Suppose that P is associated with two different maximal ℓ-ideals, say M and N. Then $R = N + M$. By Lemma 5.1, $1 = a + b + c$ with $a \in N, b \in M$, $c \in M \cap N$, and $ab = 0$. Take $x \in P$ and $y \notin P$. Let

$$z = a(x - 1) + b(y + 1).$$

In R/M,

$$z + M = (a + M)((x - 1) + M) + (b + M)((y + 1) + M) = (x - 1) + M,$$

since $a + M = 1 + M$ and $b + M = 0$. Similarly in R/N, $z + N = (y + 1) + N$. Thus $(z - x)^- + M = 1 + M$ implies that $(z - x)^- \notin M$, so $z \in P$. Then $(y - z)^- + N = 1 + N$ implies that $(y - z)^- \notin N$, so $y \in P$, which is a contradiction. Thus P is associated with at most one maximal ℓ-ideal.

(II) *Two lattice-prime ideals P_1, P_2 are associated with the same maximal ℓ-ideal if and only if $P_1 \cap P_2$ contains a lattice-prime ideal.*

Suppose that P_1 and P_2 are associated with the same maximal ℓ-ideal M. Choose $x \in P_1$ and $y \in P_2$. Write $a = (x \wedge y) - 1$ and define

$$P = \{w \in R \mid (w - a)^+ \in M\}.$$

We leave it to the reader to check that P is a lattice-prime ideal associated with M (Exercise 6). Let $r \in P$. Since

$$(r - a)^+ = (r - (x \wedge y) + 1)^+ = (r - x + 1)^+ \vee (r - y + 1)^+,$$

$(r - x + 1)^+, (r - y + 1)^+ \in M$. Hence we must have $(r - x)^- \notin M$ and $(r - y)^- \notin M$ (Exercise 7). Therefore $r \in P_1 \cap P_2$, and hence $P \subseteq P_1 \cap P_2$.

Conversely suppose that P, P_1, P_2 are lattice-prime ideals associated with M, M_1, M_2, respectively, and $P \subseteq P_1 \cap P_2$. We show that $M = M_1 = M_2$. If $M \neq M_1$, then similar to the proof in (I), for $a \in P$ and $b \notin P_1$, there exists c such that

$$c + M = (a - 1) + M \text{ and } c + M_1 = (b + 1) + M_1.$$

Thus $c \in P \subseteq P_1$ and then $b \in P_1$, which is a contradiction. Therefore $M = M_1$. Similarly $M = M_2$.

(III) *Let a be fixed element in R, and let $\mathcal{K} = \{M_\alpha \mid \alpha \in \Gamma\}$ be any nonempty set in $Max_\ell(R)$. Then a maximal ℓ-ideal M belongs to the closure of $\mathcal{K}$ if and only if there exists a lattice-prime ideal P associated with M such that P contains $A(\mathcal{K})$, which is the intersection of all lattice-prime ideals which contain a and are also associated with any member in $\mathcal{K}$.*

Suppose that M is in the closure of $\mathcal{K}$. Then

$$P = \{x \in R \mid (x - a)^+ \in M\} \text{ and } P_\alpha = \{x \in R \mid (x - a)^+ \in M_\alpha\}$$

are lattice-prime ideals associated with M and M_α, respectively (Exercise 6). Clearly $a \in P_\alpha$ for each $\alpha \in \Gamma$. Let $r \in A(\mathcal{K})$, then $r \in P_\alpha$ for each $\alpha \in \Gamma$. Thus $(r - a)^+ \in M_\alpha$ for each $M_\alpha \in \mathcal{K}$, so $(r - a)^+ \in M$ since M is in the closure of $\mathcal{K}$. Therefore $r \in P$, that is, $A(\mathcal{K}) \subseteq P$.

Conversely, suppose that M is not in the closure of $\mathcal{K}$. Then

$$I = \bigcap_{\alpha \in \Gamma} M_\alpha \nsubseteq M,$$

so $R = I + M$. Let P be any lattice-prime ideal associated with M. Take $b \notin P$. By Lemma 5.1 and a similar argument as in (I), there exists $c \in R$ such that

$$c + M_\alpha = (a - 1) + M_\alpha, \forall \alpha \in \Gamma \text{ and } c + M = (b + 1) + M.$$

Thus $(c - a)^- \notin M_\alpha$ for any $\alpha \in \Gamma$, so $c \in A(\mathcal{K})$ by the definition of $A(\mathcal{K})$. Then $(b - c)^- \notin M$ and $b \notin P$ imply that $c \notin P$, so $A(\mathcal{K})$ is not contained in any lattice-prime ideal associated with M.

We are ready to give the final proof of Theorem 5.4. Suppose that φ is a lattice isomorphism from R to S. Then clearly a subset P of R is a lattice-prime ideal of R if and only if $\varphi(P)$ is a lattice-prime ideal of S (Exercise 8). Two lattice-prime ideals are called *equivalent* if they are associated with the same maximal ℓ-ideal. Then this is an equivalence relation on the set of all lattice-prime ideals (Exercise 9). For a lattice-prime ideal P, we use $[P]$ to denote the equivalence class containing P. Define $\psi \colon Max_\ell(R) \to Max_\ell(S)$ by

$$M \mapsto [P] \mapsto [\varphi(P)] \mapsto N,$$

where M is a maximal ℓ-ideal of R, P is a lattice-prime ideal of R associated with M, and $\varphi(P)$ is the lattice-prime ideal of S associated with the maximal ℓ-ideal N of S. By (I) and (II), ψ is well-defined, one-to-one and onto (Exercise 10).

Let $b \in S$ and $b = \varphi(a)$ for some $a \in R$. For a nonempty subset $\mathcal{K}$ in $\text{Max}_\ell(R)$, $A(\mathcal{K})$ is the intersection of all the lattice-prime ideals of R that contain a and are associated with any member in $\mathcal{K}$, and $A(\psi(\mathcal{K}))$ is the intersection of all the lattice-prime ideals of S that contain b and are associated with any member in $\psi(\mathcal{K})$. We claim that

$$\varphi(A(\mathcal{K})) = A(\psi(\mathcal{K})).$$

Let $y \in \varphi(A(\mathcal{K}))$. Then $y = \varphi(x)$, where $x \in A(\mathcal{K})$. Let I be a lattice-prime ideal of S which contains b and is associated with some $N \in \psi(\mathcal{K})$. Let $I = \varphi(P)$ and $N = \psi(M)$. Then $a \in P$, P is a lattice-prime ideal of R and $\psi(M) = N$ imply that P is associated with M. Thus $A(\mathcal{K}) \subseteq P$, so $x \in P$ and $y = \varphi(x) \in I = \varphi(P)$. Therefore $y \in A(\psi(\mathcal{K}))$, that is, $\varphi(A(\mathcal{K})) \subseteq A(\psi(\mathcal{K}))$. Similarly we can show that $A(\psi(\mathcal{K})) \subseteq \varphi(A(\mathcal{K}))$ (Exercise 11).

Now let $\mathcal{K}$ be a closed set in $\text{Max}_\ell(R)$. We show that $\psi(\mathcal{K})$ is closed in $\text{Max}_\ell(S)$. Let N be in the closure of $\psi(\mathcal{K})$. By (III), there exists a lattice-prime ideal I of S associated with N such that I contains $A(\psi(\mathcal{K})) = \varphi(A(\mathcal{K}))$. Let $I = \varphi(P)$ and $N = \psi(M)$. Then P is a lattice-prime ideal of R associated with M and P contains $A(\mathcal{K})$. Thus by (III), M is in $\mathcal{K}$ since it is closed, and hence $N \in \psi(\mathcal{K})$. So $\psi(\mathcal{K})$ is closed. By a similar argument, it can be shown that if $\mathcal{J}$ is a closed subset of $\text{Max}_\ell(S)$ and $\mathcal{J} = \psi(\mathcal{K})$ for some $\mathcal{K} \subseteq \text{Max}_\ell(R)$, then $\mathcal{K}$ is also closed. Therefore ψ is a homeomorphism between $\text{Max}_\ell(R)$ and $\text{Max}_\ell(S)$. This completes the proof. □

In the following, we consider how to generalize Theorem 5.4 to ℓ-unital ℓ-reduced ℓ-semisimple ℓ-rings. Suppose that R is an ℓ-unital ℓ-reduced ℓ-semisimple ℓ-ring. A lattice-prime ideal P of R is called *dominated*, if for any $x, y \in R$, $x \in P$ and $(y - x)^+ \wedge 1 = 0$ imply $y \in P$. Another way to say it is that if $x \in P$ and $y \leq x + z$ with $z \wedge 1 = 0$, then $y \in P$. A lattice-prime ideal P of R is called *associated* with a maximal ℓ-ideal M of R if $x \in P$ and $(y - x)^+ \wedge 1 \notin M$ imply $y \in P$.

We notice that if R is an f-ring, then the identity element 1 is a weak unit, and hence $(y - x)^+ \wedge 1 = 0$ if and only if $(y - x)^+ = 0$, that is, $y \leq x$. Thus dominated lattice-prime ideals are just lattice-prime ideals when R is an f-ring.

By similar proofs used in Theorem 5.4, we have the following facts whose proofs are omitted and the reader is referred to [Ma, Wojciechowski (2002)] for more details.

(I') *Each __dominated__ lattice-prime ideal is associated with exactly one maximal ℓ-ideal.*

(II') *Two __dominated__ lattice-prime ideals P_1, P_2 are associated with the same maximal ℓ-ideal if and only if $P_1 \cap P_2$ contains a __dominated__ lattice-prime ideal.*

(III') *Let a be fixed element in R, and let $\mathcal{K} = \{M_\alpha \mid \alpha \in \Gamma\}$ be any nonempty set in $Max_\ell(R)$. Then a maximal ℓ-ideal M belongs to the closure of $\mathcal{K}$ if and only if there exists a __dominated__ lattice-prime ideal P associated with M such that P contains $A(\mathcal{K})$, which is the intersection of all __dominated__ lattice-prime ideals which contain a and are also associated with any member in $\mathcal{K}$.*

Theorem 5.5. *Let R and S be two ℓ-unital ℓ-reduced ℓ-semisimple ℓ-rings. If there exists an ℓ-isomorphism between two additive ℓ-groups of R and S which preserves identity element, then $Max_\ell(R)$ and $Max_\ell(S)$ are homeomorphic.*

Proof. Suppose that $\varphi : R \to S$ is an ℓ-isomorphism between the additive ℓ-group of R and the additive ℓ-group of S with $\varphi(1) = 1$.

We first show that for a subset P of R, P is a dominated lattice-prime ideal of R if and only if $\varphi(P)$ is a dominated lattice-prime ideal of S. Suppose that P is a dominated lattice-prime ideal of R. Then $\varphi(P)$ is a lattice-prime ideal of S. Suppose $x, y \in S$ such that $x \in \varphi(P)$ and $(y - x)^+ \wedge 1 = 0$. Let $x = \varphi(a)$ and $y = \varphi(b)$, where $a \in P$ and $b \in R$. Since

$$\varphi[(b - a)^+ \wedge 1] = \varphi[(b - a)^+] \wedge \varphi(1) = (y - x)^+ \wedge 1 = 0,$$

$(b - a)^+ \wedge 1 = 0$, so $b \in P$ and $y = \varphi(b) \in \varphi(P)$. Thus $\varphi(P)$ is dominated. Similarly, if $\varphi(P)$ is a dominated lattice-prime ideal of S, then P is also a dominated lattice-prime ideal of R. Therefore φ induces a one-to-one correspondence $P \to \varphi(P)$ between the set of all dominated lattice-prime ideals of R and the set of all dominated lattice-prime ideals of S.

Similarly to the proof of Theorem 5.4, two dominated lattice-prime ideals are called equivalent if they are associated with the same maximal ℓ-ideal. Let $[P]$ denote the equivalence class containing the dominated lattice-prime ideal P. Define $Max_\ell(R) \to Max_\ell(S)$ by

$$M \mapsto [P] \mapsto [\varphi(P)] \mapsto N,$$

where $M \in \text{Max}_\ell(R)$, P is a dominated lattice-prime ideal of R associated with M, and φP is associated with $N \in \text{Max}_\ell(S)$. Then by the same argument in Theorem 5.4, $\text{Max}_\ell(R)$ and $\text{Max}_\ell(S)$ are homeomorphic. $\square$

The conditions that R, S are ℓ-reduced and $\varphi(1) = 1$ in Theorem 5.5 cannot be dropped as shown in the following examples.

Example 5.2.

(1) Let R be the direct sum of two copies of the ℓ-field $\mathbb{Q}[\sqrt{2}]$ with the entrywise order. Then R is ℓ-unital ℓ-reduced ℓ-semisimple ℓ-ring. Clearly $\text{Max}_\ell(R)$ has two elements. Let S be the ℓ-ring $M_2(\mathbb{Q})$ with the entrywise order. Then S is an ℓ-unital ℓ-ring, however S is not ℓ-reduced. Since S is a simple ring, $\text{Max}_\ell(S)$ contains one element. Thus $\text{Max}_\ell(R)$ and $\text{Max}_\ell(S)$ cannot be homeomorphic.
Define $\varphi : R \to S$ by
$$\varphi(a + b\sqrt{2}, d + c\sqrt{2}) = \begin{pmatrix} a & b \\ c & d \end{pmatrix}.$$
Then φ is an ℓ-isomorphism between additive ℓ-groups of R, S and $\varphi(1) = 1$ (Exercise 12).

(2) Let R be the direct sum of two copies of $\mathbb{Q}$ and $S = \mathbb{Q}[\sqrt{2}]$ be the ℓ-field with the coordinatewise order. Define $f : R \to S$ by $f((a,b)) = a + b\sqrt{2}$. Then f is an ℓ-isomorphism of the additive ℓ-groups, however $f((1,1)) = 1 + \sqrt{2}$ is not the identity element in S.

A subset in a topological space is called a *clopen* set if it is closed and also open. We characterize clopen sets in $\text{Max}_\ell(R)$. Firs we consider f-rings.

Lemma 5.4. *Let A be a unital f-ring. $\mathcal{K} \subseteq \text{Max}_\ell(A)$ is clopen if and only if $\mathcal{K} = s(x)$, where $x \in A$ is an idempotent element.*

Proof. "$\Leftarrow$" Since $\mathcal{K} = s(x)$, $\mathcal{K}$ is open. Now as $x(1 - x) = 0$, for each $M \in \text{Max}_\ell(A)$, $x \in M$ or $(1 - x) \in M$, but not both. Thus $x \notin M$ if and only if $(1 - x) \in M$, so $\mathcal{K} = h(1 - x)$ is also closed.

"$\Rightarrow$" Since $\mathcal{K}$ is open, $\mathcal{K} = \cup_{x \in B} s(x)$ for some $B \subseteq A^+$. Since $\mathcal{K}$ is also closed, it is compact, so $\mathcal{K} = \cup_{i=1}^n s(x_i) = s(x)$, where $x = \vee_{i=1}^n x_i \geq 0$. Similarly $\text{Max}_\ell(A) \setminus \mathcal{K} = s(y)$, for some $0 \leq y \in A$. Thus $s(x) \cup s(y) = \text{Max}_\ell(A)$ and $s(x) \cap s(y) = \emptyset$. Thus $x \wedge y$ is contained in every maximal ℓ-ideal of A. Let $u = x - x \wedge y$ and $v = y - x \wedge y$. Then $u \wedge v = 0$, and $s(x) = s(u)$ and $s(y) = s(v)$. Let $\langle u + v \rangle$ be the ℓ-ideal of A generated by $u + v$. Since $u + v$ is not contained in any maximal ℓ-ideal of A,
$$A = \langle u + v \rangle = \{x \mid |x| \leq r(u+v)s, \text{ where } r, s \in A^+\},$$

and hence $1 \leq r(u+v)s = rus + rvs$ for some $r, s \in A^+$. Hence $1 = a + b$, where $0 \leq a \leq rus$ and $0 \leq b \leq rvs$. Since $u \wedge v = 0$, $rus \wedge rvs = 0$, so $a \wedge b = 0$. However because A is an f-ring, $a \wedge b = 0$ implies $ab = 0$. Hence $a^2 = a$, and $b^2 = b$. Finally we show that $s(u) = s(a)$. It is clear that if $M \in \mathrm{Max}_\ell(A)$ and $a \notin M$, then $u \notin M$, so $s(a) \subseteq s(u)$. Similarly, $s(b) \subseteq s(v)$. If $M \in s(u) \setminus s(a)$, then $b \notin M$ since $1 = a + b$, and hence $M \in s(b) \subseteq s(v)$, which contradicts with that $s(u) \cap s(v) = \emptyset$. Thus $s(u) = s(a)$. Therefore, $\mathcal{K} = s(a)$, where $a \in A$ is an idempotent. □

Corollary 5.2. *Let R be an ℓ-unital ℓ-reduced ℓ-ring. A set $\mathcal{K} \subseteq \mathrm{Max}_\ell(R)$ is clopen if and only if $\mathcal{K} = s(x)$, where $x \in f(R)$ is an idempotent.*

Proof. "⇐" Since $\mathcal{K} = s(x)$, $\mathcal{K}$ is open. Since $x(1-x) = 0$, and $x, (1-x) \in f(R)$, by Lemma 5.1, for each $M \in \mathrm{Max}_\ell(R)$, $x \in M$ or $(1-x) \in M$, but not both. Thus $x \in M$ if and only if $(1-x) \notin M$, and hence $\mathcal{K} = h(1-x)$. Hence $\mathcal{K}$ is closed.

"⇒" By Theorem 5.3, $\mathrm{Max}_\ell(R)$ and $\mathrm{Max}_\ell(f(R))$ are homeomorphic under the mapping $\varphi : M \to M_f$, where M_f is the unique maximal ℓ-ideal of $f(R)$ that contains $M \cap f(R)$. Let $\mathcal{K}$ be a clopen set in $\mathrm{Max}_\ell(R)$. Then $\varphi(\mathcal{K})$ is clopen in $\mathrm{Max}_\ell(f(R))$. By Lemma 5.4, there exists an idempotent element $x \in f(R)$ such that

$$\varphi(\mathcal{K}) = \{I \in \mathrm{Max}_\ell(f(R)) \mid x \notin I\}.$$

We show that $\mathcal{K} = s(x)$. Let $M \in \mathcal{K}$. Then $\varphi(M) \in \varphi(\mathcal{K})$, so $x \notin \varphi(M)$. Since $M \cap f(R) \subseteq \varphi(M)$, $x \notin M$, and hence $M \in s(x)$. Thus $\mathcal{K} \subseteq s(x)$. Now let $N \in s(x)$. Then $1 - x \in N \cap f(R) \subseteq \varphi(N)$, and hence $x \notin \varphi(N)$. Thus $\varphi(N) \in \varphi(\mathcal{K})$, so $N \in \mathcal{K}$. Therefore $s(x) \subseteq \mathcal{K}$. This completes the proof. □

5.2 ℓ-ideals in commutative ℓ-unital ℓ-rings

In this section, R denotes a commutative ℓ-unital ℓ-reduced ℓ-ring. Recall that an ℓ-prime ℓ-ideal is called a minimal ℓ-prime ℓ-ideal if it contains no smaller ℓ-prime ℓ-ideal. Let $\mathrm{Min}_\ell(R)$ denote the set of all minimal ℓ-prime ℓ-ideals of R endowed with the hull-kernel topology, that is, open sets in $\mathrm{Min}_\ell(R)$ are

$$S(X) = \{P \in \mathrm{Min}_\ell(R) \mid X \nsubseteq P\},$$

and closed sets in $\mathrm{Min}_\ell(R)$ are

$$H(X) = \{P \in \mathrm{Min}_\ell(R) \mid X \subseteq P\},$$

where $X \subseteq R$. To reduce possible confusion to the reader, for a subset X of R, we use $S(X)$ $(H(X))$ to denote the open (closed) sets in $\text{Min}_\ell(R)$ and use $s(X)$ $(h(X))$ to denote the open (closed) sets in $\text{Max}_\ell(R)$.

Recall that for $x \in R$, $\ell(x) = \{a \in R \mid |a||x| = 0\}$ is called the ℓ-annihilator of x, which is an ℓ-ideal of R. As a direct consequence of Theorem 1.30, we have the following result. We leave the proof as an exercise (Exercise 13).

Lemma 5.5. *For each element* $a \in R$, $H(\ell(a)) = S(a)$ *and* $S(\ell(a)) = H(a)$.

Theorem 5.6. *Let R be a commutative ℓ-unital ℓ-reduced ℓ-ring. Then $\text{Min}_\ell(R)$ is a Hausdorff space with a topological space basis consisting of clopen sets.*

Proof. For $P_1 \neq P_2$ in $\text{Min}_\ell(R)$, take $x \in P_1 \setminus P_2$. Then $P_1 \in H(x)$ and $P_2 \in H(\ell(x))$. By Lemma 5.5, $H(x)$ and $H(\ell(x))$ are both open, and $H(x) \cap H(\ell(x)) = \emptyset$. Therefore $\text{Min}_\ell(R)$ is a Hausdorff space. We know that $\{S(a) \mid a \in R\}$ is a base for the open sets and each $S(a)$ is clopen. $\square$

Theorem 5.7. *Let R be a commutative ℓ-unital ℓ-reduced ℓ-ring. Then $\text{Min}_\ell(R)$ is compact if and only if for each $x \in R^+$ there exists $y \in R^+$ such that $xy = 0$ and $\ell(x) \cap \ell(y) = \{0\}$.*

Proof. First assume that $\text{Min}_\ell(R)$ is compact. For $x \in R^+$, if $P \in \text{Min}_\ell(R) \setminus S(x)$, then by Theorem 1.30, $|x| \in P$ implies that there is $0 \leq z \in R$ such that $|x|z = 0$ and $z \notin P$, and hence $P \in S(z)$. It follows that

$$\text{Min}_\ell(R) = S(x) \bigcup (\cup_{0 \leq z \in \ell(x)} S(z)).$$

Then $\text{Min}_\ell(R)$ is compact implies that

$$\text{Min}_\ell(R) = S(x) \cup S(y_1) \cup \cdots \cup S(y_n)$$

for some $0 \leq y_1, \cdots, y_n \in \ell(x)$. Let $y = y_1 + \cdots + y_n$. Then $S(y) = S(y_1) \cup \cdots \cup S(y_n)$ and $y \in \ell(x)$ (Exercise 14), so $xy = 0$ and $S(y) \cup S(x) = \text{Min}_\ell(R)$. Hence $\ell(x) \cap \ell(y)$ is contained in each minimal ℓ-prime ℓ-ideal of R, and hence $\ell(x) \cap \ell(y) = \{0\}$ since R is ℓ-reduced.

Conversely suppose that the given conditions are satisfied, we show that $\text{Min}_\ell(R)$ is compact. Let $\text{Min}_\ell(R) = \cup_\alpha S(I_\alpha)$ for some ℓ-ideals I_α of R. Let $I = \sum_\alpha I_\alpha$. We have $S(I) = \text{Min}_\ell(R)$, so for each $P \in \text{Min}_\ell(R)$, $I \not\subseteq P$. We claim that there exists $0 \leq x \in I$ such that $\ell(x) = \{0\}$. Suppose for each $0 \leq x \in I$, $\ell(x) \neq \{0\}$. We derive a contradiction. Let

$$M = \{a \in R^+ \mid \ell(a) = 0\}.$$

Then M is closed under the multiplication of R and $I \cap M = \emptyset$. Thus M is an m-system. By Theorem 1.26, $I \subseteq P$ for some ℓ-prime ℓ-ideal P and $P \cap M = \emptyset$. We shall prove that P is minimal. Given $0 \leq z \in P$, there exists $w \in R^+$ such that $zw = 0$ and $\ell(z) \cap \ell(w) = \{0\}$. Since $\ell(z + w) \subseteq \ell(z) \cap \ell(w)$, $\ell(z + w) = \{0\}$, and hence $z + w \in M$. It follows that $z + w \notin P$, so $w \notin P$. By Theorem 1.30, P is a minimal ℓ-prime ℓ-ideal of R. This contradicts with the fact that I is not contained in any minimal ℓ-prime ℓ-ideal. Therefore there exists $0 \leq x \in I$ such that $\ell(x) = \{0\}$.

Suppose that $x \in I_{\alpha_1} + \cdots + I_{\alpha_k}$. For $P \in \text{Min}_\ell(R)$, if $I_{\alpha_1}, \cdots, I_{\alpha_k} \subseteq P$, then $x \in P$ implies that $xy = 0$ for some $0 \leq y \notin P$ by Theorem 1.30, so $\ell(x) \neq \{0\}$, which is a contradiction. Hence $\text{Min}_\ell(R) = S(I_{\alpha_1}) \cup \cdots \cup S(I_{\alpha_k})$. Therefore $\text{Min}_\ell(R)$ is compact. □

For an ℓ-prime ℓ-ideal P, we define $O_P = \{a \in R \mid \ell(a) \not\subseteq P\}$. Clearly O_P is an ℓ-ideal and $O_P \subseteq P$ (Exercises 15). By Theorem 1.30, an ℓ-prime ℓ-ideal P is minimal if and only if $O_P = P$.

Theorem 5.8. *Let R be a commutative ℓ-unital ℓ-reduced ℓ-ring and M be a maximal ℓ-ideal of R.*

(1) For $x \in R$, $x \in O_M$ if and only if there exists $0 \leq e \in M \cap f(R)$ such that $xe = x$.

(2) For $x \in R$, if $x \in O_M$, then $h(x)$ is a neighborhood of M in $\text{Max}_\ell(R)$. If R is ℓ-semisimple, then the converse is also true.

(3) M is the only maximal ℓ-ideal containing O_M.

(4) Every ℓ-prime ℓ-ideal of R lies between O_N and N for a unique maximal ℓ-ideal N of R.

Proof. (1) Suppose that $xe = x$ for an element $e \in M \cap f(R)$. Then $x(e - 1) = 0$, and hence $x(e-1)^+ = x(e-1)^-$. So we have $|x(e-1)^+| = |x(e-1)^-|$. Because $(e-1)^+$ and $(e-1)^-$ are both f-elements, $|x|(e-1)^+ = |x|(e-1)^-$. Since $(e - 1)^+ \wedge (e - 1)^- = 0$ implies that

$$(e - 1)^+(e - 1)^- = (e - 1)^-(e - 1)^+ = 0,$$

we have

$$|x|[(e - 1)^+]^2 = |x|[(e - 1)^-]^2 = 0,$$

that is, $[(e - 1)^+]^2 \in \ell(x)$ and $[(e - 1)^-]^2 \in \ell(x)$. Now since $e \in M$, $(e - 1) \notin M$, so

$$(e - 1)^+ \notin M \text{ or } (e - 1)^- \notin M,$$

and hence by Lemma 5.1

$$[(e-1)^+]^2 \notin M \text{ or } [(e-1)^-]^2 \notin M.$$

Thus $\ell(x) \nsubseteq M$, and hence $x \in O_M$.

Now let $x \in O_M$. Then there exists $y \in R \setminus M$ such that $|x||y| = 0$ since $\ell(x) \nsubseteq M$. Let $\langle y \rangle$ be the ℓ-ideal of R generated by y. Since $y \notin M$, $R = M + \langle y \rangle$, and hence $1 = e + z$, where $0 \le e \in M$ and $0 \le z \le r|y|$ for some $r \in R^+$. Thus $|x|z = 0$, and hence $|x| = |x|e$. Since $0 \le e \le 1$, e is an f-element, and $|x| = |x|e$ implies that $x = xe$ (Exercise 29).

(2) Let $x \in O_M$. So there exists $y \notin M$ such that $|x||y| = 0$. Then $M \in s(y) \subseteq h(x)$. Thus $h(x)$ is a neighborhood of M. Suppose that R is ℓ-semisimple and $M \in s(y) \subseteq h(x)$ for some $y \in R$. Since $s(y) = s(|y|)$ and $h(x) = h(|x|)$, $M \in s(|y|) \subseteq h(|x|)$. Let $N \in \text{Max}_\ell(R)$. If $|y| \notin N$, then $N \in s(|y|) \subseteq h(|x|)$, so $|x| \in N$. Thus $|x||y| \in N$ for each $N \in \text{Max}_\ell(R)$. Since R is ℓ-semisimple, $|x||y| = 0$ and $0 < |y| \notin M$. Thus $\ell(x) \nsubseteq M$, and so $x \in O_M$.

(3) Let L be a maximal ℓ-ideal of R such that $O_M \subseteq L$ and $L \ne M$. Then $R = L + M$. By Lemma 5.1, there exist $0 \le a \in M \setminus L$ and $0 \le b \in L \setminus M$ such that $ab = 0$. Since $a \notin L$, $a \notin O_M$, so $\ell(a) \subseteq M$. Hence $b \in M$, which is a contradiction. Thus $L = M$.

(4) Let P be an ℓ-prime ℓ-ideal of R. By Corollary 5.1, P is contained in a unique maximal ℓ-ideal N of R. Let $x \in O_N$. Then $\ell(x) \nsubseteq N$, so $\ell(x) \nsubseteq P$, and hence $x \in P$. Thus $O_N \subseteq P$. $\qquad\square$

The following example shows that the condition that R is ℓ-semisimple cannot be dropped in Theorem 5.8(2).

Example 5.3. Let $A = \mathbb{R}[x]$ be the polynomial ring in one variable over $\mathbb{R}$ with the coordinatewise order. Then A is a commutative ℓ-unital ℓ-ring with the unique maximal ℓ-ideal $M = xA$. Consider the direct sum $A \oplus A$ of two copies of A. Then $M \oplus A$ is a maximal ℓ-ideal of $A \oplus A$ and $O_{M \oplus A} = 0 \oplus A$. Clearly $M \oplus A \in s((1,0)) \subseteq h((x,1))$, but $(x,1) \notin O_{M \oplus A}$.

An ℓ-ideal I is called a *pure* ℓ-ideal if $R = I + \ell(x)$ for each $x \in I$. For an ℓ-ideal I, define $m(I) = \{a \in R \mid R = I + \ell(a)\}$. Then $m(I)$ is an ℓ-ideal and I is pure if and only if $m(I) = I$ (Exercise 16).

Theorem 5.9. *Let R be a commutative ℓ-unital ℓ-reduced ℓ-ring and I, J be ℓ-ideals.*

(1) For each maximal ℓ-ideal M of R, O_M is a pure ℓ-ideal.

(2) $m(I)$ *is a pure ℓ-ideal, and*

$$m(I) = \bigcup_{g \in I} \ell(1 - g).$$

(3) $m(I) = \{x \in R \mid x = ax \text{ for some } 0 \le a \in I \cap f(R)\}$. *In particular,* $m(I) \subseteq I$.

(4) $h(I) = h(m(I))$.

(5)

$$m(I) = \bigcap O_M, \text{ where } M \in h(I).$$

(6) $m(I) + m(J) = m(I + J)$.

(7) For an ℓ-prime ℓ-ideal P, $m(P) = O_M$ for some maximal ℓ-ideal M and O_P is a pure ℓ-ideal if and only if $O_P = O_M$.

Proof.

(1) Let $x \in O_M$. Then $\ell(x) \not\subseteq M$, so $R = M + \ell(x)$. By Lemma 5.1, $1 = a + b + c$, where $0 \le a \in f(R) \cap M$, $0 \le b \in f(R) \cap \ell(x)$, $c \in f(R) \cap M \cap \ell(x)$, and $ab = 0$. We have $b \in \ell(a) \setminus M$, and hence $a \in O_M$. Thus $1 \in O_M + \ell(x)$, so $R = O_M + \ell(x)$.

(2) First we show that

$$m(I) = \bigcup_{g \in I} \ell(1 - g).$$

Let

$$x \in \bigcup_{g \in I} \ell(1 - g).$$

Then $|x||1 - g| = 0$ for some $g \in I$. So $1 - g \in \ell(x)$, and hence $R = I + \ell(x)$. Thus $x \in m(I)$. Conversely, if $x \in m(I)$, then $R = I + \ell(x)$, and hence $1 = c + d$, where $0 \le c \in I, 0 \le d \in \ell(x)$. Therefore, $|x| = c|x|$, and hence $x \in \ell(1 - c)$ with $c \in I$. So

$$x \in \bigcup_{g \in I} \ell(1 - g).$$

To see that $m(I)$ is an ℓ-ideal we just have to show that $m(I)$ is closed under the addition of R. Let $x, y \in m(I)$. Then

$$R = I + \ell(x) = I + \ell(y).$$

Let

$$1 = a + b = a' + b',$$

where $0 \le a, a' \in I$, $0 \le b \in \ell(x)$, and $0 \le b' \in \ell(y)$. Then

$$1 = (a + b)(a' + b') = aa' + ab' + ba' + bb' \in I + \ell(x + y),$$

so $R = I + \ell(x + y)$. Thus $x + y \in m(I)$.

Finally, to see that $m(I)$ is a pure ℓ-ideal, let $a \in m(I)$. Then $R = I + \ell(a)$, and hence, by Lemma 5.1, $1 = x + y + z$, where $0 \le x \in I$, $0 \le y \in \ell(a)$, $0 \le z \in I \cap \ell(a)$, and $xy = 0$. Since $y \in \ell(x)$, we have $1 \in I + \ell(x)$, and hence $R = I + \ell(x)$. Thus $x \in m(I)$. This implies $1 = x + y + z \in m(I) + \ell(a)$, and hence $R = m(I) + \ell(a)$. Hence $m(I)$ is a pure ℓ-ideal.

(3) Let $x \in m(I)$. Then $R = I + \ell(x)$, and hence $1 = a + b$ for some $0 \le a \in I, 0 \le b \in \ell(x)$. Thus $x = ax$ and $0 \le a \in I \cap f(R)$. Conversely, let $x = ax$ for some $0 \le a \in I \cap f(R)$. Then $x(1 - a) = 0$. Since $1 - a \in f(R)$, $(1 - a)^2 \ge 0$ is an f-element, and hence $x(1 - a)^2 = 0$ implies that $|x|(1 - a)^2 = 0$. Hence

$$(1 - a)^2 = 1 - 2a + a^2 \in \ell(x).$$

Since $a \in I$, we have $R = I + \ell(x)$. Thus $x \in m(I)$.

(4) Since $m(I) \subseteq I$, $h(I) \subseteq h(m(I))$. Now let M be a maximal ℓ-ideal of R and $m(I) \subseteq M$. If $I \not\subseteq M$, then $R = I + M$. By Lemma 5.1, $1 = a + b + c$, where $0 \le a \in I$, $0 \le b \in M$, $0 \le c \in I \cap M$, and $ab = 0$. Thus $b \in \ell(a)$, so $R = I + \ell(a)$, and hence $a \in m(I)$. So $a \in M$, which implies $1 = a + b + c \in M$, which is a contradiction. Thus $I \subseteq M$.

(5) If $x \in m(I)$, then $R = I + \ell(x)$, so $\ell(x) \not\subseteq M$ for each $M \in h(I)$, and hence $x \in O_M$ for each $M \in h(I)$. Conversely, suppose $x \in O_M$ for each $M \in h(I)$. If $I + \ell(x) \ne R$, then there exists a maximal ℓ-ideal N such that $I + \ell(x) \subseteq N$. Since $I \subseteq N$ and $x \in O_N$, $\ell(x) \not\subseteq N$, which contradicts with that $\ell(x) \subseteq N$. Thus $I + \ell(x) = R$, so $x \in m(I)$.

(6) From (4), we have

$$h(m(I) + m(J)) = h(m(I)) \bigcap h(m(J))$$
$$= h(I) \bigcap h(J)$$
$$= h(I + J).$$

Then from (5), we have

$$m(m(I) + m(J)) = \bigcap O_N, \text{ where } N \in h(m(I) + m(J))$$
$$= \bigcap O_N, \text{ where } N \in h(I + J)$$
$$= m(I + J).$$

Thus $m(I+J) \subseteq m(I)+m(J)$ by (3). Since clearly $m(I)+m(J) \subseteq m(I+J)$, we have $m(I+J) = m(I) + m(J)$.

(7) Let $P \subseteq M$, where M is the unique maximal ℓ-ideal of R containing P. By (5), $m(P) = O_M$. Since O_M is the largest pure ℓ-ideal contained in M by (5), if O_P is a pure ℓ-ideal, then $O_P \subseteq O_M$. Clearly $O_M \subseteq O_P \subseteq M$ is always true. Thus $O_P = O_M$. $\qquad\square$

The following result characterizes those ℓ-rings R for which each principal ℓ-ideal is a pure ℓ-ideal. An ℓ-ideal I of R is called a *direct summand* if there exists an ℓ-ideal J such that R is a direct sum of I and J as ℓ-ideals, that is, $R = I \oplus J$.

Theorem 5.10. *Let R be a commutative ℓ-unital ℓ-reduced ℓ-ring. Then the following statements are equivalent.*

(1) Every ℓ-prime ℓ-ideal of R is maximal.
(2) Every principal ℓ-ideal of R is a pure ℓ-ideal.
(3) Every principal ℓ-ideal of R is a direct summand.

Proof. (1) $\Rightarrow$ (2). First we notice that (1) implies that every ℓ-prime ℓ-ideal of R is minimal (Exercise 17). Let $a \in R$ and $b \in \langle a \rangle$. If $R \neq \langle a \rangle + \ell(b)$, then $\langle a \rangle + \ell(b) \subseteq M$ for some maximal ℓ-ideal M. Since M is a minimal ℓ-prime ℓ-ideal and $b \in M$, $\ell(b) \not\subseteq M$ by Theorem 1.30, which is a contradiction. Therefore, $R = \langle a \rangle + \ell(b)$, so $\langle a \rangle$ is a pure ℓ-ideal.

(2) $\Rightarrow$ (3). Let $a \in R$. From (2), we have $R = \langle a \rangle + \ell(a)$. Let $b \in \langle a \rangle \cap \ell(a)$. Then $|b| \leq r|a|$ for some $r \in R^+$ and $|b||a| = 0$, so $|b|^2 = 0$, and hence $|b| = 0$ since R is ℓ-reduced. Thus $b = 0$, so $\langle a \rangle \cap \ell(a) = \{0\}$, and hence $R = \langle a \rangle \oplus \ell(a)$.

(3) $\Rightarrow$ (1). Suppose that P is an ℓ-prime ℓ-ideal. Let $P \subseteq I$ for some ℓ-ideal I of R. If $P \neq I$, then there exists an element $a \in I \setminus P$. Since $R = \langle a \rangle \oplus J$ for some ℓ-ideal J of R, $J \subseteq \ell(a) \subseteq P \subseteq I$, and hence $R = \langle a \rangle + J \subseteq I$. Thus P is a maximal ℓ-ideal of R. $\qquad\square$

An ℓ-ideal $I \neq R$ is called ℓ-*pseudoprime* if $ab = 0$ for $a, b \in R^+$ implies $a \in I$ or $b \in I$. An ℓ-prime ℓ-ideal is certainly ℓ-pseudoprime. However the converse is not true. For instance, let $R = \mathbb{R}[x, y]$ be the polynomial ℓ-ring in two variables over $\mathbb{R}$ with the coordinatewise order. Since R is a domain, $xR \cap yR$ is an ℓ-pseudoprime ℓ-ideal, however $xR \cap yR$ is not ℓ-prime. An ℓ-ideal $I \neq R$ is called ℓ-*semiprime* if for any $a \in R^+$, $a^2 \in I$ implies that $a \in I$. We leave it as an exercise to the reader to check that an ℓ-ideal is ℓ-semiprime if and only if it is the intersection of ℓ-prime ℓ-ideals containing it (Exercise 18).

Theorem 5.11. *Let R be a commutative ℓ-unital ℓ-reduced ℓ-ring. Let I and J be pure ℓ-ideals of R.*

(1) I is ℓ-semiprime.

(2) If I is ℓ-pseudoprime, then I is ℓ-prime.

(3) If I and J are ℓ-prime, then either $I + J = R$ or $I = J$.

(4) For an ℓ-prime ℓ-ideal P, O_P is ℓ-prime if and only if O_P is ℓ-pseudoprime.

Proof.   (1) Let $x^2 \in I$ for some $x \in R^+$. Then $R = I + \ell(x^2)$, and hence $x = a + b$, where $0 \leq a \in I$ and $0 \leq b \in \ell(x^2)$. Since $b \leq x$, and $bx^2 = 0$, we have $b^3 \leq bx^2 = 0$. Hence $b^3 = 0$, and so $b = 0$ since R is ℓ-reduced. Thus $x = a \in I$.

(2) Let $ab \in I$ for some $a, b \in R^+$ and $a \notin I$. Since $R = I + \ell(ab)$, $1 = u + v$, where $0 \leq u \in I$, $0 \leq v \in \ell(ab)$. Since $abv = 0$ and I is ℓ-pseudoprime, we have $bv \in I$. Thus $b = bu + bv \in I$.

(3) Since I and J are ℓ-prime, by Theorem 5.9, there exist maximal ℓ-ideals M and N such that $I = O_M$ and $J = O_N$. If $I + J \neq R$, then $I + J$ is contained in some maximal ℓ-ideal, so by Theorem 5.8 $M = N$, and hence $I = J$.

(4) Suppose that O_P is ℓ-pseudoprime, and let $ab \in O_P$ for some $a, b \in R^+$. Suppose that $a \notin O_P$ and $b \notin O_P$. We get a contradiction as follows. Since $ab \in O_P$, $\ell(ab) \not\subseteq P$, and so there exists $0 \leq c \notin P$ such that $abc = 0$. Since O_P is ℓ-pseudoprime and $a \notin O_P$, $bc \in O_P$. Hence there exists $0 \leq d \notin P$ such that $bcd = 0$, so $cd \in O_P \subseteq P$ since $b \notin O_P$. Now $cd \in P$ implies $c \in P$ or $d \in P$, which is a contradiction.   □

For a commutative ℓ-unital ℓ-reduced ℓ-ring R if $R = \ell(a) + \ell(b)$ whenever $ab = 0$ for some $a, b \in R^+$, then R is called *normal*. Clearly if R is an ℓ-domain, then R is normal.

Theorem 5.12. *Let R be a commutative ℓ-unital ℓ-reduced ℓ-ring. The following are equivalent.*

(1) R is normal.

(2) O_P is ℓ-prime for each ℓ-prime ℓ-ideal P.

(3) O_M is ℓ-prime for each maximal ℓ-ideal M.

(4) O_M is minimal ℓ-prime for each maximal ℓ-ideal M.

(5) $R = P + Q$ for any two distinct minimal ℓ-prime ℓ-ideals P and Q.

(6) Each maximal ℓ-ideal contains a unique minimal ℓ-prime ℓ-ideal.

Proof. (1) ⇒ (2). Since O_P is ℓ-prime if and only if O_P is ℓ-pseudoprime by Theorem 5.11, we just need to show that O_P is ℓ-pseudoprime. Let $ab = 0$ for some $a, b \in R^+$. Then $R = \ell(a) + \ell(b)$. If $a \notin O_P$ and $b \notin O_P$, then $\ell(a) \subseteq P$ and $\ell(b) \subseteq P$, and hence $R = \ell(a) + \ell(b) \subseteq P$, which is a contradiction.

(2) implies (3) is clear.

(3) ⇒ (4). Let P be an ℓ-prime ℓ-ideal and $P \subseteq O_M$. Then $P \subseteq M$ implies $O_M \subseteq O_P \subseteq P$, so $P = O_M$. Thus O_M is a minimal ℓ-prime ℓ-ideal.

(4) ⇒ (5). Given a minimal ℓ-prime ℓ-ideal J of R, let M be the unique maximal ℓ-ideal containing J. Then $O_M \subseteq J \subseteq M$ by Theorem 5.8(4). Since O_M is ℓ-prime, $J = O_M$. Thus, minimal ℓ-prime ℓ-ideals of R are O_M, where $M \in \text{Max}_\ell(R)$. By Theorem 5.11(3), $R = O_M + O_N$ if $O_M \neq O_N$, where M, N are maximal ℓ-ideals.

(5) ⇒ (6). Obvious.

(6) ⇒ (1). Let $ab = 0$ for some $a, b \in R^+$. If $R \neq \ell(a) + \ell(b)$, then there exists a maximal ℓ-ideal M such that $\ell(a) + \ell(b) \subseteq M$, so $a \notin O_M$ and $b \notin O_M$. Now let P be the unique minimal ℓ-prime ℓ-ideal contained in M. Then $O_M \subseteq P$. Since O_M is ℓ-semiprime by Theorem 5.11, O_M is an intersection of ℓ-prime ℓ-ideals. Since each ℓ-prime ℓ-ideal containing O_M is contained in M by Theorem 5.8, each ℓ-prime ℓ-ideal containing O_M contains P. Thus $O_M = P$, so O_M is ℓ-prime, and hence $a \in O_M$ or $b \in O_M$, which is a contradiction. Thus we have $R = \ell(a) + \ell(b)$. □

For a commutative ℓ-unital ℓ-reduced ℓ-ring R, if R is normal, from Theorem 5.12, we have

$$\text{Min}_\ell(R) = \{O_M \mid M \in \text{Max}_\ell(R)\}.$$

Thus maximal ℓ-ideals of R and minimal ℓ-ideals of R are in one-to-one correspondence. This is not true if R is not normal as shown in the following example.

Example 5.4. Let $A = \mathbb{R}[x]$ be the polynomial ring over $\mathbb{R}$ with the coordinatewise order. Let R be the ℓ-subring of $A \times A$ defined as follows.

$$R = \{(f, g) \in A \times A \mid f(0) = g(0)\}.$$

Then R is commutative without any nilpotent element (Exercise 19). Since $(x, 0)(0, x) = (0, 0)$ and $R \neq \ell((x, 0)) + \ell((0, x))$, R is not normal. R has only one maximal ℓ-ideal $M = \{(f, g) \mid f(0) = g(0) = 0\}$ with $O_M = \{0\}$. Clearly, $\ell((x, 0))$ and $\ell((0, x))$ are minimal ℓ-prime ℓ-ideals with $O_{\ell((x,0))} = \ell((x, 0))$ and $O_{\ell((0,x))} = \ell((0, x))$. We leave it to the reader to verify this fact (Exercise 20).

Theorem 5.13. *Let R be normal. Then $Max_\ell(R)$ and $Min_\ell(R)$ are home-omorphic if and only if for each $x \in R^+$ there exists $y \in R^+$ such that $xy = 0$ and $\ell(x) \cap \ell(y) = \{0\}$ (or equivalently, $Min_\ell(R)$ is compact).*

Proof. Since $Max_\ell(R)$ is compact, if $Min_\ell(R)$ and $Max_\ell(R)$ are homeomorphic then $Min_\ell(R)$ is compact, so R has the desired property by Theorem 5.7.

Conversely, suppose that $Min_\ell(R)$ is compact. We show that $Min_\ell(R)$ and $Max_\ell(R)$ are homeomorphic. Since R is normal, we have

$$Min_\ell(R) = \{O_M \mid M \in Max_\ell(R)\}.$$

The mapping $M \mapsto O_M$ is clearly a one-to-one and onto mapping from $Max_\ell(R)$ to $Min_\ell(R)$.

Let $\{M_\alpha \mid \alpha \in \Gamma\}$ be a closed set in $Max_\ell(R)$. We show that $\{O_{M_\alpha} \mid \alpha \in \Gamma\}$ is closed in $Min_\ell(R)$. Let P be a minimal ℓ-prime ℓ-ideal of R and

$$\bigcap_{\alpha \in \Gamma} O_{M_\alpha} \subseteq P.$$

Let M be the unique maximal ℓ-ideal of R containing P. If $\cap_{\alpha \in \Gamma} M_\alpha \not\subseteq M$, then

$$R = (\bigcap_{\alpha \in \Gamma} M_\alpha) + M.$$

By Lemma 5.1, $1 = a + b + c$, where $0 \le a \in \cap_{\alpha \in \Gamma} M_\alpha$, $0 \le b \in M$, $0 \le c \in (\cap_{\alpha \in \Gamma} M_\alpha) \cap M$, and $ab = 0$. Since $b \notin M_\alpha$ for each $\alpha \in \Gamma$, $a \in O_{M_\alpha}$ for each $\alpha \in \Gamma$. Thus

$$a \in \bigcap_{\alpha \in \Gamma} O_{M_\alpha} \subseteq P \subseteq M,$$

so $1 \in M$, which is a contradiction. Therefore, $\cap_{\alpha \in \Gamma} M_\alpha \subseteq M$, and hence $M \in \{M_\alpha \mid \alpha \in \Gamma\}$ since $\{M_\alpha \mid \alpha \in \Gamma\}$ is closed. So $P = O_M \in \{O_{M_\alpha} \mid \alpha \in \Gamma\}$. Hence $\{O_{M_\alpha} \mid \alpha \in \Gamma\}$ is closed.

Now, suppose that $\{O_{M_\alpha} \mid \alpha \in \Gamma\}$ is a closed set in $Min_\ell(R)$. We show that $\{M_\alpha \mid \alpha \in \Gamma\}$ is closed in $Max_\ell(R)$. Let $M \in Max_\ell(R)$ and

$$\bigcap_{\alpha \in \Gamma} M_\alpha \subseteq M.$$

If $M \notin \{M_\alpha \mid \alpha \in \Gamma\}$, then $R = M_\alpha + M$ for each $\alpha \in \Gamma$. Hence, by Lemma 5.1, $1 = x_\alpha + y_\alpha + z_\alpha$, where $0 \le x_\alpha \in M_\alpha, 0 \le y_\alpha \in M, 0 \le z_\alpha \in M_\alpha \cap M$, and $x_\alpha y_\alpha = 0$ for each $\alpha \in \Gamma$. Since $x_\alpha \notin M$, $y_\alpha \in O_M$ for each $\alpha \in \Gamma$, and since $y_\alpha \notin M_\alpha$, $x_\alpha \in O_{M_\alpha}$. But then $y_\alpha \notin O_{M_\alpha}$ for each $\alpha \in \Gamma$. For each α

$$S(y_\alpha) = \{P \in Min_\ell(R) \mid y_\alpha \notin P\}$$

is open and $\{S(y_\alpha) \mid \alpha \in \Gamma\}$ is an open cover for the set $\{O_{M_\alpha} \mid \alpha \in \Gamma\}$, and hence a finite subcover $S(y_i)$, $i = 1, ..., n$, can be extracted out of this cover because of the compactness of $\text{Min}_\ell(R)$ implies that any closed set of it is compact. Now for each O_{M_α}, there exists $y_j \notin O_{M_\alpha}$ for some $1 \leq j \leq n$, so $x_j \in O_{M_\alpha}$. Therefore

$$x_1 \cdots x_n \in \bigcap_{\alpha \in \Gamma} O_{M_\alpha} \subseteq \bigcap_{\alpha \in \Gamma} M_\alpha \subseteq M.$$

Thus $x_k \in M$ for some $k \in \Gamma$, and so $1 = x_k + y_k + z_k \in M$, which is a contradiction. Therefore $M \in \{M_\alpha \mid \alpha \in \Gamma\}$, and hence $\{M_\alpha \mid \alpha \in \Gamma\}$ is closed. $\quad\square$

Corollary 5.3. *Suppose R is normal and for each $x \in R^+$ there exists $y \in R^+$ such that $xy = 0$ and $\ell(x) \cap \ell(y) = \{0\}$. Then $\text{Max}_\ell(R)$, $\text{Max}_\ell(f(R))$, $\text{Min}_\ell(R)$, and $\text{Min}_\ell(f(R))$ are all homeomorphic.*

Proof. If R is normal with the property that for each $x \in R^+$ there exists $y \in R^+$ such that $xy = 0$ and $\ell(x) \cap \ell(y) = \{0\}$, then it is easy to check that $f(R)$ is normal with the same property. Now the conclusion follows from Theorems 5.3 and 5.13. $\quad\square$

We provide a characterization for normal ℓ-rings under certain conditions.

Lemma 5.6. *Suppose R is normal with the property that for each $x \in R^+$ there exists $y \in R^+$ such that $xy = 0$ and $\ell(x) \cap \ell(y) = \{0\}$. If R is not an ℓ-domain, then there exists an idempotent element $a \in f(R)$, $0 < a < 1$, such that $R = Ra \oplus R(1 - a)$ as ℓ-ideals.*

Proof. Let $wz = 0$ and $0 < w \in R$, $0 < z \in R$. Then there exists $u \in R^+$ such that $wu = 0$ and $\ell(w) \cap \ell(u) = \{0\}$. Now $u \neq 0$, as otherwise $0 < z \in \ell(w) = \ell(w) \cap \ell(u) = \{0\}$, contradicting with $z > 0$. So $u > 0$. Consider the following sets in $\text{Min}_\ell(R)$.

$$\mathcal{K} = \{P \in \text{Min}_\ell(R) \mid \ell(w) \subseteq P\},$$

$$\mathcal{J} = \{P \in \text{Min}_\ell(R) \mid \ell(u) \subseteq P\}.$$

For each $P \in \text{Min}_\ell(R)$, $w \in P$ or $u \in P$ implies that $\ell(w) \nsubseteq P$ or $\ell(u) \nsubseteq P$, and hence $\mathcal{K} \cap \mathcal{J} = \emptyset$. Since $\ell(w) \cap \ell(u) = \{0\}$, for each $P \in \text{Min}_\ell(R)$, $\ell(w) \subseteq P$ or $\ell(u) \subseteq P$, and hence $\text{Min}_\ell(R) = \mathcal{K} \cup \mathcal{J}$. Since R is ℓ-reduced and $\ell(w) \neq \{0\}$ and $\ell(u) \neq \{0\}$, $\mathcal{K} \neq \emptyset$ and $\mathcal{J} \neq \emptyset$. Clearly $\mathcal{K}, \mathcal{J}$ are closed sets. Since P is a minimal ℓ-prime ℓ-ideal, we have

$$\mathcal{K} = \text{Min}_\ell(R) \setminus \{P \in \text{Min}_\ell(R) \mid w \in P\},$$

$$\mathcal{J} = \text{Min}_\ell(R) \setminus \{P \in \text{Min}_\ell(R) \mid u \in P\}.$$

They are thus both open.

Now since $\text{Max}_\ell(R)$ and $\text{Min}_\ell(R)$ are homeomorphic by Theorem 5.13, there exist clopen sets $\mathcal{N} \neq \emptyset, \mathcal{M} \neq \emptyset$ in $\text{Max}_\ell(R)$ such that

$$\text{Max}_\ell(R) = \mathcal{N} \cup \mathcal{M} \quad \text{and} \quad \mathcal{N} \cap \mathcal{M} = \emptyset.$$

By Corollary 5.2, $\mathcal{N} = s(e)$ and $\mathcal{M} = s(f)$, where $e, f \in f(R)$ are idempotent elements. Let $a = 1 - e$ and $b = 1 - f$, we have $\mathcal{N} = h(a)$, $\mathcal{M} = h(b)$, and $a, b \in f(R)$ are idempotent elements. Since $1 = a + (1 - a)$, we have $R = Ra + R(1 - a)$, and since $\mathcal{N}$ and $\mathcal{M}$ are not empty, $0 < a < 1$. From $(a \wedge (1 - a))^2 \leq a(1 - a) = 0$, $a \wedge (1 - a) = 0$ since R is ℓ-reduced. Then Ra and $R(1 - a)$ are ℓ-ideals of R with $Ra \cap R(1 - a) = \{0\}$. Therefore, we have $R = Ra \oplus R(1 - a)$ as ℓ-ideals of R. □

Theorem 5.14. *Let R be a commutative ℓ-unital ℓ-reduced ℓ-ring R. Suppose that the identity element 1 is greater than only a finite number of disjoint elements. Then R is normal and satisfies that*

$$(\star) \ \forall x \in R^+, \exists y \in R^+ \ \text{such that} \ xy = 0, \ell(x) \cap \ell(y) = \{0\}$$

if and only if R is a finite direct sum of commutative ℓ-unital ℓ-domains.

Proof. If R is a direct sum of commutative ℓ-unital ℓ-domains, then R is normal and has the given condition (Exercise 21).

Now suppose that R is normal and for each $x \in R^+$ there exists $y \in R^+$ such that $xy = 0$ and $\ell(x) \cap \ell(y) = \{0\}$. By Lemma 5.6, if R is not a domain, then $R = Ra \oplus R(1-a)$ as ℓ-ideals and $0 < a < 1$ is an idempotent element. Then Ra and $R(1 - a)$ are both commutative ℓ-unital ℓ-reduced normal ℓ-ring satisfying the given condition $(\star)$. Thus if Ra or $R(1 - a)$ is not an ℓ-domain, we may repeat using Lemma 5.6 to direct summand Ra or $R(1 - a)$. Since 1 is greater than only a finite number of disjoint elements, R is a direct sum of commutative ℓ-unital ℓ-domains. □

An ℓ-pseudoprime ℓ-ideal in a commutative ℓ-unital ℓ-reduced normal ℓ-ring may be contained in two ℓ-prime ℓ-ideals which are not comparable. For example, let $R = \mathbb{R}[x, y]$ be the polynomial ring in two variables over $\mathbb{R}$ with the coordinatewise order. Then xR and yR are ℓ-prime ℓ-ideals of R. Since R is a domain, $\{0\}$ is ℓ-pseudoprime, but $xR \not\subseteq yR$ and $yR \not\subseteq xR$.

However if an ℓ-ideal I is contained in a unique maximal ℓ-ideal, then I must be ℓ-pseudoprime. Thus if any two ℓ-prime ℓ-ideals containing I are comparable, then I is ℓ-pseudoprime.

Theorem 5.15. *Let R be a commutative ℓ-unital ℓ-reduced normal ℓ-ring. For an ℓ-ideal I of R, if I is contained in a unique maximal ℓ-ideal, then I is ℓ-pseudoprime.*

Proof. Since I is contained in a unique maximal ℓ-ideal M, by Theorem 5.9(5), $m(I) = O_M \subseteq I$. Since O_M is ℓ-prime by Theorem 5.12, I is ℓ-pseudoprime. □

If R is an f-ring, then the situation is different.

Theorem 5.16. *Let R be a commutative unital ℓ-reduced normal f-ring and I be an ℓ-ideal of R. Then I is ℓ-pseudoprime if and only if the ℓ-prime ℓ-ideals containing I form a chain.*

Proof. Suppose that I is an ℓ-pseudoprime ℓ-ideal and P, Q are ℓ-prime ℓ-ideals containing I. If $0 \leq a \in O_P$, then there exists $b \in R^+$ such that $ab = 0$ and $b \notin P$. Since I is ℓ-pseudoprime, $a \in I$. Thus $O_P \subseteq I$. By Lemma 5.12, O_P is ℓ-prime. Since R is an f-ring, by Theorem 4.10, P and Q are comparable since they both contain ℓ-prime ℓ-ideal O_P. □

For an ℓ-ideal I of R, define

$$\sqrt{I} = \{a \in R \mid |a|^n \in I \text{ for some positive integer } n\}.$$

Let R be a commutative ℓ-unital ℓ-reduced ℓ-ring and I be an ℓ-ideal of R. Then $\sqrt{I}$ is the smallest ℓ-semiprime ℓ-ideal containing I. We leave the verification of this fact to the reader (Exercise 22).

Theorem 5.17. *Let R be a commutative ℓ-unital ℓ-reduced normal ℓ-ring and I be an ℓ-ideal of R. If $\sqrt{I}$ is ℓ-prime, then I is ℓ-pseudoprime.*

Proof. By Corollary 5.1, there exists a unique maximal ℓ-ideal M such that $\sqrt{I} \subseteq M$. So $I \subseteq M$. Let N be a maximal ℓ-ideal of R and $I \subseteq N$. Then $\sqrt{I} \subseteq N$, and hence $N = M$. Thus M is the unique maximal ℓ-ideal containing I. By Theorem 5.9(5), $m(I) = O_M \subseteq I$. Since R is normal, O_M is ℓ-prime, and hence I is ℓ-pseudoprime. □

In the ℓ-ring $R = \mathbb{R}[x, y]$ with the entrywise order, if $I = xR \cap yR$, then $\sqrt{I} = I$. It is clear that I is ℓ-pseudoprime and $\sqrt{I}$ is not ℓ-prime. However for f-rings, the situation is changed.

Theorem 5.18. *Let R be a commutative unital ℓ-reduced normal f-ring and I be a proper ℓ-ideal of R. Then I is ℓ-pseudoprime if and only if $\sqrt{I}$ is ℓ-prime.*

Proof. Suppose that I is ℓ-pseudoprime. Let M be a maximal ℓ-ideal and $I \subseteq M$. If $a \in O_M$, then there exists $b \in R$ such that $|a||b| = 0$ and $b \notin M$. Since I is ℓ-pseudoprime, $a \in I$. Thus $O_M \subseteq I$. By Theorem

5.12, O_M is ℓ-prime, and hence R is an f-ring implies that any two ℓ-ideals containing I are comparable. Thus $\sqrt{I}$ is ℓ-prime (Exercise 23). $\square$

We refer the reader to [Larson (1988)] for an example showing the hypothesis of normality in Theorem 5.18 cannot be dropped.

We notice that in a commutative ring R, an ideal I is called *prime* (*semiprime*) if for any $a, b \in R$, $ab \in I$ implies that $a \in I$ or $b \in I$ (for any $a \in R$, $a^2 \in I$ implies that $a \in I$), and I is called *pseudoprime* if for any $a, b \in R$, $ab = 0$ implies that $a \in I$ or $b \in I$. In general ℓ-rings, an ℓ-prime (ℓ-semiprime, ℓ-pseudoprime) ℓ-ideal may not be prime (semiprime, pseudoprime). However in an f-ring, since for any two elements a and b, $|ab| = |a||b|$, an ℓ-prime (ℓ-semiprime, ℓ-pseudoprime) ℓ-ideal must be prime (semiprime, pseudoprime).

At the end of this section, we consider some properties of commutative ℓ-unital ℓ-rings in which each maximal ideal is an ℓ-ideal. For an f-ring, a prime ideal P must be a sublattice. In fact, for any $a \in P$, $a^+ a^- = 0$ implies that $a^+ \in P$ or $a^- \in P$. Therefore for a maximal ideal M of a commutative unital f-ring to be an ℓ-ideal, it just needs to be a convex set.

A commutative ℓ-unital ℓ-ring R is said to have *bounded inversion property* if whenever $a \geq 1$ for $a \in R$, then a is a unit.

Theorem 5.19. *Let R be a commutative ℓ-unital ℓ-ring. Each maximal ideal of R is convex if and only if R has bounded inversion property.*

Proof. "$\Rightarrow$" Suppose that $a \in R$ and $a \geq 1$. If Ra is contained in a maximal ideal M, then M is convex and $1 \leq a$ implies that $1 \in M$, which is a contradiction. Thus $Ra = R$ and a is invertible.

"$\Leftarrow$" Let M be a maximal ideal and $0 \leq a \leq b \in M$ and $a \in R$. If $a \notin M$, then $R = Ra + M$ and $1 = ra + m$ for some $r \in R$ and $m \in M$. Then

$$1 = ra + m \leq |r|a + m \leq |r|b + m$$

implies that $1 = (|r|b + m)s$ for some $s \in R$, so $1 \in M$, which is a contradiction. Thus we must have $a \in M$ and M is convex. $\square$

As a direct consequence of Theorem 5.19, in a commutative unital f-ring R each maximal ideal is an ℓ-ideal if and only if R has bounded inversion property.

By Theorem 5.19, for a general commutative ℓ-unital ℓ-ring, if each maximal ideal is an ℓ-ideal, then it has bounded inversion property.

Theorem 5.20. *Let R be an ℓ-unital commutative ℓ-ring. If R has bounded inversion property, then $f(R)$ also has bounded inversion property.*

Proof. Let $a \in f(R)$ and $a \geq 1$. Then a^{-1} exists in R. Since a is an invertible f-element, by Theorem 1.20(2), $a^{-1} > 0$, so by Theorem 1.20(2) again, a^{-1} is a d-element. For $x, y \in R$ with $x \wedge y = 0$,

$$a^{-1}x \wedge a^{-1}y = 0 \implies aa^{-1}x \wedge a^{-1}y = 0 \implies x \wedge a^{-1}y = 0.$$

Thus $a^{-1} \in f(R)$. Hence $f(R)$ has bounded inversion property. $\square$

For a ring R, $\mathrm{Max}(R)$ denotes the set of all maximal ideals equipped with the hull-kernel topology. For any subset $X \subseteq R$, define

$$U(X) = \{M \in \mathrm{Max}(R) \mid X \nsubseteq M\},$$

and

$$V(X) = \{M \in \mathrm{Max}(R) \mid X \subseteq M\}.$$

Then $U(X)$ are open sets in $\mathrm{Max}(R)$ and $\{U(a), a \in R\}$ forms a basis for open sets.

A topological space is called *zero-dimensional* if it contains a topological space basis consisting of clopen sets. A ring is called *clean* if each element in it is a sum of a unit and an idempotent.

Theorem 5.21. *Let A be a commutative ℓ-unital ℓ-reduced ℓ-ring in which each maximal ideal is an ℓ-ideal, that is, $\mathrm{Max}(A) = \mathrm{Max}_\ell(A)$. Then $\mathrm{Max}(A)$ is zero-dimensional if and only if each element in A is a sum of a unit and an idempotent element in $f(R)$.*

Proof. "⇒" Take $a \in A$, if $V(a-1) = \emptyset$, then $(a-1)R = R$ implies that $a - 1$ is a unite and $a = (a-1) + 1$.

For the following, assume that $V(a-1) \neq \emptyset$. Then $V(a-1)$ and $V(a)$ are disjoint closed sets. Since $\mathrm{Max}(A)$ is compact and zero-dimensional, there is clopen set $\mathcal{K}$ such that $V(a) \subseteq \mathcal{K}$ and $V(a-1) \cap \mathcal{K} = \emptyset$ (Exercise 30). Since $\mathrm{Max}(A) = \mathrm{Max}_\ell(A)$, by Corollary 5.2, $\mathcal{K} = U(e)$ for some idempotent element $e \in f(A)$.

Define $g = e(a-1)$ and $f = (1-e)a$. Then

$$(g+f) + e = ea - e + a - ea + e = a.$$

We show that $g + f$ is not contained in any maximal ideal of A, so $g + f$ is a unit of A. Suppose that $g + f \in M$ for some $M \in \mathrm{Max}(A)$. If $e \in M$, then $a \in M$. On the other hand, $M \notin \mathcal{K} = U(e)$, and $V(a) \subseteq U(e)$ implies

$M \notin V(a)$, that is, $a \notin M$, which is a contradiction. If $1 - e \in M$, then $f \in M$, and hence $g \in M$. On the other hand, $e \notin M$ implies that $M \in \mathcal{K}$, so $M \notin V(a-1)$. Thus $a - 1 \notin M$ and $e \notin M$ imply that $g = e(a-1) \notin M$, which is a contradiction. Therefore $g + f$ is not contained in any maximal ideal of A, so $g + f$ is a unit.

"$\Leftarrow$" We show that clopen sets consist of a topological space basis for open sets of $\text{Max}(R)$. Let $a \in A$ and $M \in U(a)$. Then A/M is an ℓ-field implies that there is an element $b \in A$ such that $ab + M = 1 + M$ in A/M. By assumption, $ab = u + e$, where u is a unit of A and $e \in f(A)$ is an idempotent. If $e \notin M$, then $(e + M)^2 = e + M$ implies that $e + M = 1 + M$, and hence

$$1 + M = ab + M = (u + e) + M = (u + M) + (1 + M)$$

implies that $u + M = 0$, that is, $u \in M$, which is a contradiction. Thus we must have $e \in M$, so $1 - e \notin M$, that is, $M \in U(1 - e)$. Suppose that $N \in U(1 - e)$. Then $e \in N$, so $ab \notin N$ since $u \notin N$. Thus $N \in U(ab)$. Therefore $U(1 - e) \subseteq U(ab) \subseteq U(a)$ and $U(1 - e)$ is clopen by Corollary 5.2. $\qquad\square$

For a commutative unital semiprime f-ring A, each maximal ideal of A is an ℓ-ideal if and only if A has bounded inversion property. Thus we have the following consequence of Theorem 5.12.

Corollary 5.4. *For a commutative unital semiprime f-ring A with bounded inversion property, $\text{Max}(A)$ is zero-dimensional if and only if A is clean.*

Exercises

(1) Prove that $s(a), a \in R^+$, form a basis for the open sets of $\text{Max}_\ell(R)$.

(2) Prove that the ring $C(X)$ of real-valued continuous functions on X is an f-ring.

(3) Check $R[x; \delta]$ defined in Example 5.1 is a ring, that is, the multiplication is associative and distributive over the addition.

(4) Let R be a totally ordered integral domain. Prove that if R is ℓ-simple, then for any $0 < a \in R$, there exists $b \in R^+$ such that $1 \leq ba$.

(5) Show that $R[x; \delta]x$ defined in Example 5.1 is a maximal left ℓ-ideal of $R[x; \delta]$.

In Exercises 6-11, R is assumed to be a unital ℓ-semisimple f-ring.

(6) Let M be a maximal ℓ-ideal of R and $a \in R$. Define $P = \{w \in R \mid (w-a)^+ \in M\}$. Prove that P is a lattice-prime ideal of R associated with M.

(7) Let M be a maximal ℓ-ideal of R and $x, y, r \in R$. Prove that if $(r - x + 1)^+, (r - y + 1)^+ \in M$, then $(r - x)^-, (r - y)^- \notin M$.

(8) Let φ be a lattice isomorphism between lattices L_1 and L_2. Prove that a subset P of L_1 is a lattice-prime ideal if and only if $\varphi(P)$ is a lattice-prime ideal of L_2.

(9) Two lattice-prime ideals are called equivalent if they are associated with the same maximal ℓ-ideal. Prove the relation is an equivalence relation.

(10) Prove the ψ defined in Theorem 5.4 is well-defined, one-to-one and onto.

(11) Prove that $A(\psi(\mathcal{K})) \subseteq \varphi(A(\mathcal{K}))$ in Theorem 5.4.

(12) Verify that φ in Example 5.2(1) is an ℓ-isomorphism between additive ℓ-groups of R and S.

(13) Prove Lemma 5.5.

(14) Prove that in Theorem 5.6, if $y = y_1 + \cdots + y_n$, then $S(y) = S(y_1) \cup \cdots \cup S(y_n)$.

(15) Let R be a commutative ℓ-unital ℓ-reduced ℓ-ring and P be an ℓ-prime ℓ-ideal. Prove that $O_P = \{a \in R \mid \ell(a) \not\subseteq P\}$ is an ℓ-ideal and $O_P \subseteq P$.

(16) Suppose that R is a commutative ℓ-unital ℓ-reduced ℓ-ring and I is an ℓ-ideal of R. Prove that $m(I) = \{a \in R \mid R = I + \ell(a)\}$ is an ℓ-ideal of R and I is a pure ℓ-ideal if and only if $m(I) = I$.

(17) Suppose that R is a commutative ℓ-unital ℓ-reduced ℓ-ring in which every ℓ-prime ℓ-ideal is maximal. Prove that every ℓ-prime ℓ-ideal of R is a minimal ℓ-prime ℓ-ideal.

(18) Prove that an ℓ-ideal I is ℓ-semiprime if and only if I is the intersection of ℓ-prime ℓ-ideals containing I.

(19) Verify that R defined in Example 5.4 is a commutative ℓ-unital reduced ℓ-subring of $A \times A$, where A is the polynomial ℓ-ring $\mathbb{R}[x]$ with the entrywise order.

(20) Verify that $\ell((x, 0))$ in Example 5.4 is a minimal ℓ-prime ℓ-ideal with $\ell((x, 0)) = O_{\ell((x,0))}$.

(21) Suppose that an ℓ-ring R is a direct sum of commutative ℓ-unital ℓ-domains. Prove that R is normal and for each $x \in R^+$ there exists $y \in R^+$ such that $xy = 0$ and $\ell(x) \cap \ell(y) = \{0\}$.

(22) Let R be a commutative ℓ-unital ℓ-reduced ℓ-ring and I be an ℓ-ideal of R. Prove that $\sqrt{I}$ is the smallest ℓ-semiprime ℓ-ideal containing I.

(23) Let R be a commutative ℓ-unital ℓ-reduced normal ℓ-ring and I be a proper ℓ-ideal of R. Prove that if any two ℓ-prime ℓ-ideals containing I are comparable, then $\sqrt{I}$ is ℓ-prime.

(24) Let $A = \mathbb{R}[x]$ be the totally ordered domain in which a polynomial is positive if the coefficient of its lowest power is positive. Define

$$R = \{(a, b) \in A \times A \mid a - b \in xA\}.$$

With respect to the coordinatewise operations and order, R is a commutative unital ℓ-reduced f-ring. Prove that R is not normal.

(25) Let R be a commutative ℓ-semisimple ℓ-unital ℓ-ring. Prove that if I is a minimal nonzero ℓ-ideal, then

$$I = \bigcap \{M \in \text{Max}_\ell(R) \mid I \subseteq M\}.$$

(26) Let R be a unital commutative ℓ-ring with squares positive. Prove that if R has bounded inversion property, then R is an almost f-ring.

(27) Suppose that R is an ℓ-unital Archimedean ℓ-domain in which $f(R)$ is a totally ordered field and $f(R)^\perp$ is a subring of R. Prove that each maximal ideal of R is an ℓ-ideal if and only if for any $0 \neq a \in f(R)$ and $b \in f(R)^\perp$, $a + b$ is a unit.

(28) Find a commutative ℓ-unital ℓ-ring with bounded inversion property that contains a maximal ideal which is not an ℓ-ideal.

(29) Let R be an ℓ-ring and $0 \le e$ be an f-element. Prove that for any $x \in R$, if $|x| = |x|e$, then $x = xe$.

(30) Prove that in Theorem 5.21, $V(a) \subseteq \mathcal{K}$ and $V(a - 1) \cap \mathcal{K} = \emptyset$ for some clopen set $\mathcal{K}$.

List of Symbols

$\subseteq, \supseteq$ set inclusion

$\subsetneqq, \supsetneqq$ proper set inclusion

$a \vee b$ least upper bound of $\{a, b\}$

$a \vee_{\geq} b$ least upper bound of $\{a, b\}$ with respect to $\geq$

$a \wedge b$ greatest lower bound of $\{a, b\}$

$a \wedge_{\geq} b$ greatest lower bound of $\{a, b\}$ with respect to $\geq$

$a \leq b, b \geq a$ a is less than or equal to b

$a < b, b > a$ a is strictly less than b

$U_A(B)$ set of upper bounds of B in A

$L_A(B)$ set of lower bounds of B in A

P_A power set of a set A

$\emptyset$ empty set

G^+ positive cone

$-G^+$ negative cone

$\in$ belongs to

$\cup$ set union

$\cap$ set intersection

$\mathbb{Z}$ ring of integers

$\mathbb{Z}^+$ set of positive integers

$\mathbb{Q}$ field of rational numbers

$\mathbb{R}$ totally ordered field of real numbers

$\mathbb{C}$ field of complex numbers

$\mathbb{R} \times \mathbb{R}$ direct product of two $\mathbb{R}$

g^+ positive part of g

g^- negative part of g

$|g|$ absolute value of g

$C_G(X)$ convex ℓ-subgroup generated by X in G

$X^\perp$ polar of X

$X^{\perp\perp}$ double polar of X

$\mathcal{C}(G)$ lattice of all convex ℓ-subgroups of G

$\oplus_{i\in I} C_i$ direct sum of convex ℓ-subgroups C_i

$\oplus_{i\in I} V_i$ direct sum of convex vector sublattices V_i

$\oplus_{i\in I} R_i$ direct sum of ℓ-rings R_i

G/N quotient ℓ-group

$G \cong H$ ℓ-isomorphic ℓ-groups

$\varphi : G \to G/N$ projection

$i = \sqrt{-1}$ imaginary unit

$Ker(\varphi)$ kernel of φ

$R \cong S$ ℓ-isomorphic ℓ-rings R and S

$M_n(R)$ $n \times n$ matrix ring over an ℓ-ring R

$T_n(R)$ $n \times n$ upper triangular matrix ring over an ℓ-ring R

e_{ij} standard matrix units

$F[G]$ group (semigroup) ℓ-algebra

$F[x]$ polynomial ring

$d(R)$ set of positive d-elements of R

$\overline{d}(R)$ set of d-elements of R

$f(R)$ set of all elements whose absolute value is an f-element of R

$\overline{f}(R)$ set of all f-elements of R

π_k canonical epimorphism

$\langle X \rangle$ ℓ-ideal generated by X

$\langle a \rangle$ ℓ-ideal generated by a

$I_1 + \cdots + I_n$ sum of ℓ-ideals

$\ell\text{-}N(R)$ ℓ-radical of an ℓ-ring R

$\ell\text{-}P(R)$ p-radical of an ℓ-ring R

$i(A)$ i-ideal of an ℓ-algebra A

$F[[x]]$ ring of formal power series

$F((x))$ formal Laurent series field

$G \times G$ Cartesian product

$F \setminus \{0\}$ the set of nonzero elements in F

$F^t[G]$ twisted group ℓ-algebra

$|S|$ cardinality of a set S

$Orth(R)$ orthomorphism of R

$u(R)$ band generated by units in an ℓ-ring R

$ab(R)$ set of almost bounded elements in an ℓ-ring R

$\vee x_i$ sup of x_i

$r(a)$ right ℓ-annihilator of a

$\ell(a)$ left ℓ-annihilator of a

$f'(x)$ derivative of $f(x)$

$A \setminus B$ different of sets A and B

R/I quotient ℓ-ring of R to an ℓ-ideal I

I_0 set of element r in an f-ring such that $\mathbb{Z}r$ is bounded

D_a inner derivation induced by a

(G, P) partially ordered group G with positive cone P

(R, P) partially ordered ring R with positive cone

$[u, v]$ commutator $uv - vu$

δ_{jk} Kronecker delta

R_R ℓ-ring R as right ℓ-module over R

$End_R(aR, aR)$ ring of endmorphisms of right R-module aR

ℓ_x mapping by left multiplication of x

$S(a, f)$ semigroup generated by a and f

$i(x)$ set $\{a \in R \mid ax = xa = a\}$ in an ℓ-ring R

$dim_F V$ dimension of vector space V over F

$(i_1 i_2 \cdots i_n)$ n-cycle

$U_f(R)$ set of upper bounds of $f(R)$ in R

$F[x; \sigma]$ skew polynomial ring

$tr f$ trace of a matrix f

$det(a)$ determinant of a

$gcd(a, b)$ greatest common divisor of a and b

F^n n-dimensional column space over F

$cone_F(K)$ the cone generated by a subset K over F

$|v|$ the length of vector v in $\mathbb{R}^n$

$G \underset{\rightarrow}{\oplus} H$ lexicographic order of two totally ordered groups

$Max_\ell(R)$ space of maximal ℓ-ideals of an ℓ-ring R

$Min_\ell(R)$ space of minimal ℓ-prime ℓ-ideals of an ℓ-ring R

$Max(R)$ space of maximal ideals of R

$s(X)$ set of maximal ℓ-ideals not containing $X \subseteq R$

$S(X)$ set of minimal ℓ-prime ℓ-ideals not containing $X \subseteq R$

$h(X)$ set of maximal ℓ-ideals containing $X \subseteq R$

$H(X)$ set of minimal ℓ-prime ℓ-ideals containing $X \subseteq R$

$C(X)$ ring of real-valued continuous functions on X

$\forall$ for all, for any

$\exists$ there exists

$A(\mathcal{K})$ intersection of lattice-prime ideals that contain a fixed point and are associated with some maximal ℓ-ideal in $\mathcal{K}$

$\sqrt{I}$ intersection of all ℓ-prime ℓ-ideals containing ℓ-ideal I

$a|b$ a divides b

$(I : R)$ largest ℓ-ideal contained a left ℓ-ideal I of R

$U(X)$ set of maximal ideals not containing $X \subseteq R$

$V(X)$ set of maximal ideals containing $X \subseteq R$

O_P set $\{a \in R \mid \ell(a) \nsubseteq P\}$

$m(I)$ set $\{a \in R \mid R = I + \ell(a)\}$

$\Rightarrow$ implication

Bibliography

Agnarsson, G., Amitsur, S. A., and Robson, J. C. (1996). Recognition of matrix rings II, *Isreal J. Math.*, **96**, pp. 1–13.

Anderson, F. W. (1962). On *f*-rings with the ascending chain condition, *Proc. Amer. Math. Soc.*, **13**, pp. 715–721.

Anderson, F. W. (1965). Lattice-ordered rings of quotients, *Canad. J. Math.*, **17**, pp. 434–448.

Bernau, S. J. and Huijsmans, C. B. (1990). Almost *f*-algebras and *d*-algebras, *Math. Proc. Camb. Phil. Soc.*, **107**, pp. 287–308.

Bigard, A. and Keimel, K. (1969). Sur les endomorphismes conservant les polaires d'un groupe reticule archimedean, *Bull. Soc. Math. France*, **97**, pp. 381–398.

Birkhoff, G. and Pierce, R. S. (1956) Lattice-ordered rings, *An. Acad. Brasil. Cienc.*, **28**, pp. 41–69.

Coelho, S. and Milies, C. (1993) Derivations of upper triangular matrix rings, *Linear Algebra Appl.*, **187**, pp. 263–267

Colville, P., (1975). Characterizing *f*-rings, *Glasgow Math. J.*, **16**, pp. 88–90.

Colville, P., Davis, G. and Keimel, K. (1977). Positive derivations on *f*-rings, *J. Austral. Math. Soc.*, **23**, pp. 371–375.

Conrad, P. (1961). Some structure theorems for lattice-ordered groups, *Trans. Amer. Math. Soc.* **99**, pp. 212–240.

Dai, T. Y. and Demarr, R. (1978). Positive derivations on partially ordered linear algebra with an order unit, *Proc. Amer. Math. Soc.*, **72**, pp. 21–26.

Dauns, J. (1989). Lattice-ordered division rings exist, *Ordered Algebraic Structures*, Kluwer Acad. Publ., 229–234.

Diem, J. E. (1968). A radical for lattice-ordered rings, *Pacific J. Math.*, **25**, 71–82.

Eowen, L. (1988). Ring theory, Vol. 1, Academic Press.

Fuchs, L. (1963). Partially ordered algebraic systems, Pergamon Press.

Fuchs, R. (1991). A characterization result for matrix rings, *Bull. Austral. Math. Soc.*, **43**, 265–267.

Hansen, D. J. (1984). Positive derivations on partially ordered strongly regular rings, *J. Austral. Math. Soc.*, **37**, pp. 178–180.

Hayes, A. (1964). A characterization of *f*-rings without non-zero nilpotent, *Jour-*

nal London Math. Soc., **39**, pp. 706–707.

Henriksen, M. (1977). Semiprime ideals of f-rings, *Symposia Mathematica*, **21**, pp. 401–409.

Henriksen, M. (1995). A survey of f-rings and some of their generalizations, Ordered algebraic structures (Curaçao, 1995), 1–26, Kluwer Acad. Publ., Dordrecht, 1997.

Henriksen, M. (2002). Old and new unsolved problems in lattice-ordered rings that need not be f-rings, Ordered algebraic structures, 183–815, Kluwer Acad. Publ., Dordrecht, 2002.

Henriksen, M. and Isrell, J (1962). Lattice-ordered rings and function rings, *Pacific J. Math.*, **12**, pp. 533–565.

Henriksen, M., Isrell, J and Johnson, D., (1961). Residue class fields of lattice-ordered algebras, *Fund. Math.*, **50**, pp. 107–117.

Henriksen, M. and Jerison, M (1965). The space of minimal prime ideals of a commutative ring, *Trans. Amer. Math. Soc.*, **115**, pp. 110–130.

Henriksen, M. and Smith, F. A. (1982). Some properties of positive derivations on f-rings, *Contemporary Math.*, **8**, pp. 175–184.

Herstein, I. N. (1978). A note on derivations, *Canad. Math. Bull.*, **21**, pp. 369–370.

Herstein, I. N. (1979). A note on derivations II, *Canad. Math. Bull.*, **22**, pp. 509–511.

Hungerford, T. W. (1974). Algebra, Springer.

Jacobson, N. (1980). Basic Algebra II, Freeman, San Francisco.

Johnson, D. G. (1960). A structure theory for a class of lattice-ordered rings, *Acta Math.*, **104**, pp. 533–565.

Kaplansky, I. (1947). Lattices of continuous functions, *Bull. Amer. Math. Soc.*, **53**, pp. 617–623.

Kaplansky, I. (1958). Projective modules, *Math. Ann.*, **68**, pp. 372–377.

Keimel, K. (1973). Radicals in lattice-ordered rings, *Rings, modules and radicals* (Proc. Colloq., Keszthely, 1971), 237–253. Colloq. Math. Janos Bolyai, Vol. 6, North-Holland, Amsterdam.

Kreinovich, V. and Wojciechowski, P. (1997). On lattice extensions of partial orders of rings, *Communications in Algebra*, **25**, pp. 935–941.

Lam, T. Y. (1999) Modules with isomorphic multiples, and rings with isomorphic matrix rings, a survey, *L'Enseignement Mathematique*, No. 35.

Lam, T. Y. (1999) Lectures on modules and rings, *Graduate Texas in Mathematics*, No. 189, Springer.

Lam, T. Y. (2001). A first course in noncommutative rings, *Graduate Texas in Mathematics*, No. 131, Second Edition, Springer.

Lam, T. Y. and Leroy, A. (1996). Recognition and computations of matrix rings, *Isbel J. of Math.*, **96**, pp. 379–397.

Larson, S. (1988). Pseudoprime ℓ-ideals in a class of f-rings, *Proc. Amer. Math. Soc.*, **104**, pp. 685–692.

Larson, S. (1997). Quasi-normal f-rings, *Ordered Algebraic Structures*, pp. 261–275, Kluwer Academic Publishers.

Li, F., Bai, X. and Qiu, D. (2013). Lattice-ordered matrix algebras over real gcd-domains, *Comm. Algebra*, **41**, pp. 2109–2113.

Ma, J. (2011). Finite-dimensional ℓ-simple ℓ-algebras with a d-basis, *Algebra Univers.*, **65**, pp. 341–351.

Ma, J. (2013). Recognition of lattice-ordered matrix rings, *Order*, **30**, pp. 617–623.

Ma, J., Wojciechowski, P. (2002). A proof of Weinberg's conjecture on lattice-ordered matrix algebras, *Proc. Amer. Math. Soc.*, **130**, pp. 2845–2851.

Ma, J., Wojciechowski, P. (2002). Structure spaces of maximal l-ideals of lattice-ordered rings, Ord. Alg. Struct. (Gainesville, 2001), J. Martinez (Ed), Kluwer Acad. Publ., (2002), 261–274.

Ma, J., Zhang, Y. (2013). Lattice-ordered matrix algebras over totally ordered integral domains, *Order* (9 March 2013 online first).

Mason, G. (1973). z-ideals and prime ideals, *J. Algebra*, **26**, pp. 280–297.

McGovern, W. (2003). Clean semiprime f-rings with bounded inversion, *Comm. Algebra*, **31**, pp. 3295–3304.

McHaffey, R. (1962). A proof that the quaternions do not form a lattice-ordered algebra, *Proc. of Iraqi Scientific Societies*, **5**, pp. 70–71.

Pajoohesh, H. (2007). Positive derivations on lattice ordered rings of matrices, *Quaestiones Mathematicae*, **30**, pp. 275–284.

Passman, D. S. (2011). *The algebraic structure of group rings*, Dover.

Robson, J. C. (1991). Recognition of matrix rings, *Communications in Algebra*, **19**, pp. 2113–2124.

Redfield, R. (1989). Constructing lattice-ordered power series fields and division rings, *Bull. Austral. Math. Soc.*, **40**, 365–369.

Schwartz, N. (1986). Lattice-ordered fields, *Order*, **3**, pp. 179–194.

Schwartz, N. and Yang, Y. (2011). Field with directed partial orders, *J. Algebra*, **336**, pp. 342–348.

Semrl, P. (2006). Maps on matrix spaces, *Linear Algebra Appl.*, **413**, pp. 364–393.

Steinberg, A. S. (1970). *Lattice-ordered rings and modules*, Ph.D. Thesis, University of Illinois at Urbana-Champaign.

Steinberg, A. S. (1983). *Unital ℓ-prime lattice-ordered rings with polynomial constraints*, Trans. Amer. Math. Soc. **276**, pp. 145–164.

Steinberg, A. S. (2010). *Lattice-ordered rings and modules*, Springer.

Subramanian, H. (1967). ℓ-prime ideals in f-rings, *Bull. Soc. Math. France*, **95**, pp. 193–203.

Subramanian, H. (1968). Kaplansky's Theorem for f-rings, *Math. Ann.*, **179**, pp. 70–73.

Weinberg, E. C. (1966). On the scarcity of lattice-ordered matrix rings, *Pacific J. Math.*, **19**, pp. 561–571.

Woodward, S. (1993). A characterization of local-global f-rings, *Ordered Algebraic Structures* (Gainesville, FL, 1991); Kluwer Acad. Publ. ; Dordrecht; pp. 235–249.

Index

245

Printed in the United States
By Bookmasters